Van der Waals and Molecular Science

Van der Waals in his middle thirties. (Photograph from his grandson.)

Van der Waals and Molecular Science

by

A. Ya. KIPNIS

B. E. YAVELOV

and

J. S. ROWLINSON

CLARENDON PRESS · OXFORD

1996

This book has been printed digitally in order to ensure its continuing availability

OXFORD
UNIVERSITY PRESS

Great Clarendon Street, Oxford OX2 6DP

Oxford University Press is a department of the University of Oxford.
It furthers the University's objective of excellence in research, scholarship,
and education by publishing worldwide in

Oxford New York

Auckland Bangkok Buenos Aires Cape Town Chennai
Dar es Salaam Delhi Hong Kong Istanbul Karachi Kolkata
Kuala Lumpur Madrid Melbourne Mexico City Mumbai Nairobi
São Paulo Shanghai Singapore Taipei Tokyo Toronto

with an associated company in Berlin

Oxford is a registered trade mark of Oxford University Press
in the UK and in certain other countries

Published in the United States
by Oxford University Press Inc., New York

Reprinted 2001

A catalogue record for this book is available from the British Library

Library of Congress Cataloging in Publication Data
Kipnis, Aleksandr IAkovlevich.
[Iogannes Diderik Van-der-Vaal's, 1837-1923. English]
Van der Waals and molecular sciences / by A. Ya. Kipnis, B. E.
Yavelov and J. S. Rowlinson.
Includes bibliographical references and index.
1. Waals, J. D. van der (Johannes Diderik), 1837-1923.
2. Physics—history. 3. Chemistry, Physical and theoretical-
-History. 4. Physicists—Netherlands—Biography. I. IAvelov, B.
E. (Boris Efimovich) II. Rowlinson, J. S. (John Shipley), 1926-
III. Title.
QC16.W33K5613 1996 530'.092—dc20 95-48336

ISBN 0-19-855210-6

Preface

The development of molecular physics and physical chemistry cannot be understood without a knowledge of the work of the Dutch physicist Johannes Diderik van der Waals. His doctoral thesis of 1873 was the first theory of liquids and gases in which the essential differences and similarities of these two phases were interpreted in terms of the properties of the constituent molecules. For forty years he and his school held fast to the view that atoms, molecules and their interactions were the basis of much of physics and chemistry and, by example rather than by direct attack, rebutted the anti-atomistic views of Mach, Ostwald, Duhem and the other 'energeticists'. The Dutch contribution to physical science had never been greater, nor has it ever again had the influence that it manifested in the forty years before the outbreak of the First World War in 1914. Van der Waals was the senior member of the group of physicists responsible for this surge and his life and work are an essential component of any attempt to understand the scientific history of the period.

Work that is important in its day does not always seem so relevant to the scientists of the next generation and this was, for a time, the fate of the work of van der Waals. There has, however, been a re-appraisal in the years since the end of the Second World War and we have come to realise not only the historic importance of his work but also its essential correctness (in our eyes) and the far-sighted originality of many of his contributions. After reading it for the first time a theoretical physicist wrote recently that 'to me, his thesis read as if it had been written 20 years ago, not 120.' In this scientific life we have tried to bring out all these facets of his work and so put van der Waals into his historic place as one of the founders of molecular science.

In preparing both the original Russian edition of 1985, *Johannes Diderik van der Waals, 1837–1923*, by A.Ya. Kipnis and B. E. Yavelov, and this revised and enlarged English edition we have incurred many debts. Some of these are acknowledged in the notes at the end of each chapter but others are more wide-ranging. We thank for their help the following friends, colleagues and correspondents:

in Russia, L. P. Danyushevskaya, V. Ya. Frenkel', A. T. Grigoryan, N. F. Mikhailova, G. I. Milovidova, Z. K. Sokolovskaya, Yu. I. Soloviev, and N. V. Vdovichenko; in the Netherlands and Belgium, H. Bordewijk (Municipal Archives, The Hague), W. Downer (Municipal Archives, Leiden), E. P. van Emmerik (Amsterdam), A. J. F. Gogelein (Museum Boerhaave, Leiden), A. Hoff (University of Leiden), G. A. Kohnstamm (Rosendaal), M. Kohnstamm (Brussels), A. J. Kox (University of Amsterdam), H. J. F. M. Leechburch Auwers (Museum Boerhaave, Leiden), J. van

Meurs (University Library, Amsterdam), J. Tabak (Municipal Archives, Deventer), G. M. van Thienen (Society for the Advancement of the Natural Sciences, Medicine, and Surgery, Amsterdam), L. J. W. Waaijers (Library of Delft University of Technology), the late J. C. van der Waals (grandson of J. D. van der Waals), Jan van der Waals (great-grandson), J. H. van der Waals (University of Leiden), J. Valenbreder (State Archives of North-Holland, The Hague), G. A. C. Veeneman (Museum Boerhaave, Leiden) and M. L. Wurfbain (Museum 'de Lakenhaal', Leiden); in Germany, P. L. Butzer (Technical University, Aachen); and in the United Kingdom, M. Kelly (Library of Queen's University, Belfast) and A. E. B. Owen (Cambridge University Library), and the Leverhulme Trust for an Emeritus Fellowship for J. S. R.

For the preparation, checking and publication of the manuscript we are indebted to the staff of Oxford University Press, to Nancy Rowlinson, and particularly to Marjorie Wallace for the seven years she has worked on this book.

Mannheim A. Ya. K.
Moscow B. E. Ya.
Oxford J. S. R.
August 1995

Contents

Note on Conventions

Superscript numbers in the text refer to the Notes at the end of each chapter. Upright numbers in square braces refer to the entries in the Bibliography of works by and about van der Waals at the end of the book. Italic numbers in square braces denote the references at the end of each chapter. Abbreviations of journal titles and a further explanation of notations used in references are to be found at the start of the Bibliography.

Modern Dutch usage is to spell names such as van der Waals with a capital V if they stand alone but with a small v if they are preceded by initials or forenames. We have followed the older convention of writers in English, including many Dutch writers of English, of using the small v throughout.

CHAPTER 1

Leiden

The first record of the forebears of the great Dutch physicist Johannes Diderik van der Waals is dated 20 April, 1639. On that day the birth of Michiel Bartholomeesz van der Waal from Brielle, then a little fort at the confluence of the Maas and Waal, was registered at Leiden Town Hall. His father Bartholomeus – not yet van der Waal – was born between 1580 and 1590, probably also in Brielle[1]. The name van der Waal seems, however, to suggest that the family came originally from the banks of the Waal, the southern arm of the lower Rhine, which flows into the sea close to Brielle.

From 1718 the family started to call themselves van der Waals, perhaps to distinguish themselves from many others with the name van der Waal[2]. It was in Leiden that all the paternal and maternal ancestors of J. D. van der Waals were born, lived, worked, produced children, and died. The only exception was his maternal grandmother Jannetje Montfoort (1787–1854), who died at Delft. The men were all industrious Leiden workers, usually builders or clothiers.

At the time when Michiel van der Waal settled in Leiden, the United Provinces of the Netherlands had overthrown Spanish rule and, at the end of the sixteenth century, had established the first bourgeois republic. They were entering on the period of their history that the Dutch with justifiable pride call the *Gouden Eeuw*, the Golden Age. This state of two to three million people grew into a world power, outstripping other countries in economic, trading and even military strength. They made outstanding advances in the sciences and humanities, and in technologies such as dyke building, land reclamation, the making of optical instruments, and diamond cutting. At sea their ships were the first to reach Australia, Tasmania and New Zealand. Dutch painting rose to new heights and literature entered its classical period under the leadership of Jacob Cats (1577–1660), Pieter Hooft (1581–1647), Joost van den Vondel (1587–1679), and Constantijn Huygens (1596–1687), the father of the physicist Christiaan Huygens (1629–95). The country hospitably opened its doors to prominent European scholars and men in public life who had been persecuted elsewhere by absolutist regimes. Free from the constraints of religious orthodoxy they were able to publish in Holland the fruits of their free-thinking in large handsome editions in a way that was undreamt of in France or Italy.

The old town of Leiden was prominent both in the struggle for independence and in the subsequent flowering of national culture. In 1574 it was besieged for 131 days by the Spanish troops of the Duke of Alba; many thousands died of hunger and disease but the town held out. As a reward for its courage William the Silent, Prince of Orange (1533–84) granted a special charter to Leiden to enable it to found the first university in Holland. On 8 February 1575, when the war of independence was still raging, the 'Batavian Academy' was inaugurated in the former nunnery of St Barbara. Religious toleration, the great scope for scientific work and its publication (the house of Elsevier, then the Leiden university publishers, was well known by 1580), a well-endowed library and, not least, generous salaries, attracted to Leiden many of the most brilliant European intellectuals. The 'Athens of the North' came to international fame through the work of such scholars and scientists as the botanist Carolus Clusius (1526–1609), the historian Joseph Scaliger (1540–1609), 'the bottomless well of erudition' as his contemporaries called him, the lawyer Hugo Grotius (1583–1645), the astronomer Willebrord Snel (1580–1626) and the mathematician Simon Stevin (1548–1620) who did much to establish the decimal notation. The great blossoming of Dutch painting also owed much to Leiden. It was the home town not only of Rembrandt (1606–69) but also of Gerrit (or Gerard) Dou (1613–75), Gabriël Metsu (1629–67), and of Heine's favourite painter, the good-humoured Leiden innkeeper, Jan Steen (1626–79).

The population of the town grew rapidly from about 12 000 early in the sixteenth century to 100 000 by 1620–50, mainly because of the great influx of Protestant refugees from France and Flanders. As Leiden grew its prosperity nourished the famous scholars of its University, its painters, its merchants, and its many thousands of workmen including the ancestors of van der Waals. In the *Gouden Eeuw* Leiden became the world centre of clothmaking, and such terms as 'fine Leiden cloth', 'Leiden baize' and 'Leiden camlet' were as well known as the term 'Leyden jar' became in the next century, after its invention by the Leiden professor Pieter van Musschenbroek (1692–1761).

Slowly, however, the Golden Age drifted into a period first of stagnation and then of decline. The incessant wars of the late seventeenth and early eighteenth century exhausted the country. The grand bourgeoisie and aristocracy of the towns and the provinces opposed all attempts to establish some central rule, thus preventing the pursuit of a consistent domestic and foreign policy that met the national needs. Dutch commerce and industry gradually declined in the face of foreign competition, particularly from Britain which rapidly overtook the Netherlands as the centre of entrepreneurial progress. The rate of reclamation of land, perhaps the most significant economic indicator for Holland, declined sharply.

Dutch painting, however, passed its peak of brilliance several decades before the economic decline set in, and its literature fell similarly into a derivative lifelessness. Leiden did not escape these trends. The town's trade died away, cloth factories were closed at an incredible rate; by 1710–17 the population of 100 000 had fallen to a third of that number and, once more, the town had become agrarian. Only Dutch science kept its pre-eminent position and, indeed, was only in its growing phase at the middle of the seventeenth century when economic activity and the cultivation of the fine arts were at their zenith.

Among the leading figures of this 'phase-shifted' golden age were Christiaan Huygens, Benedict Spinoza (1632–77), Anthonie van Leeuwenhoek (1632–1723), Jan Swammerdam (1637–80), and the Leiden professors Herman Boerhaave (1668–1738), Willem 's Gravesande (1688–1742), Pieter van Musschenbroek, and Bernard Albinus (1697–1770). All but the first two and the last one of these men graduated at Leiden, and Huygens, although not a graduate, attended lectures there.

Leiden reached the height of its European fame during the time when Boerhaave, 'the Batavian Hippocrates' and 'the teacher of all Europe', was professor there. He was appointed in 1709, and taught medicine, botany and chemistry. In his time the number of students rose to two thousand; they came from all over Europe and included such future celebrities as the Dutch physician Gerard van Swieten (1700–72), the Swede Carl Linné (1707–78), and the Swiss Albrecht von Haller (1708–77). Among the 120 medical students in 1726 there were forty Britons, twenty Germans, about ten Russians, and many Frenchmen, Greeks and Arabs. The Tsar, Peter the Great, attended a few of Boerhaave's lectures, and Lavrentii Blumentrost (1692–1755), a pupil of Boerhaave's, became later the first President of the St Petersburg Academy of Sciences.

By the middle of the eighteenth century Leiden's claim to first place among European universities could no longer be sustained, although it was still pre-eminent in the Netherlands, where there were now five universities altogether – at Leiden, Franeker, Harderwijk, Groningen and Utrecht. Nevertheless there were now no brilliant names among its staff, and the numbers of both foreign and Dutch students were falling rapidly. Half a century after Boerhaave's time there were only 500–600, of whom two-thirds were studying law or theology; science was almost totally neglected. In the town more and more houses fell empty, markets were closed, the life on the canals disappeared, the cheerful noise and music of the students fell silent, and Leiden gradually took on the melancholic and bewitched appearance of a sleeping town that is to be found in so many of the descriptions of the leisured travellers of the last century.

'Arriving at Leiden one feels the chill of a dead town . . . , in the broad streets and squares, and at the cross-roads there is no one to be seen apart from a few scattered passers-by who look like the survivors of some epidemic. One walks up lanes with weeds underfoot, and on each side closed shutters; the silence is complete and it seems that the inhabitants have fallen into a supernatural sleep. . . . One can wander for hours from bridge to bridge, from canal to canal, and from isle to isle searching in vain for the life and clamour of old Leiden, but all one finds is desolation, silence and reflections in the water of the sombre splendour of the town' (E. de Amicis, 1878 [*1*, p. 214]).

'The town . . . spends the rest of its life on the banks of the dying Rhine. . . . There are no ships, barques or even small boats. Only clouds of gnats and flocks of ducks float over the undisturbed surface of the water. One can walk from one end of the town to the other without meeting a soul. . . . It seems as if the town were bewitched and its inhabitants condemned to eternal sleep.

The weeds growing over the roads are witness to the absence of all traffic in Leiden' (M. L. Werner, 1884 [*2*, p. 165–168]).

'It were as if the faint smile of death were imprinted on the whole town — on its buildings and on its streets' (Vas. I. Nemirovich-Danchenko, 1892 [*3*, p. 452]).

The decline was at its deepest in 1810 when the country collapsed politically and economically after its incorporation by Napoleon into the First Empire. French domination brought the Dutch distress and humiliation, but it had also the effect of bringing together the provinces and towns of the Netherlands, which had resisted consolidation, into an effective state under a firm federal rule. This favourable effect became apparent when, after Napoleon's fall and the re-establishment of national sovereignty in 1813, there was a slow but steady economic recovery.

The year 1824 saw the founding of the powerful *Nederlandsche Handel-Maatschappij* (Netherlands Trading Society) which undertook a systematic exploitation of the colonies. The next year the North-Holland Canal was opened — the largest in the world until the completion of the Suez Canal in 1869. The first railway followed in 1837, and in that year also appeared the first issue of the intellectual monthly *De Gids* (The Guide), which later became a rallying point for liberal views. In the 1840s the Dutch astounded the whole world by a spectacular project — by means of pumps driven by gigantic steam engines they dried up Haarlemmer Meer, as if to punish it for the disastrous flooding of Leiden it had produced in 1836.

During the French occupation it seems that Leiden fared worse than other Dutch towns. This was not only because of predatory extortions and the wrecking of the economic life of the town, but also because of the disaster on 12 January 1807 when a French ship laden with gunpowder blew up in the Rapenburg canal in the middle of Leiden. The terrible explosion, which was heard as far away as Amsterdam, razed entire quarters on both sides of the canal, killing and wounding hundreds. This part of the town became known as the Ruins.

Leiden University escaped comparatively unscathed in these troubled times; indeed its importance in the Netherlands increased following the closure of Utrecht, Franeker and Harderwijk during the French occupation. The last two never opened again. A few years later a large university building was erected at the Ruins which later became the famous Kamerlingh Onnes Cryogenic Laboratory. The business and intellectual life of Leiden slowly revived after the country regained its independence in 1815, encouraged by the Literary and Mathematics Society of the Netherlands, founded at Leiden in the eighteenth century, and known as *Mathesis Scientiarum Genetrix* or 'Mathematics, the ancestor of the sciences'.

In 1820 there graduated from Leiden University, Johan Rudolf Thorbecke (1798–1872), a man who played an outstanding rôle in his country's history. In 1831 he became Professor of Law at Leiden, and became the brilliant leader of the liberal forces in the country, who in 1848 curbed the royal power and established a more democratic form of government. For the next twenty years he led the long struggle against conservative opinion and religious factions that launched the Netherlands on a path of dynamic entrepreneurial development.

Johannes Diderik van der Waals was born in Leiden on 23 November 1837. His home at *No. 972 Jan Vossensteeg* in the sixth quarter still stands[3], but the number is now 30. He was the son of a carpenter, Jacobus van der Waals (1812–98) and his wife Elisabeth, née van den Burg (1816–96). It appears that he was named after his maternal grandfather Johannes Diederik van den Burg (1790–1838), a painter and glazier. The future scientist was the eldest of the van der Waals's ten children – four daughters and six sons. Apart from Johannes Diderik, several of these acquired some reputation; Gerardus (1851–?) became a senior colonial officer in the East Indies, and the youngest, Hendrik Gerardus (1857–?), a translator of Sanskrit. The second child Elisabeth (1840) lived only two weeks, and the same name was given to the third child who was born the next year. The eighth child, Cornelis, was an able mathematician and chess player, but took to drink and ran into financial trouble[4].

No records have survived from Johannes Diderik's childhood and adolescence; all that is known is that the family's scanty means allowed him only a primary education, and then attendance for some time at the so-called MULO, an advanced primary school with a three-year course [231]. It should be said, however, that in those times there was no public secondary education in the Netherlands or in other countries. As to university, a young man of the working class could not even dream of it, since attendance there would have prevented him from earning his daily bread until he was 23–25 years old, apart, of course, from the cost of finding lecture and examination fees. Moreover a knowledge of the classical languages was a *sine qua non* for entering a university, a restriction that was not abolished in the Netherlands until 1917. Thus the route to a higher education had to pass through the gymnasia, latin schools, athenaeums or other similar establishments which was quite incompatible with earning one's living. At that time primary education in the Netherlands, as in other countries, was neither free nor compulsory, although it was strongly encouraged by the communal authorities. On the other hand it is worth noting that the teaching standards in Dutch elementary schools were then believed to be very high [*4*, pp. 45–111; *5*, p. 165, 169]. This level was due, to a considerable extent, to the law of 1806 which removed the primary schools from the control of the churches, and to the adoption of a systematic inspection of schools and of teachers' qualifications. An important role in the development of Dutch public education was played by many independent public associations, from the powerful *Maatschappij tot Nut van 't Algemeen* (Society for the General Good) to more modest ones like the Mathesis, mentioned above. The latter promoted in Leiden a three-year industrial evening school (*Burger-avondschool*) for working-class youths. This was founded as early as 1785 and was supported mainly by the Leiden community [*6*, p. 7].

By our standards, a Dutch elementary school of those days was a very modest establishment. There was only one teacher to impart all subjects – reading, writing, geography, singing, and the rudiments of natural science. All children studied in one crowded room which was often divided into 'classes' by means of light curtains. If there were more than seventy pupils the teacher had the help of an apprentice, the so-called 'pupil-teacher'. If the numbers went beyond a hundred a teacher's assistant could be appointed and, finally, if they were over a hundred and fifty then they were taught by a team of three, the teacher, his assistant and the pupil-teacher [5, p. 254].

Van der Waals entered on his pedagogical career at an early age, and by fourteen or fifteen he was already a pupil-teacher. In 1856, at the age of nineteen, he started on a course of examinations which occupied the next seventeen years of his life. The first of these many examinations was to obtain the 'Aktes' (in full, *Akte van Bekwaamheid*, or Certificate of Ability) which testified to their owner's command of various subjects, and which allowed him to climb up the rungs of the ladder of primary and, later, secondary education. Van der Waals's Aktes for German, French, and the 'Teacher's Akte of the 3rd Degree' have been preserved[5]. He obtained also Aktes for the rudiments of mathematics and for the natural sciences. By the end of the 1850s he was working as a teacher's assistant in a primary school [231, p. 149]. On 12 October 1860, at The Hague, the South Holland provincial commission for primary school teaching awarded him a pass in the next examination, the Akte 'of teacher's qualification' which entitled him to teach mathematics. The next year he obtained the 'principal teacher's' Akte which allowed him to hold the post of director of a primary school.

There can be little doubt that van der Waals showed at an early age great ability in natural science, and skill as a teacher. Without these abilities it is hard to believe that he would have broken away from the family tradition of working at a trade and, in a way that was rare in Dutch society at that time, have climbed steadily up the ladder of the teaching profession.

In 1862 the twenty-five-year-old teacher first set foot in Leiden University when he enrolled for classes in mathematics. His lack of knowledge of the classical languages naturally prevented him from enrolling as a regular student and, in any case, it would have been a foolhardy step to have given up his regular and much sought-after job as a teacher. There was, however, provision for students to enrol for up to four lecture courses a year, at a cost of fl. 40 per course, which was about a tenth of his annual salary as a primary school teacher.

His first professor was Gideon Jan Verdam (1802–66), a Leiden graduate and a leading Dutch mathematician of the first half of the nineteenth century. He had been teaching at Leiden since 1839, and earlier had become well known throughout the Netherlands as the young winner of mathematical competitions. He wrote many textbooks on elementary and higher mathematics and mechanics. These ran to only a few editions, but some were translated into German. He was a member of the Royal Academy of Sciences in Amsterdam, the Holland Society of Sciences of Haarlem (*Hollandsche Maatschappij der*

Wetenschappen, henceforth abbreviated as HMW), and many other learned societies including that of Prague. He covered all branches of mathematics in his lectures but showed a particular fondness for geometry, a partiality that may have been related to his remarkable gift for drawing. He drew all the figures for his textbooks, and professional engravers could not match his skill in making complicated diagrams. It may well be that these lectures, the first that van der Waals attended at the University, played an important rôle in developing the strong geometric cast of his scientific thought.

Verdam's health deteriorated in the early 1860s, and his place was often taken by David Bierens de Haan (1822–95), who was appointed an assistant professor in 1863, and succeeded to the full professorship on Verdam's death three years later. Bierens de Haan had taken his doctorate, with distinction, under Verdam in 1847, and by the 1860s he had acquired a considerable reputation in the Netherlands and abroad. He was a member of the Academy and of the HMW, a corresponding member of the British Association, had an honorary doctorate from Edinburgh University, and was an honorary member of the Kazan Mathematical Society. His reputation rested on work in two fields, first, his comprehensive and extremely useful compilation of tables of integrals – a monumental volume that was revised and re-issued as late as 1939 – and second, his work on the history of Dutch mathematics. He edited previously unknown mathematical memoirs by Stevin, Spinoza and Huygens. In Leiden he was a well-known public figure; for many years he was dean of the local Mennonite community, and he was chairman of the Leiden Mathesis.

He had, apparently, no gift for teaching and it is perhaps worth noting that when Hendrik Antoon Lorentz (1853–1928), a good-natured man who had attended his lectures in the 1870s, had to praise his former professor on a formal occasion he said that he was an example to the younger generation for his 'untameable and strenuous work', but he passed over his lectures in silence [7, p. 328].

Van der Waals may also have been dissatisfied with Verdam's successor, for he decided to supplement his studies with a course in mathematical astronomy, although he had originally been interested only in mathematics. It is possible, also, that he needed astronomy to enrich his collection of Aktes or, again, that he was simply attracted by the brilliancy and personal charm of Friedrich (or Frederik) Kaiser (1808–72) the Professor of Astronomy. It is known that Kaiser's lectures were the principal attraction of Leiden University for the young Lorentz [7, p. 329].

Kaiser had at an early age acquired a great reputation all over the Netherlands for his astronomical observations and calculations. He was awarded an honorary doctorate by Leiden at the age of twenty-seven. His achievements aroused an interest in astronomy throughout the country, and this grew to real enthusiasm after the appearance in the 1840s of the two volumes of his popular encyclopaedia of astronomy, *The Starry Sky*. He exploited this enthusiasm to the full in his campaign for the construction of a new observatory at Leiden.

The government, headed by the economical Thorbecke, refused financial support and so Kaiser opened a subscription list. The large sum of fl. 13 000 was collected and building began in 1858. Inspired by their professor, the Leiden students freely gave their services, and the observatory was ready for use by 1861. At first the staff was absurdly small, but again student volunteers came to the rescue. The first research work was of as high a standard as any in the world at that time, and Kaiser's prestige rose accordingly. He became a member of the Royal Institution in London, of the Royal Astronomical Society, of the Berlin Academy of Sciences, etc.

Kaiser became the pride of Leiden, and one of the local sights; for most Dutchmen the Leiden Faculty of Physics and Mathematics was Kaiser's Faculty. Students of all subjects flocked to his lectures, even the philologists and the theologians. It is, however, noteworthy that this energetic and optimistic scientist, who did so much for his university and for Dutch science, never led an easy life. He had to earn his own living from the age of thirteen, and was often in poor health, suffering from headaches and colds. Three years after his death the Rector asked in his address at the 300th anniversary of the university, 'What was it about Kaiser that was irresistibly attractive? His science, his great learning, his penetrating mind, his wonderful gift for teaching? Yes, all these, but there was in him also an absolute and boundless love of science burning within him, so that he knew how to inspire his students' hearts' [*8*, p. 25].

The young Lorentz was strongly influenced by Kaiser's teaching. Throughout his student years he was, he relates [*7*, p. 330], none too successful in astronomy but it was Kaiser's influence that led him to become a physicist. As will be seen later, this remarkable scientist and teacher had also a strong influence on van der Waals, although for the decisive impulse that turned the young school teacher into a physicist we must be grateful to another Leiden professor, Pieter Leonhard Rijke (1812–99). Van der Waals had, of course, a general idea of physics before meeting Rijke, but this idea was a distorted one, and he had little sympathy with this branch of science. Heike Kamerlingh Onnes (1853–1926), in reporting a private conversation with van der Waals, wrote that he 'arrived at physics in his own particular way. He had originally shown a strong inclination towards mathematics, whose truths appealed most strongly to his critical spirit and his need for a deep faith. He had learnt his physics from books, and took its premises to be arbitrary, so that this science totally repelled him. He was told, however, that those wanting to study physics had to take Professor Rijke's experimental lectures. From these he quickly learnt that, if physics could not match the certainties given to mathematics by its deductive proofs, this shortcoming was compensated for by the reliability provided by experiment and observation' ([233], letter of 8/x/1898).

Rijke had graduated from Leiden in 1835 and occupied the chair of physics from 1854 until he reached the age of compulsory retirement in 1882. He was a member of the Academy and of the HMW, although much of his research was published abroad, often in the *Annalen der Physik*. He was purely an empiricist, apparently devoid of any imagination, but observant and extremely meticulous. He worked principally in the field of electricity and, in particular, on the so-called 'sounding flames' and on electric arcs. It is of some interest that it was he who first observed the magnetic suppression of electric arcs that arises from the switching on and off of high currents. The effect is nowadays widely used in electrotechnology.

His experiments were mainly of the kind suitable for lecture demonstration, as was natural since the first Leiden physics laboratory was intended for just such experiments; indeed it was little more than a roomful of scientific instruments some of which went back to the *Gouden Eeuw*. Moreover teaching was his life's work. Lorentz recalled that, 'We, his students, admired best his scrupulous care. Rijke aimed constantly at developing the students' practical skills, and, since he would not tolerate the slightest failure of his demonstrations, he prepared them with great thoroughness and inexhaustible patience' [7, p. 328]. Nowadays, demonstrations at physics lectures are regarded as something quite secondary; a faith in physics is almost inborn so there is no need to demonstrate its raison d'être. That, however, was not the case in the past and, according to Kamerlingh Onnes, it was the perfection and reliability of Rijke's experiments that made van der Waals believe that there was a firm basis to this science.

Many students and postgraduates passed through Rijke's hands in the course of his long life; van der Waals and Lorentz took their doctorates under his supervision, Jacobus Henricus van 't Hoff (1852–1911) and Hugo de Vries (1848–1935) attended his lectures. He contributed much to the secondary level of education in the Netherlands, but his name is now almost totally forgotten. He is ignored by the encyclopaedias and treatises of the history of science, and even Dutch biographical dictionaries give him only a line or two. Perhaps surprisingly, such oblivion was what he wanted. When he died no traditional obituaries were published. The Proceedings (*Verslagen*) of the Academy, of which he had been a member for thirty-six years, published only a short notice in a rectangle surrounded by a black border which said that since the deceased wanted no description of his life and work, the Academy and the country could only gratefully honour his memory [9]. His contemporaries knew him only as a pedantic unsmiling figure, dressed invariably in black, but Lorentz wrote 'My Promotor, Professor Rijke, who appeared to the students to be quite inaccessible, had, nevertheless, a kind heart as I came to know well in later years' [7, p. 328].

Van der Waals's first contacts with professional science occurred at a time of important changes in the Dutch educational system – changes that were to affect his career very directly. In February 1862 the Liberal government of Thorbecke came to power. This obstinate minister had been head of government from 1845 to 1853 and had then compelled the King to grant some substantial democratic concessions and to improve primary instruction in the schools. He had no time, however, to bring about a corresponding improvement in secondary education before he was overthrown by the Conservative opposition. Having become Prime Minister again, he introduced in the States General in June 1862 a bill to establish a system of secondary education. This was signed by the King, and came into force on 2 May 1863.

The main educational goal of this legislation was to give 'to many middle-class children, who cannot manage only with a primary level of instruction, some general knowledge that was necessary for their careers and to prepare them to some degree for a commercial or industrial life' [*6*, p. 3]. Thorbecke did not intend that secondary education should be narrowly vocational, but that it should enable youths to develop their 'general culture in the broadest sense of the term, a culture that would enable them to work most effectively in all kinds of fields' [*6*, p. 3].

Two kinds of secondary school were introduced, burgher schools of a lower rank for the children of workmen and farm labourers, and advanced burgher schools (*Hoogere Burgerscholen*, or HBS) for young men whose families were 'sufficiently middle-class for them to have enough time to be able to develop their abilities to a greater degree than the children of artisans or husbandmen' [*6*, p. 4]. The HBS were divided into five-year and three-year schools, and there were also evening schools of both kinds. Time was to show the wisdom of Thorbecke and his followers, for the HBS contributed strongly to the great national revival that started in the 1870s, and were the foundation of the remarkable advances of Dutch science at the turn of the century. Almost all Dutch scientists passed through the HBS, either as students or as teachers, or both. For twelve years van der Waals was closely associated with the HBS that offered the five-year course.

What kind of instruction did these schools offer? The closest analogy is the *Realgymnasia* of Germany and Russia, which had started as early as the eighteenth century, but which became widespread only about the time that the Dutch HBS were founded. The Realgymnasia and the HBS differed from the classical gymnasia in that the centre of gravity of their teaching was in the subjects of the 'real cycle' – physics, mathematics, mechanics and chemistry. The Dutch 'realists' studied, however, for only five years, compared with the six or seven of their German and Russian counterparts. The lesson in religion, that was compulsory in Germany and Russia, was absent from the curriculum of the HBS – a happy consequence of the mutual neutralization of the claims of the Protestant and Catholic halves of the population. Finally, the students in the HBS learnt three, not two, modern languages – English, German and French – and so, in compensation, were freed from the cramming of Latin that still persisted in an attenuated form in the German and Russian schools. This release from the yoke of the dead languages, however, blocked the direct route to the universities.

Pupils of twelve to fourteen years old were admitted to the HBS on passing an entrance examination, and they paid an annual fee of fl. 30–60. Nine HBS, with about 600 pupils, had been opened by 1864, and in the next decade the number of schools rose to thirty-nine and their pupils totalled 3500. The new schools needed new teachers, but despite the shortage, particularly of those able to teach the 'real cycle', the law of 2 May 1863 laid down high standards for applicants for teachers' posts at the HBS. These posts were highly respected and carried a substantial annual salary of up to fl. 3000. For the main subjects teachers normally had to have the degree of doctor or *candidaat*, that is, to have graduated from a university without having taken a doctorate. It was, however, possible to obtain such a post without these degrees, but then only by passing difficult examinations, that is, by accumulating the necessary Aktes.

Van der Waals was among the first applicants for a job in the HBS. Early in 1864 he set about preparing for the Akte A examination, which would qualify him to compete on equal terms with the *candidaats* for a post in mathematics or mechanics. It seems that he had no hesitation in choosing to specialize in teaching these subjects, notwithstanding the fact that the only failure he had met so far in his examination odyssey had been in mathematics [206, p. 8]. The Akte A examinations comprised a series of tests spread over several days in the various branches of science chosen by the candidate, and accompanied by a test of wider erudition. The examinations, all of which were written, took twelve hours. All went well, however, and on 20 July 1864, van der Waals returned home as the happy owner of the Akte A. His enjoyment quickly turned to disappointment when it turned out that there had been an annoying misunderstanding. But, who knows, it may be that it was because of this misunderstanding that van der Waals became a physicist, and not a mathematician.

What happened was this; van der Waals was looking forward to being appointed to the post of mathematics teacher at the HBS at Deventer, a town some 100 km east of Leiden. The influential Rijke had told him in advance that he was guaranteed the offer of the post when he had acquired his Akte A. Moreover Kaiser, who had no doubts about his student's ability, had written the following reference for the authorities at Deventer.

> The undersigned Professor of the Physics and Mathematics Faculty of Leiden University affirms that during the session 1863/64 this gentleman, Mr J. D. van der Waals, attended the courses in mathematics and astronomy most punctiliously, and exhibited great ability, learning and diligence. The undersigned is much impressed by the modesty and composure that Mr van der Waals has shown, and regards it as his great good fortune to have among his students so outstanding a young man. He gives it as his opinion that Mr van der Waals is an extremely competent secondary school teacher.
>
> F. Kaiser
>
> Leiden, 11 June, 1864.

There can be no doubt that such a recommendation from the distinguished astronomer would count for much in the small provincial town of Deventer.

Thus all would have been well had not Rijke muddled up the whole affair. He had promised van der Waals the post of mathematics teacher at Deventer and, at the same time, had told the Ministry that he was sending a physicist. So it was a physicist who was expected at Deventer, and the mathematics post had already been filled[7]. This was a painful blow, but such difficulties never stopped van der Waals. He put in one more year of hard work – teaching in school, lectures at the University, and preparations for the next round of examinations filled his time. He was already Director of the Leiden evening primary school, mentioned above as established under the auspices of *Mathesis Scientiarum Genetrix*[8]. On 12 August 1865, he obtained the Aktes B and C which gave him equal status as an applicant for a HBS post with those who held doctorates in mathematics and physics respectively. He had passed the examinations brilliantly. Kaiser, one of the examiners, wrote that 'he succeeded in doing something that seems to me to be beyond human power'[9]. Kamerlingh Onnes spoke in the same way about the physics examination, 'I believe that the last examination was passed *only*[10] by van der Waals; at least it was a rare occurrence' [233]. Rijke, whose agreement with Deventer still held good, sent a supplementary letter of recommendation saying that he 'presented the most convincing evidence of outstanding talent and extraordinary assiduity'[11]. On 21 September 1865, van der Waals was appointed physics teacher at the HBS in Deventer [237].

Six days later he married a charming eighteen-year-old Leiden girl, Anna Magdalena Smit (1847–81). She was the youngest of the three children of a Leiden hatter, Johannes Smit (1805–72) and his wife Anna Pierrot Magnez (c.1810–?)[12]. Her elder brother, Joseph Smit (1836–1929) moved to London in 1866 where he became known as one of the best lithographers and illustrators. He specialized in animal and, more particularly, in ornithological drawings, contributing much to the journal *Nature* [*10*, p. 64].

Van der Waals met his future wife when she was still a young girl. For her thirteenth birthday he had sent a poem in which the little Anna Magdalena was compared, with little originality, to a rosebud[13]. The van der Waals and Smit families lived at opposite ends of Leiden and belonged to different congregations, so it seems plausible[14] to guess that he and his future wife met first at primary school, as teacher and pupil.

On 1 October 1865, he wrote to the authorities at Deventer to accept the appointment, and on 24 October the newly married couple moved to that town [237].

Notes

1. These particulars of van der Waals's ancestry come from the notice by N. B. van der Waals on p. 272 of *Gens Nostra XVII*, 1962, and from the information on p. 226–227 of the same volume. For copies of these pages the authors are indebted to Professor J. H. van der Waals of Leiden, and Dr H. Bordewijk, the Director of the Municipal Archives of The Hague.
2. Professor E. M. Bruins of Amsterdam University is inclined to see in this 's' a suggestion of belonging to a group.
3. This information comes from Dr W. Downer, the Municipal Archivist of Leiden [236].
4. The information on J.D. van der Waals's brothers and sisters came to the authors

from Mr J. C. van der Waals, his grandson, who died on 5 February 1984, shortly before his 72nd birthday. The letter, which was unfinished, contained also several interesting pieces of family lore, and was sent by his son Mr Jan van der Waals (1947-) through the good offices of Professor A. J. Kox of Amsterdam University [242].

5. Communicated by Dr W. Downer [236].
6. From direct and reliable observations by Ph. Kohnstamm [206, p. 8] and H. Kamerlingh Onnes [233].
7. This anecdote is quoted by Kamerlingh Onnes in 1898 [233] and by Kohnstamm [206, p. 8]; there can be little doubt that both heard it directly from van der Waals.
8. Professor A. J. Kox told the authors of a letter of recommendation of 16 June 1865 written by Professor Verdam. In this letter, which is now in the Van der Waals Archive of the Van der Waals Laboratory at Amsterdam, he is described as 'Directeur der Burger-avondschool van het Genootschap Mathesis Scientiarum Genetrix te Leiden'.
9. Cited in [205, p. 110].
10. Underlined by Kamerlingh Onnes.
11. Cited in [231, p. 149].
12. This information is from Mr J. C. van der Waals and from Dr M. L. Wurfbain, Director of the Museum de Lakenhaal in Leiden.
13. This information is from Mr J. C. van der Waals, and is also mentioned in [232, p. 9].
14. From Mr J. C. van der Waals.

References

1. Amicis, E. de, *La Hollande*, Paris, 1878, 413 p.

2. Werner, M. L. (М. Л. Вернер), *Strana plotin (The country of dykes)*, Moscow, 1884, 230 p. (in Russian).

3. Nemirovich-Danchenko Vas.I. (Вас.Ив.Немирович-Данченко), *Po Germanii i Gollandii (Around Germany and Holland)*, St Petersburg, 1892, 474 p. (in Russian).

4. Sagra, H. de la, *Voyage en Hollande et en Belgique*, Paris, 1839, t. 1, *Hollande*, 366 p.

5. Laveleye, E. de, *L'Instruction du Peuple*, Paris, 1872, 489 p.

6. Loos, D. de, *Organisation de l'enseignement secondaire dans le royaume des Pays-Bas*, Leeuwarden, 1894, 130 p.

7. Lorentz, H. A., Rede bij de aanvaarding van het doctoraat in de geneeskunde h.c. te Leiden. In *Collected papers*, v. 9, Den Haag, 1939, p. 326–334.

8. Heynsius, A., *Discours solemnel prononcé le 8 fevr. 1875 pour le jubilé de l'Université de Leiden par le Recteur de l'Université A. Heynsius*, Leiden, 1875, 42 p.

9. van de Sande Bakhuyzen, H. G., Petrus Leonardus Rijke, *Versl.*, 1899, d. 7, b. 532.

10. Nissen, C., *Die Illustrierten Vogelbücher. Ihre Geschichte und Bibliographie*, Stuttgart, 1953, 223 S.

CHAPTER 2

The HBS teacher

Two points of interest never failed to be mentioned by the nineteenth century guide-books to Deventer, 'a clean and prosperous town', as the omniscient Baedeker put it [*1*, p. 306]. These were, first, the famous *Deventer koeken*, or honey-cakes, which were sold all over the country and even exported to the East and West Indies and, secondly, the *Deventer Athenaeum Illustre* which was founded as early as 1630.

According to the *Guide des Voyageurs* of 1863 this once-flourishing school had been reduced to a mere fifteen pupils housed within its fine buildings with their magnificent assembly-hall, excellent laboratory and vast library [*2*, p. 146]. The fall of numbers occurred in spite of the high standard of the teaching staff. Thus the humanities were taught by A. J. Witringa (1827–1901), a well-known writer of the 1870s and 1880s. The exact sciences were in the hands of V. S. M. van der Willigen (1822–78), a former student of Rijke's, and a member both of the Academy (from 1857) and the HMW (from 1858). He was one of the leading Dutch physicists of his time, and his name is familiar to historians of optics for his early work on the spectra of gas discharges.

'It is said', the guide-book of 1863 confided, 'that the Athenaeum will be transformed into an industrial school' [*2*, p. 146]. The author was clearly well informed for in March 1864 the Deventer Magistrate passed an order closing the Athenaeum and establishing a HBS in its buildings. The next year the order was cancelled, but the old school eventually died a natural death in 1878, when it seems that there were no pupils left [*3*].

The Deventer HBS, one of the first to be established, was opened on 8 September 1864. As mentioned in Chapter 1, Bierens de Haan was partly responsible for its establishment. Having taken his degree at Leiden, he had come to Deventer to teach mathematics at the local gymnasium and, after his election to the City Council, had become involved in wider educational matters. It was at his insistence that in 1862, the year before the law setting up the HBS was passed, the Council acknowledged the need for a secondary school [*4*, p. 6]. Bierens de Haan would probably have been its first mathematics teacher if he had not been called to Leiden to help Verdam [*5*].

The post of Director of the HBS was filled by M. J. Cop (1818–76), a Leiden doctor and a former Athenaeum professor, who also took over the teaching of chemistry[1]. Instruction in mathematics and physics was given initially by van der Willigen, but he moved to Haarlem in the autumn of 1864 on his appointment as Director of Physics at the Teyler Foundation. Fifty years later this post was held by Lorentz on his retirement from Leiden; it was the one dream of every Dutch physicist, since it carried a generous salary, had no teaching duties, and the holder had at his disposal a splendid collection of scientific instruments and even a scientific journal, the *Archives Musée Teyler*.

In 1864 van der Waals missed the first vacancy in mathematics at the HBS, as has been described in the previous chapter, but a year later he arrived in Deventer to take up the physics post.

It seems that the choice of lodging was approached by van der Waals as earnestly as everything else in his long and hard-working life. The newly married couple settled in the prosperous street *Groote Poot*, now *De Hereeniging*, [237], in a house of which the ground floor was occupied by a fine bookshop, a great asset to a keen reader like van der Waals. Next door was an ironmonger's shop which was of equal value to his young wife just starting out as a housekeeper. Finally they were only a few steps from van der Waals's new place of work, the Deventer HBS[2] [*6*, p. 51].

He started to teach physics in October 1865 and, on 14 December, also took on the physics teaching at the evening classes of the HBS. At the same time Cop was taken ill and he had to take over the chemistry classes also. The HBS opened with thirty-two schoolboys[3], the next year it had seventy-five, and in 1870 had to move to another more spacious building. It was evidently the growing popularity of the HBS that hastened the Athenaeum's decline. After further removals and transformations the HBS has now become the *Alexander Hegius School*[4].

In the summer of 1866, after the pupils had left for their holidays (only one had graduated that year [*7*, p. 9]), van der Waals and his wife returned to Leiden; their first child was expected soon. They settled temporarily in the house of the Smits on the *Koepoortsgracht* [234][5] as apparently the van der Waals family home was too overcrowded at that time. While in Deventer he had applied unsuccessfully for a position at the gymnasium in Rotterdam [231, p. 149]. Probably his aim was to leave the remote town of Deventer and settle somewhere within easy reach of Leiden and its University. When he arrived at Leiden, Kaiser gave him the welcome news that there was a good chance of a post at The Hague.

In 1865 the gymnasium at The Hague had been split into two independent institutions under one roof – the gymnasium proper and a HBS. Mathematics and physics were taught in both schools by H. G. van de Sande Bakhuyzen (1838–1923), Kaiser's favourite pupil who had taken his doctorate at Leiden in 1863 [*8*]. In July 1866 he asked The Hague municipal authorities if he might resign to take a post in the new HBS in the university city of Utrecht. The young astronomer could not resist the opportunity of again being close to academic research. Kaiser, who no doubt had a hand in all of this, at once told van der Waals of the vacancy at The Hague. The matter needed prompt and careful action, particularly in view of the missed opportunity at Deventer. Van der Waals sent his letter of application to the Magistrate at The Hague on 20 July, enclosing his Aktes A, B and C [238]. Kaiser wrote the next day[6].

Mr J. D. van der Waals, whose performance in the most advanced examinations prescribed by the law of secondary instruction caused his examiners to make public their highest commendations, needs no praise or recommendation from me. I am happy to say, however, that I shall always regard it an honour and a privilege to count Mr J. D. van der Waals amoung my pupils, and to have had a hand in his education. Mr J. D. van der Waals is distinguished by extraordinary gifts and he possesses industry and perseverance which have led to his success in a way that seems to me to be beyond human abilities. With his unusual talents, especially for mathematics, physics and astronomy, Mr J. D. van der Waals combines the strictest morals and, since he is also a highly experienced teacher, I believe it will be the good fortune of secondary education if such a man is appointed.

F. Kaiser
Professor of the Physics and Mathematics Faculty of Leiden University.

Leiden
21 July 1866

There were seven other applicants, some with doctor's degrees, but on 26 July van der Waals was one of the two recommended by the Magistrate to the City Council, and on 31 July he was their choice [234].

It can be inferred from van der Waals's letters to the Magistrate at The Hague that there was some resistance in Deventer to his leaving [234]. The Deventer HBS had, in fact, a problem because of the coincidence of van der Waals's departure with Cop's retirement because of illness. Moreover, The Hague HBS tried, this time unsuccessfully, to get from Deventer the chemist they had chosen as Cop's successor. However on 12 September van der Waals was honourably discharged by the Deventer City Council.

It may be noted that his successor at Deventer was H.W. Schroeder van der Kolk (1836–67)[7] who, before his premature death, earned a place in the history of physical chemistry by being among the first who tried to apply the mechanical theory of heat to the treatment of chemical reactions [*9*].

On 14 September 1866 van der Waals became a father; his wife gave birth to their first child Anne Madeleine (1866–1937). Time pressed, however, and having left his wife and daughter in the care of the Smits, he went at once to The Hague where, on 17 September, he gave his first lesson [234].

The Hague, the seat of the monarch, the national government and the States General, had only acquired the status of a town early in the nineteenth century. It was one of the first Dutch cities to throw off the lethargy of the early years of that century and by the 1850s its bourgeois prosperity was growing rapidly. From 1855 to 1878 its population grew from 66 000 to 100 000. It was the city of the aristocracy, the wealthy merchants, the rentiers and the highest ranks of the civil service; it housed the richest museum and art galleries, fashionable shops, luxury hotels, magnificent private houses and elaborate gardens [*10*].

The Hague Gymnasium, and the HBS to which it gave birth (now the *J. R. Thorbecke School*) occupied a large three-storied house of dull exterior in the *Westeindestraat* in the centre of the town. It had previously been the Royal

Music School and then a Fine Arts Society. The interior was decorated with panels of exotic African woods, now partly preserved in the municipal museum. The schoolboys nicknamed the building the *Hok* [12, p. 30], that is, a shed or larder, possibly because the classrooms were high and old-fashioned[8]. Van der Waals settled at first on the nearby *Prinsengracht* ('the best street is beautifully built' wrote a Russian traveller of the mid-nineteenth century [*11*, p. 127]). On 8 January 1867 he formally took himself and his wife off the Deventer list [227], and the next day they were registered as citizens of The Hague, living in *Wagenstraat*, a winding crowded street that led from the main railway station to the centre of the town. There was a games hall in this street where, perhaps, he would call for a game of billiards of which he was very fond — 'like several other pioneers of the kinetic theory of gases', as S. G. Brush drily remarks[9] [170].

In June 1868 a second daughter was born, Jacqueline Elisabeth, and three years later, a third, named Johanna Diderika after her father who had perhaps by then given up hope of a son and heir. In 1870 or 1871 the family moved to their final home in The Hague in *Zuidwal.* In 1874 the HBS moved also, to a mansion at *Bleyenburg 38* [234] not far from the central square, the *Plein*, the major junction for the trams, at first horse-drawn, then steam and finally electric. By then the number of pupils had quadrupled from the initial forty-four [*7*, p. 87].

According to the time-tables that have survived, van der Waals taught mathematics, physics and cosmography to Forms 3, 4 and 5 of the HBS from 1866–7 to 1868–69, and also algebra to Form 2 in 1867–68. From 1869–70 until 1877 he taught physics and cosmography there, but during the years 1867–68 to 1872–73 he also taught physics to the first four forms of the Gymnasium. The HBS worked six days a week starting at 9 a.m.; from Monday to Friday they had six lessons a day with a break for lunch from 1 to 2 p.m., and on Saturday four lessons and no lunch break. Van der Waals was usually teaching from fourteen to sixteen hours a week, although his load was at first considerably heavier. He had no day completely free of lessons during his first three years (except of course Sunday) [234]. We do not know his duties from 1872–76, but it is safe to infer that by 1876–77 he had Saturdays free.

The high standard of the teaching at the HBS can be deduced from the fact that, apart from van der Waals, there were several other figures who were well known, at least within the Netherlands. Van de Sande Bakhuyzen has already been mentioned; he succeeded Kaiser at Leiden, became a member of the Academy, and was later Chairman of its Mathematics and Physics Section. C. Bellaar Spruyt (1842–1901) had a doctorate in physics, and taught chemistry and natural science until his transfer to Utrecht with van de Sande Bakhuyzen. He later became Professor of Philosophy at Amsterdam, a member of the Academy, and acquired the reputation of a brilliant and open-minded thinker. In 1874 P. H. Schoute (1846–1913) took the post of teacher of mechanics, geometry and drawing. He was later Professor of Mathematics at Groningen, a member of the Academy, and an internationally recognized authority on multi-dimensional geometry. J. ten Brink (1834–1901), a prolific writer and critic, taught history, geography, civic affairs and Dutch literature in both the Gymnasium and the HBS until he was invited to Leiden in 1884 as Professor of the History of Dutch Literature, when he also became a member of the Academy.

The new type of school needed a new type of textbook, and a number of these were written by the first teachers in the HBS. Thus in 1866–67 the course for Form 5 of the HBS at The Hague recommended the textbook of geometry by A. J. van Pesch (1837–1916), a colleague of van der Waals at Deventer, the *Rudiments of Mechanics* by van de Sande Bakhuyzen, and a chemistry manual written by W. F. Koppeschaar (1831–1909), who was a substitute for Bellaar Spruyt at the HBS. Koppeschaar's life resembled that of van der Waals and may throw some light on the obscurities of the latter's youth. The son of a textile dyer, Koppeschaar had to be content with the elementary training of the Leiden Mathesis evening school, in which he became a teacher's assistant at the age of seventeen. Four years later he became a teacher there and soon after joined the governing body of the Mathesis. He managed to take some lecture courses at Leiden University and in 1865 enrolled there as a student after a dispensation from the examinations in classical languages given him by Thorbecke in person[10]. In 1867 he obtained his doctorate in chemistry. The course of instruction in physics that he had to take was written by J. Bosscha, a man who greatly influenced van der Waals's entry into science; his life and work will be described later.

Van der Waals was a conscientious and methodical teacher. The Municipal Annual Report of 1866 stated that the new teacher 'fulfils from the start our favourable expectations of both his knowledge and his teaching. If only his health will stand up to the strain imposed by his zeal! The very many hours that the programme allots to his lessons need changing.' The next year the report ran, 'As to van der Waals's teaching we can only reiterate the praises in last year's report based on the first impressions of his lessons.' The curators of the Gymnasium were more fulsome in their praise; 'This teacher is a master in the art of organising his lessons, he is a pedagogue who can hold the attention by a clear exposition, by laying a firm grounding and by having no fear of repeating the work until a clear understanding has been achieved, whether or not this impairs the consistency of the lesson. The teacher does not merely make a speech but talks directly to the pupils, making them realize for themselves the degree to which they have understood everything he has told them. We should unhesitatingly accept his proposal to revise his time-table so that each form can listen to his exposition of one of the branches of physics: heat for the 1st Form, magnetism and electrostatics for the 2nd Form, current electricity for the 3rd Form, and light and sound for the 4th Form' [234].

Van der Waals was less enthusiastic about his lessons at the Gymnasium than he was about those at the HBS with its 'real' bias. He spoke out against the practice at the Gymnasium of attaching so much importance to physics that its marks became the criteria for promotion to the next form. Nevertheless he thought that an interest in heat, magnetism, electricity, light and sound should be aroused in every way possible, not only by high-flown theories but also by numerous examples of useful and interesting applications. The mathematical standards should not be set too high otherwise the exact sciences would become mere addenda to the basic mathematical programme [234].

Whilst teaching at The Hague, van der Waals had been immersing himself more and more deeply in science. It might be supposed that he was the kind of schoolmaster of whom the pupils are rather afraid, but this was probably not so, for he aimed to enlighten his pupils rather than to try to turn them all into 'little scientists', or to cram them with bookish learning. Evidence on this point, and on his liberal pedagogical views, comes from the fact that in late 1871 and early 1872 he and his colleague ten Brink started to give lectures on the exact sciences and literature to girls of 'civilised standing' [*8*]. He would certainly have continued with these when, in October 1873, a girls' department offering a three-year course was affiliated to the HBS. Thirty-five girls were enrolled in the department while the HBS and the Gymnasium had 108 and 54 boys respectively [*12*, p. 96].

Van der Waals's participation in the Hague Physical Society called the *Diligentia* is further evidence of his sympathy with the cause of enlightenment. The society had been founded in 1793 as a club of local enthusiasts for 'studies in experimental philosophy' [*13*]. The Diligentia was housed in a stately building with a large auditorium and several rooms containing valuable scientific collections. In the 1870s there was a meeting on every other Friday evening when a member or an invited scientist of repute lectured on some problem of science or advanced technology, and among the list of visiting lecturers can be found such names as H. A. Lorentz and Hugo de Vries. In 1873 the first collection was published in which reports of these lecturers for 1872–73 were placed under the title *Natuurkundige Voordrachten*[11]. (The word *Natuurkundige* (physical) has in this context the old meaning of *Natuurwetenschappelijke* (natural scientific), in the sense that it includes also 'medical and technical'.) For fifty years the reports were compiled by a journalist, P. A. Haaxman Jr. They appeared first in the daily newspaper *Dagblad van Zuidholland en 's Gravenhage* and then, after completion of a yearly series, they were collected and published in one volume.

Van der Waals was admitted to the Diligentia in 1868; only one of his colleagues at the HBS — the first director — was a member. (In the mid-1870s there were about 250 members, many of whom were of high social standing.) It can be inferred from the Society's publications that van der Waals gave two lectures there, on Friday 5 February 1875 and Friday 3 December 1880 [2,*13*]. (Haaxman's reports were published first in the newspaper on 7–8 February 1875 and 5–6 December 1880.) Both of these lectures are purely popular, but they are interesting as evidence that is hard to come by on how van der Waals considered the phenomena of electricity; they reflect also some of his views on other subjects.

The lecture of 1880 belongs to the time when he was living in Amsterdam and will be considered in Chapter 4, but the first, 'On the influence of magnets upon electric currents and especially upon electric light' [2], deserves to be considered here. It begins with a remark on the rapid progress in recent years in this domain of science. He reminded his listeners of discoveries due to Oersted and Ampère, and then demonstrated how a thin platinum wire moved between the poles of a horseshoe magnet when an electric current from a Bunsen cell was passed through it. The installation was designed in such a way that the wire glowed and was easily observed by the audience. The movement and the equilibrium form of the wire, for example an S-shape, depended on the mutual disposition of the magnet and the wire.

Then he turned to experiments with rarefied gases. Glowing discharges in tubes were obtained by means of Ruhmkorff's induction coil. 'We see here,' he said, 'how the substance shines owing to the heat evolved by the electric current overcoming the resistance of the gas. The gas is in a glowing state; thus there is no proper electrical light here' [2, p. 89]. He demonstrated tubes with hydrogen, nitrogen, carbon dioxide and sulphur dioxide, all of which showed different colours. He emphasized especially that the 'three sorts of light that are characteristic of glowing hydrogen were later discovered in the Sun's atmospheric spectrum and investigations into the constitution of that great heavenly body conducted on this basis have already given us so many fruitful and enjoyable results that the scientific world can expect to have a more profound knowledge of the Sun than of the child fed by it, the Earth' [2, p. 90].

The analogy between a white-hot platinum wire and glowing gas in a tube was then demonstrated by the deformation of a glowing beam of gas when the tube was inserted between poles of the magnet. Some beautiful experiments by Plücker and Gassiot were then reproduced, in which the course of the electric current was rather complicated and also susceptible to magnetic influence. He concluded with a demonstration of how the Earth's magnetism can affect electric currents of the atmosphere travelling from the equator to the poles — a possible way of seeking an explanation of the northern lights.

There are some interesting details here. Firstly, this report shows that van der Waals had at his disposal modern equipment for physical research and that he could carry out experiments successfully; one can see that the laboratory resources of the HBSs were excellent. Secondly, he considered electric discharge in a rarefied gas to be like the usual conductive heating by a current — an erroneous explanation, but evidence that he avoided introducing new hypotheses until known theory is exhausted. This position was sensible because of the lack of understanding of the whole of this field in those days. Finally, he shows here his interest in astronomical problems[12].

The Diligentia comprised the city's intellectual elite. Scientific activities, unless there was some particular stir, never occupied in the public life of The Hague the position they held in Leiden, which, as Paul Ehrenfest (1880–1933) put it, 'consisted mainly of professors' [*14*, p. 89]. There were, nevertheless, a few scientists at The Hague whose senior positions made them no less influential than the Olympians at Leiden. In the Diligentia van der Waals first met a scientific community, not as a junior but as an equal, and it was here that he first came into regular contact with Bosscha, an active and competent physicist who was a member of the governing body of the Diligentia.

Johannes Bosscha (1831–1911) came from a family that had long been known in the Netherlands for its learning. His great-grandfather and his grandfather were classical scholars, his father Johannes senior (1797–1874) was an academic philologist who was the Minister responsible for relations between church and state in two governments, neither of them Thorbecke's.

It is hardly surprising that, coming from such a background and being himself a man of great abilities, Bosscha had gained a higher position in science and in public life than van der Waals (his junior by six years) by the time that they met at The Hague. In 1863 he had been elected to the Academy and in the same year, only two months after the law setting up the HBS had come into force, he became Inspector of Secondary Education for the two provinces of Holland. (The first province where van der Waals taught was outside his jurisdiction; Deventer is in Overijssel, but The Hague and Leiden are in South Holland.)

In 1844–48 Bosscha had attended a Latin school in Amsterdam where physics was taught by van der Willigen. His instruction in physics continued under van der Willigen at the Amsterdam Athenaeum Illustre. When van der Willigen moved to the Athenaeum at Deventer, Bosscha followed him there too; he had already taken a liking to physics and knew that he would not find a better teacher. He graduated from the Athenaeum in 1850 and went to Leiden University where, four years later, he took his degree with a thesis under Rijke's supervision, 'On the differential galvanometer'. After a short stay in Berlin he returned to Leiden where he worked as Rijke's assistant until 1860. During this time he carried out some first-rate work. In 1855 he devised the method of compensation for the measurement of electromotive forces. He received wide recognition for his experiments on energy conservation in the electrical circuits of galvanic cells. This law was not universally accepted in the middle of the nineteenth century, and his results did much to establish its universality [*15*].

At Leiden he inevitably fell under Kaiser's spell. As Rijke's pupil and collaborator he became the centre of the students' efforts in building the new observatory, and later helped Kaiser in his astronomical researches. Kaiser's insistence on accuracy undoubtedly influenced the metrological bias of Bosscha's later research in physics.

From 1860–63 he taught physics and mathematics at the Military Academy in his native town of Breda, and for the next ten years he was Inspector of Secondary Education at The Hague. He was one of the team entrusted by Thorbecke with the task of creating the first HBS established under the law of 1863. It was his job to select the teachers, work out the school programmes and chair the committees for the provincial examinations. It was due to his enthusiasm and perseverance that the HBS had such good physical and chemical apparatus that the teachers could even carry out research. The first part of his *Textbook of physics and its most important applications* [*16*] appeared in 1865; it was aimed specifically at the HBS. This book was revised repeatedly and went through seven editions in thirty years. It became the standard textbook throughout the Netherlands although it had initially been conceived as no more than a Dutch adaptation of a once popular French text by A. Boutan and J.Ch. d'Almeida [*17*].

Bosscha did not abandon his scientific activity while he held this administrative post. Since he had no laboratory facilities, as an enthusiast for metrology, he set out to analyse and refine the extensive thermophysical measurements on gases published in 1847 by H. V. Regnault (1810–78), which were probably the most cited at that time. Between 1869 and 1871 Bosscha conducted a rather heated argument with this celebrated Frenchman on the thermal expansion of mercury and on the accuracy of his temperature measurements with mercury and gas thermometers.

In 1873 Bosscha was appointed Professor of Physics at the Polytechnic at Delft; five years later he became its Director. It was due to his efforts that the standard of education at Delft rose to that of a Technical University and its physics laboratory grew into the equivalent of a National Bureau of Standards. As the leading Dutch metrologist he became one of the major figures in the international activity of the 1880s aimed at establishing a worldwide system of weights and measures. At the request of the international community he had carried out at Delft comparisons of several standard metres that reached new levels of accuracy. Nevertheless when, in May 1875, the International Convention of the Metre was established after long scientific and diplomatic debates, the Netherlands was, at Bosscha's insistence, one of the countries that did not join [*18*].

In 1885 he accepted the Secretaryship of the Holland Society of Sciences at Haarlem, which had been founded in 1752, and which included among its members the most important figures in Dutch science and culture, as well as many from abroad. He held this post for a quarter of a century, so becoming a patriarch of Dutch science. As Secretary he managed the Society's journal *Archives néerlandaises des sciences exactes et naturelles*. This was written in French and did much to familiarize European scientists with Dutch research, including much of that of van der Waals which Bosscha once translated himself.

When he died in 1911, Kamerlingh Onnes, who had been his assistant at Delft from 1878–82, wrote 'He combined great courage and force of mind with an almost childlike simplicity and trustfulness, and an infectious optimism. We lose in him a noble, frank character and a friend to whom one never went without being warmed by his kind affection and stimulated by his faith that the future belongs to the great ideas' [*19*].

It is likely that, after he had settled at The Hague, van der Waals resumed attending lectures at Leiden University, to the extent that this was compatible with his duties. The train journey from the administrative capital to the scientific took only three-quarters of an hour. 'The train from The Hague to Leiden goes slowly, almost without jerks; there is no talk in the carriage, one falls asleep' was the experience of an Italian author in the mid 1870s [*20*, p. 213]. It seems, however, that there was one teacher from The Hague who accounted for every minute of his time and who could hardly afford to doze peacefully in the rumbling train.

Although the train trips to Leiden may remain conjectural, there is documentary evidence of his attendance at lectures there from 1870–73 in two records in the tercentenary list of Leiden students from 1575–1875 [*21*]. The first of these is dated 16 February 1871 and the second, which has the addition 'a. s. 3' or *anno studiorum 3* (that is, the third year of study) is dated 22 April 1873. It is not known what lectures he attended but it is safe to conjecture that they comprised the familiar trinity of mathematics, physics and mathematical astronomy. Physics was still in Rijke's hands, but there were changes in the chairs of astronomy and mathematics. After Kaiser's death in 1872 the astronomy chair was taken by van der Sande Bakhuyzen, van der Waals's predecessor at The Hague, and the mathematics chair passed after Verdam's death in 1866 to Bierens de Haan, who was assisted a year later by P. van Geer (1841–1919), another pupil of Verdam.

Van Geer was not an original mathematician of the first rank but made some contributions to the mathematical problems of social insurance and pensions. He was, however, a remarkable teacher who wrote several textbooks and took an active part in advising school teachers. His lectures, mainly on geometry and mechanics, were praised for their clarity and liveliness [*22*]. Lorentz, who heard his lectures on analytic geometry, recalled in 1925 that 'Van Geer gave us in his absorbing lectures a feel for the poetry of mathematics' [*23*, p. 329]. He was a permanent member of committees of the province of South Holland that examined HBS graduates and candidates for Aktes in mathematics and physics, and so would certainly have played a part in van der Waals's examination odyssey.

Van der Waals would undoubtedly have been in the company of Lorentz and J. H. van 't Hoff at the lectures given by Kaiser, Rijke, Bierens de Haan and van Geer. Lorentz studied at Leiden from October 1870 until February 1872 [*24*, p. 20], and van 't Hoff from October 1871 to June 1872 [*25*, p. 88]. He did not, however, make the acquaintance of his future colleagues; he was far from sociable, he came to Leiden only to attend the lectures, and he was nearly twice their age.

With whom then could van der Waals discuss the scientific problems in which he was interested, and so test his own ideas which he had probably started to formulate in the mid 1860s? Nowadays scientific life is almost unthinkable without a ceaseless stream of seminars, schools and symposia; it is hard to imagine how restricted scientific intercourse was in the past, and particularly in the Netherlands. 'There are no contacts between the physicists!' Ehrenfest had exclaimed in astonishment when he arrived in Leiden in 1912 as Lorentz's successor; he had come from St Petersburg, itself hardly a centre of lively scientific discussion [*14*, p. 97]. It was Ehrenfest who had just introduced to Dutch physicists the tradition of regular physics seminars open to informal discussion. 'Until now', he wrote to A. F. Joffe, his friend in St Petersburg, 'scientific intercourse of such a kind was absolutely unknown at Leiden' [*14*, p. 94]. The Dutch chemists were no better off; thus Arrhenius, working in van 't Hoff's laboratory in Amsterdam in 1888, complained that his colleagues saw each other only on duty and had no wish to worry about each other's work [*26*, p. 42]. And if such aloofness was the rule between colleagues of equal standing, what contacts between the students and their professors could be expected? The etiquette at Leiden reduced these to official 'tea-visits' to the professor's house, once or twice a term. Custom required that the students dressed formally, in tail-coat, with a pair of gloves, a top-hat and a walking-stick. A humorous account of his evening tea-visit to Rijke was given by van 't Hoff who had to borrow his wardrobe piece by piece from his fellow students [*25*, p. 32].

There was only one man to whom van der Waals could turn to discuss his scientific problems, and that was Bosscha, the Inspector of Secondary Education, a most knowledgeable physicist who was full of benevolence and took a lively interest in his colleagues and their scientific activities. The two men were linked by their common interest in physics, their residence in the same town, their common service to the system of secondary education, and by their membership of Diligentia. Bosscha's eldest son was a HBS student under van der Waals[13], who taught him physics from his father's textbook. In the foreword to the first volume of this work, on mechanics and heat, Bosscha had 'honoured an agreeable debt' by thanking 'Dr van der Waals, who had helped him with collecting together the vast amount of material needed for the treatment of the teaching of molecular forces and vibrations' [*27*].

The great event of his Hague period, and indeed of van der Waals's whole life, was the presentation of his famous thesis on 14 June 1873 at Leiden. It is certain that Bosscha played at least a catalytic role in this event but firm evidence is scanty. We have the witness of Kamerlingh Onnes that 'Bosscha, a well-known Dutch physicist, . . . advised van der Waals to take his doctorate and to write a dissertation. Van der Waals passed the necessary examinations and chose heat as the subject of his studies' [233].

In spite of his vast collection of Aktes, van der Waals could not be admitted to the academic examinations since he was not drilled in the ancient languages. We may note in passing, however, that Bosscha, an expert in these matters, in his report to the International Conference on Education in London in 1884, remarked that it was not really difficult to meet this requirement; 'a cursory study of a few months was sufficient to acquire so many Latin and Greek words as were necessary' [*28*]. Fortunately there was a turn for the better on 4 January 1871 when Thorbecke's party took office for the third time. Faithfully attached to the principle of laissez-faire, the Liberals at once amended the legislation governing higher education to lower a little the barrier of the classical languages.

Van der Waals followed Bosscha's advice and in March 1871, with Thorbecke's dispensation from Latin and Greek, he passed brilliantly (*magna cum laude*) the examinations in mathematics and physics [231, p. 151]. (These were, presumably, the *candidaats* examination.) In December of the same year his seventeen-year long course of examinations came at last to an end; he was through his doctor's examinations [231, p. 151]. It is of interest that he continued nevertheless to attend lectures in Leiden in 1872 and again in April 1873, only six weeks before he defended his thesis, when he signed up for a further year's study. Clearly he attended lectures to widen his education as a physicist, not to pass examinations.

The dissertation was now the next item on the agenda. Kamerlingh Onnes was not quite correct in suggesting that van der Waals first passed his examinations and then chose heat as the subject of his thesis. It will be argued below that his scientific interests had taken shape in the 1860s, and from then on he had kept his eye on the relevant papers, had performed some calculations, and collected material. Some of this he placed at Bosscha's disposal, but when?

There is little doubt that the thesis was written after the doctor's examination, and completed by the end of 1872. Two facts justify this date. First, he refers in his last chapter to a paper [*29*] 'in the last issue of the *Annales de Chimie et de Physique*' [1, p. 120; $1e_2$, p. 232]; this was the issue of October 1872[14]. Secondly there is a reference to 'the striking experiments lately made by Bosscha' [1, p. 37; $1e_2$, p. 158]. These were experiments that suggested that tobacco smoke and other aerosols possess a surface tension[15], and were reported by Bosscha at the monthly meeting of the Academy on 25 November 1871 [*30*]. He had not made a report at the previous meeting and on 30 September he had spoken of his work on capillarity but without saying anything about aerosols [*31*]. The 'lately' of van der Waals is therefore October–December 1871. He did not indicate the source of his information but it is almost certain that it came directly from Bosscha himself. The results so impressed van der Waals that he carried out similar experiments with tobacco smoke and apparently confirmed Bosscha's conclusions [1, p. 38; $1e_2$, p. 159]. One can correctly deduce from this episode that van der Waals was a smoker.

Thorbecke died on 4 July 1872, and van 't Hoff wrote at that time that after his death the Netherlands 'looks like a bare tree' [*25*, p. 30]. The Liberals could only just retain office and a new cabinet was formed by Gerrit de Vries (1818–1900), a Leiden doctor-of-law since 1839 and the father of Hugo de Vries. Several further alterations were made in the law of secondary education [*32*, p. 84] which may have eased the formalities surrounding van der Waals's submission of his thesis at Leiden. Whether or not that was the case, Saturday 14 June 1873 goes down in the history of physics as the date of the appearance of van der Waals's doctoral thesis *Over de continuiteit van den gas- en vloeistoftoestand* [1].

Notes

1. The post of Director of a HBS was given as a rule to one of the teachers in the subjects of the 'real' cycle.
2. A number of details relating to van der Waals's stay in Deventer were communicated to the authors by Mr J. Tabak, Archivist of the Municipal Archives at Deventer.
3. Girls appeared for the first time as students at the HBS at Deventer only in 1908 [*4*, p. 12]. The secondary education law of 1863 applied only to boys.
4. This information comes from Mrs M. van Brink, the Russian language teacher at the school, and from her pupils.
5. This material [234] was made available to the authors by Dr H. Bordewijk, Director of the Municipal Archives at The Hague, who also communicated much other interesting information on van der Waals's time at The Hague.
6. Cited in [205, p. 110].
7. Son of the respected Utrecht professor of physiology and anatomy, J. L. C. Schroeder van der Kolk (1797–1862), and father of the once well-known mineralogist, J. L. C. Schroeder van der Kolk (1865–1906).
8. Explanation given by Dr H. Bordewijk [239].
9. This remark is evidently based on the recollection of J. D. van der Waals, Jr.
10. The state universities were under the Ministry of the Interior. In his three governments (1849–53, 1862–66, and 1871–72) Thorbecke held both the posts of Prime Minister and Minister of the Interior.
11. Here and below the authors have used information from Mr J.van Meurs of Universiteits Bibliotheek, Amsterdam, to whom they are indebted for bibliographic investigations as to *Natuurkundige Voordrachten* and especially for the dates of van der Waals's lectures.
12. It is worth noting that two of the *Stellingen* or 'Propositions' in van der Waals's thesis (V and X, see Chapter 3) also concerned problems of the Sun. Fifteen years later he turned again to this subject when he considered the possibility of the separation into layers of the matter on the surface of the Sun under influence of gravitation (see [31, the end of Section 15]).
13. Communication from Dr H. Bordewijk [234].
14. It can be deduced from the lists of the library acquisitions of the Academy that were enclosed with its *Verslagen en Mededelingen*, that the monthly issues of the *Annales de Chimie et de Physique* were received in Holland within a month of their publication.
15. As one of the authors was informed by Professor A. A. Kirsch of the Editorial Board of the *Journal of Aerosol Science*, the question of the surface tension of uncharged aerosols is still not fully understood.

References

1. Baedeker, K., *Belgium and Holland*, 6th edn, London and Leipzig, 1881.
2. Plantenga, P. P. *Les Pays-Bas, Guide des Voyageurs*, Zutphen, 1863.
3. van der Slee, J. C., *De Illustre School te Deventer*, 's Gravenhage, 1916.
4. Bottema, O., De Hoogere Burgerschool te Deventer, 1864–1919. In *Gedenkboek te Deventer Hoogere Burgerschool, 1864–1939*, Deventer, 1939, b. 5–27.
5. Schrek, D. J. E., David Bierens de Haan, *Scripta Mathematica*, 1955, v. 21, p. 31–41.
6. Berkenvelder-Helfreich, B., *Deventer in oude ansichten*, Zaltbommel, 1968.

7. Anon, *Elementary and middle-class instruction in the Netherlands*, Leiden, 1876.

8. Went, F. A. F. C., Hendricus Gerardus van de Sande Bakhuyzen, *Versl.*, 1923, d. 32, b. 4–7.

9. Snelders H. A. M., Hendrik Willem Schroeder van der Kolk (1836–1867) en de physische chemie, *Scientiarum Historia*, 1971, Jg. 13, b. 185–197.

10. Schwencke, J., *'s Gravenhage in oude ansichten*, Zaltbommel, 1967.

11. Rykachev, V. (В. Рыкачев), *Three months in Flushing*, Morskoi sbornik, 1856, v. 24, p. 3, No. 10, p. 102–131; No. 11, p. 402–423 (in Russian).

12. Loos, D. de, *Organisation de l'enseignement secondaire dans le royaume des Pays-Bas*, Leeuwarden, 1894.

13. Schierbeck, A., Grepen uit de geschiedenis van de natuurkundige maatschappij Diligentia, 1793-1943. In *Jaarboek: Die Haghe voor 1943*.

14. Ehrenfest, P. and Joffe, A. F. (П. Эренфест, А. Иоффе), *Scientific correspondence*, ed. V.Ya. Frenkel, Leningrad, 1973 (in Russian).

15. Lorentz, H. A. Johannes Bosscha (1831-1911). In *Collected papers*, v. 9, Den Haag, 1939, p. 391-403.

16. Bosscha, J., *Leerboek der natuurkunde en van hare voornaamste toepassingen*, Leiden, 1865–1873, b. 1–4, st. 1–7.

17. Boutan, A. and d'Almeida, J.Ch., *Cours élémentaire de physique*, Paris, 1862.

18. Anon, J. Bosscha, *Rev. Gén. Sci. Pures Appl.*, 1911, t. 22, p. 433–434.

19. Kamerlingh Onnes, H., Prof. Johannes Bosscha, *Nature*, 1911, v. 86, p. 419–420.

20. Amicis, E. de, *La Hollande*, Paris, 1878.

21. *Album studiosorum Academiae Lugdono Batavae: 1575–1875*, Hagae, 1875.

22. Kluyver, J. C., Dr. Pieter van Geer. In *Almanak van het Leidsch Studenten Corps.*, Leiden, 1920, p. 287–290.

23. Lorentz, H. A., Rede bij de aanvaarding van het doctoraat in de geneeskunde honoris causa. In *Collected papers*, v. 9, Den Haag, 1939, p. 326–334.

24. Haas-Lorentz, G. L. de, Reminiscences. In *H. A. Lorentz, Impressions of his life and work*, ed. G. L. de Haas-Lorentz, Amsterdam, 1957, p. 15–47, 160–164.

25. Cohen, E., *Jacob Henricus van 't Hoff: Sein Leben und Wirken*, Leipzig, 1912.

26. *Aus dem Wissenschaftlichen Briefwechsel Wilhelm Ostwalds*, Bd 2, Berlin, 1969.

27. Ref. *16*, 2nd edn, 1875, b. viii.

28. Bosscha, J., On secondary education in the Netherlands. In *The Health Exhibition Literature*, London, 1884, v. 16, Conference on Education, Section D, Training of teachers, organisation of intermediate and higher education, p. 197–201.

29. Saint-Loup, L., Sur l'expression de la force élastique d'une vapeur saturée en fonction de la temperature, *Ann. Chim. Phys.*, 1872, t. 27, p. 211–227.

30. Bosscha, J., Over drupvorming, *P. V.*, 1871, N5, b. 2–4.

31. Bosscha, J., Eene eevoudige proef, niet zonder belang voor de theorie der capillaire werkingen, *P. V.*, 1871, N3, b. 5–7.

32. Roell, J., Hr. Gerrit de Vries Azn., *Jb. KAW*, 1900, b. 41–107.

CHAPTER 3

'An exceedingly ingenious thesis'[1]

Van der Waals entitled his thesis *On the continuity of the gaseous and liquid state*. Later, when the thesis had become famous and was used by many physicists and chemists this title was sometimes abbreviated simply to *Continuity*, and similarly in other languages – *de Continuiteit, die Kontinuität, la Continuité, Nepreryvnost*, etc.

As Hertz [*1*] said of Maxwell's electrodynamics, we can answer the question of what is van der Waals's theory, as set out in his thesis, by saying that his theory is his equation of state

$$\left(P + \frac{a}{V^2}\right)(V - b) = RT, \tag{3.1}$$

where P is the pressure, V the molar volume, R the gas constant, T the temperature, and a and b are the parameters that, respectively, account for the effects of molecular attraction and size. This differed from the long-established equation for an ideal gas,

$$PV = RT, \tag{3.2}$$

by the two correction terms. The first, a/V^2, is the so-called internal pressure, and is to be added to the actual pressure P, which is what one can measure directly. This correction term arises from the intermolecular attractions between the molecules as they move about continuously. The second term b reflects the fact that all of the space occupied by the molecules is not available to accommodate these motions, and so the effective volume is less than the measured volume V.

Van der Waals's two corrections have a wonderful effect; the simple equation (3.2), which describes only the limiting case of an ideal gas (a theoretical rather than a real substance) metamorphoses into an equation that covers the fluid states of matter from a dilute gas to a dense liquid, and accounts also for the existence of a critical point at which the distinction between these phases vanishes. How did he obtain this remarkable result?

3.1 The reality of molecules

In his Nobel Lecture in the autumn of 1910, van der Waals recalled that 'The first incentive to this my life's work came to me when, after my studies at university, I learned of a treatise by Clausius (1857) on the nature of the motion which we call heat' [103e, p. 254]. This treatise is, of course, Clausius's famous paper *Ueber die Art der Bewegung, welche wir Wärme nennen* [*2*].

Van der Waals's pupil and close associate Ph.A. Kohnstamm maintained, we believe correctly, that van der Waals's first period of university work ended in August 1865 with his physics examinations [206, p. 10]. Kohnstamm is on less certain ground in deducing from the passage above that 'We can be sure that by 1865 van der Waals had been captivated by the conclusions of Clausius and was familiar with the new ideas of molecular theory' [206, p. 10]. It is significant that all the references in *Continuity* to the work of Clausius are to the second volume of his *Abhandlungen über die mechanischen Wärmetheorie* published in 1867. Kohnstamm himself says that the original paper of 1857 was not known in the Netherlands until Bosscha discussed it in his lectures at the Delft Polytechnicum in 1873 [206, p. 8]. Bosscha was certainly familiar with the concepts of molecular kinetic theory much earlier than this. Thus in a paper of 1869 he wrote, 'According to the physicists' modern view, the temperature is nothing other than the energy of their [i.e. the molecules'] movement' [*3*, p. 175]. This 'modern view' was, however, omitted from the thermophysics chapters of his textbook of 1871. Clausius was, of course, well known and honoured in the Netherlands by then, but for his work in the 1850s on the two fundamental laws of thermodynamics [*4*], not for his work on kinetic theory. He was one of nine scientists who competed for a prize of fl. 1000 offered in 1852 by the Hollandsche Maatschappij der Wetenschappen to mark their centenary. The prize was for the most important work in the natural sciences in the four years after the celebration. It went in the end to two botanists, Alphonse de Candolle and Oswald Heer, and Clausius had to be content with election as a foreign member of the HMW in 1857. In 1870, however, they awarded him the first Huygens Gold Medal, given every four years 'for outstanding achievements in the exact sciences in the preceding twenty years' [*5*, p. 273–84].

Brush [*6*, p. 198–210] has observed that the Netherlands was rather slow to assimilate the new work on kinetic theory, notwithstanding the fact that the most perceptive and useful comment on Clausius's paper of 1857 had come from C.H.D. Buys Ballot, a professor at Utrecht. He raised the powerful objection that the rate of mixing of gases by diffusion – familiar in everyday life as the time taken for a scent to cross a room – was much slower than was consistent with the high speeds and rectilinear paths required by the new kinetic theory [*7*]. He himself had pondered since 1844, when he took his doctorate in capillarity, on a rather speculative 'unified molecular theory of matter', not dissimilar from the ideas of Rudjer Bošković (Roger Boscovich) in 1758. Buys Ballot took his atoms to be point centres of force and it was, apparently, just this difficulty over the rate of diffusion that led him to endow his 'mathematical' atoms with oscillatory rather than rectilinear motion [*8*].

It took Clausius only about two months to rebut this rather naïve objection from Utrecht, but it is curious that it is a point that had never previously occurred to him or to the other pioneers of kinetic theory. In his next paper [*9*] he had to abandon his original postulate that gas molecules are of infinitesimal size, and to assume instead 'that they have a large enough diameter, or "sphere of action" so that a molecule cannot move very far without hitting another one' [*10*, p. 53]. In this way the fundamental concept of the mean free path of gas molecules was introduced into physics. Buys Ballot later became a leading figure in meteorology, but he never again took part in the discussions on kinetic theory.

Van der Waals's recollections in his 1910 Nobel Lecture continued follows, 'Clausius' treatise was a revelation for me although it occurred to me at the same time that if a gas in the extremely dilute state, where the volume is so large that the molecules can be regarded as points, consists of small moving particles, this is obviously still so when the volume is reduced; indeed, such must still be the case down to the maximum compression and also in liquids, which can only be regarded as compressed gases at low temperature. Thus I conceived the idea that there is no essential difference between the gaseous and the liquid state of matter – that the factors which, apart from the motion of the molecules, act to determine the pressure must be regarded as quantitatively different when the density changes and perhaps also when the temperature changes, but that they must be the very factors which exercise their influence throughout. And so the idea of continuity occurred to me' [103e, p. 254–255].

Klein has suggested that the 'revelation' that van der Waals found in Clausius's paper of 1857 was his remark 'that the intermolecular attractions keeping the liquid together were the same mutual attractions that come into play in a gas only when the molecules are close enough to collide with one another' [176, p. 36]. It seems more likely that his passionate expression reflected, rather, his fascination of forty years ago for the vigour, elegance and comprehensiveness of the kinetic theory. In the introduction to this paper Clausius gave a qualitative description of the molecular constitution of all three states of matter, and this no doubt played a powerful rôle in starting van der Waals on the train of thought that he recalled so many years later in Stockholm. It seems that, contrary to the concept of continuity, Clausius sought rather to show how naturally the ideas of kinetic theory led to the differentiation of the three states, and that his discussion of molecular attraction was designed to show how important this was to the two condensed states and, by implication, that it was of little or no relevance to the gas. It is probable that van der Waals's developing ideas on continuity were stimulated but not induced by Clausius's classic paper.

His ideas on the need for a correction for the intrinsic size of the molecules were also stimulated, but probably not directly inspired, by the ideas of others. 'Van der Waals *believes* in extended molecules' wrote Kamerlingh Onnes[2] in 1898 [233]. This belief and a firm conviction of the reality of atoms and molecules were also repeated in the Nobel Lecture. 'It will be perfectly clear that in all my studies I was quite convinced of the real existence of molecules, that I never regarded them as a figment of my imagination, nor even as mere centres of force effects. I considered them to be the actual bodies, thus what we term "body" in daily speech ought better to be called "pseudo body". It is an aggregate of bodies and empty space. We do not know the nature of a molecule consisting of a single chemical atom. It would be premature to seek to answer this question but to admit this ignorance in no way impairs the belief in its real existence. When I began my studies I had the feeling that I was almost alone in holding that view. And when, as occurred already in my 1873 treatise, I determined their number in one gram-mol, their size and the nature of their action, I was strengthened in my opinion, yet still there often arose within me the question whether in the final analysis a molecule is a figment of the imagination and the entire molecular theory too' [103e, p. 264–265]. He had, apparently, little difficulty in quelling this questioning. In several places the thesis corroborates these recollections: 'Most probably the molecule must be considered as belonging to the solid state of matter' [p. 160][3]; '. . . we ought rather to view molecules as small extended bodies' [p. 168]; 'The ordinary conception[4] of ideal gases, if realized, would give us bodies deprived of the single essential property of matter – extension in space' [p. 186].

His recollection, in 1910, of the unpopularity of the idea of the reality of molecules in the late 1860s may seem to be exaggerated, but was probably correct as far as the Netherlands were concerned. The old physicist's views could well, however, have been coloured by his more recent memories of the strong wave of antagonism to the idea of the reality of atoms that occurred around the turn of the century. At the time of the writing of his thesis there was, in fact, a broad spectrum of views from a fervent belief in atoms to an unconditional and occasionally aggressive rejection. One can hardly doubt the firm belief in atoms and molecules of the founders of kinetic theory, Clausius, Maxwell and Boltzmann, whatever their epistemological views of physical reality, although it was Maxwell himself who remarked slyly, in a lecture to the British Association in 1873, 'an atom, if there is such a thing . . .' [*11*, p. 437]. In the 1860s and 1870s these three physicists and a few of their followers turned readily to the model of molecules as hard spherical bodies of microscopic dimensions. Few except van der Waals actually believed that this model was the real picture of actual gases, liquids and solids. For him, these little billiard balls were as real as everyday bodies. It was this belief that led him to exclaim that 'I had the feeling that I was almost alone in holding this view'.

Garber has written that the main motive of Clausius's research was 'to explore the nature of heat', while Maxwell used the kinetic theory in order to explore the structure of matter [*12*, p. 303]. There can be no doubt that Boltzmann followed Clausius's example, greatly enriching the theory with statistical methods as he did so. Van der Waals's motive was somewhat different again, namely to use kinetic theory and thermodynamics to arrive at an adequate description of the different states of aggregation of matter. Naturally Clausius, Maxwell and Boltzmann did not take the hard-sphere model to be a true representation of the real structure of molecules. All had tried hard to enquire into the nature of this structure, and Maxwell in particular was made uncomfortable by his inability to link thermal properties and optical spectra [*13*]. Van der Waals also knew better than to imagine molecules as having no more structure than that of undeformable hard balls. He wrote in his thesis: 'According to the most modern view of chemistry, the position which an atom occupies in a molecule is of the most far-reaching importance as causing the phenomenon of isomerism; and must in so far be determinate. . . . Inasmuch as a molecule is built up of several atoms, we cannot regard it as rigid' [p. 160,164]. He did not choose to plunge into the arguments of the three leading proponents of kinetic theory on the internal structure of molecules, but sought rather to marshall the arguments that would justify him in treating them as real small hard spheres. All 'ordinary' bodies are composed of molecules but that does not prevent us from treating them as single entities in accord with the laws of classical mechanics. In the light of his profession of faith, it is not surprising that he went on, in his Nobel Lecture, to claim that his life's work had been one of the instruments in converting physicists to a belief in the reality of molecules; '. . . my theory may have been a contributory factor. And precisely this, I feel, is a step forward' [103e, p. 265]. This view was at once endorsed by Ostwald, himself one of the leading opponents of the belief in the reality of atoms: 'He is perfectly right to ascribe this result to some extent (indeed, to a great extent) to his works' [103a].

3.2 Continuity, identity and capillarity

'The choice of the subject of van der Waals's thesis', wrote Kamerlingh Onnes, 'was determined by the fact that there were, in the science of heat, all kinds of numbers that were not related to each other (e.g. coefficients of thermal expansion and of compressibility) and, *a priori*, it seemed to van der Waals that all bodies probably follow the same law. So the chapter on heat seemed to him to be far less advanced than other branches of physics' [233]. It is reasonable to assume that Kamerlingh Onnes was describing the general motives of van der Waals's work since the opening sentence of the thesis states unambiguously: 'The choice of the subject which furnished the material for the present treatise arose out of a desire to understand a magnitude which plays a special part in the theory of capillarity as developed by Laplace' [p. 125]. This was the quantity K of Laplace's well-known expression for the 'internal' pressure inside a drop of liquid of radius R,

$$P_{\text{int}} = K + H/R \qquad (3.3)$$

The quantity H, which is equal to 2γ, where γ is the surface tension, is easily measured by any of the standard techniques of capillarity, but K is not directly related to any observable property. Van der Waals says of K that it is 'the molecular pressure' and the it is 'a measure of the cohesion of matter', and so it must be properly known if we are to understand fluids fully. It is not smaller than the experimentally observable terms, 'on the contrary, it is a million times as great', but since it never appears in the final equations for any observable property it must be found theoretically. Such considerations 'led me to establish the continuity between the gaseous and liquid condition, the existence of which, as I afterwards learned had already been suspected by others' [1, p. v–vi].

This version of how the idea of 'continuity, came to him is somewhat at variance with the account in his Nobel Lecture. The latter must seem to be the less trustworthy, but it is of interest that the word *continuiteit* in the sentence quoted above became the more colourless *Zusammenhang* in the German translation of 1881 [$1d_1'$] and *connexion* in the English translation of 1890 [$1e_1'$], both of which were prepared under the author's supervision. He then went on to clarify the point in two new paragraphs written for the translations.

'The expression "continuity of the gaseous and liquid state" is, perhaps, the most suitable, because the considerations are on the idea that we can proceed continuously from one state of aggregation to the other; geometrically expressed, both portions of the isotherm belong to *one* curve, even in the case in which these portions are connected by a part which cannot be realized.

'I have, strictly speaking, desired to prove more; that is, the *identity* of the two states of aggregation. For if the supposition, which is partly established, that in the liquid state the molecules do not merge into each other to form greater atomic complexes – if this supposition should be fully confirmed – then there would only be a difference of greater or smaller density in the two states, and thus only a quantitative difference' [p. 125]. These stronger statements from the Preface to the translations can be matched by a quotation from the body of the text of the original thesis; '. . . it is our object [in this thesis] to establish that between the liquid and gaseous states there is nothing but a quantitative distinction' [p. 159].

Thus the sequence of his ideas on continuity seems to have been, first, some vague views on 'identity' inspired by kinetic theory, which inevitably led, in their turn, to the idea of continuity, and finally to the search for a concrete connection which was achieved through his equation, which was established in the course of his attempts to understand Laplace's 'molecular pressure' K. He was too familiar with Laplace's theory to be in any doubt that it was the mutual attraction of the molecules that underlay all the phenomena of capillarity. He even noted that molecular attraction, although very weak at macroscopic distances, might affect such mobile bodies as the particles of tobacco smoke or other aerosols which, following the results of Bosscha (and himself), he believed exhibit capillarity [p. 159].

It is easy to see in the light of this commitment to attractive forces why the thesis opens with a strong rebuttal of the 'generally abandoned doctrine' of a static molecular equilibrium, and a brisk dismissal of 'the assumption of a repulsive force between molecules' [p. 127]. In a well-known paper of 1867, Maxwell had, purely for reasons of mathematical convenience, replaced his previous preference for hard spherical molecules by centres of force repelling as r^{-5}[*14,15,16*, p. 26–29]. Such repulsive forces were not acceptable to van der Waals. They are, he showed, incompatible with the experiments of Joule and Thomson in 1854 who found that the adiabatic expansion of a flowing gas led to its cooling, as is to be expected if the forces are attractive, and not to a warming, as would be found if the forces were repulsive. Much later in the thesis he comments that 'In all gases, except hydrogen, cooling was observed. It is surprising that this fact has not long ago drawn attention to the difficulty of explaining the observed phenomena, even in permanent gases, unless molecular attraction is supposed to exist' [p. 221]. Even Maxwell, a close friend of Thomson, took but little notice of it in his *Theory of Heat* in 1871, describing it merely as 'a slight effect' [*17*, p. 195]. It was, however, probably through Maxwell's book that van der Waals first learnt of this effect; Clausius seems to have ignored it. Its implications appear to have been noticed first by van der Waals and, simultaneously, by Meyer [*18*, p. 30], who incorporated it into the conceptual framework of kinetic theory. Later authors have, however, usually referred to Meyer's treatise [*18*] on kinetic theory of 1877 in which there is no mention of van der Waals's contribution. Perhaps little notice was taken generally of van der Waals's use of the Joule–Thomson effect of 1854 (on the expansion of a continuously flowing gas) because of its smallness, and perhaps because everyone was more impressed by the null effect of Joule's more famous experiment of 1845 (on expansion in a single-stage process). It was this powerful negative result that had led, in van der Waals's words, to the rout of 'the prophets of the Caloric school' [p. 225][5].

Thus the need to take account of the attractive forces was clear and, as a convinced believer in the reality of molecules, he could not ignore their intrinsic volume. As a keen student of Clausius's and (later) of Maxwell's papers, he knew that 'researches on the diffusion and viscosity of gases have shown that we ought rather to view molecules as small extended bodies' [p. 168]. Since the notion of repulsive forces was quite abhorrent to him, he had no choice but to regard the molecules as hard bodies with a defined internal volume.

3.3 Van der Waals's equation: derivation or justification?

3.3.1 The 'fundamental equation of the isotherms'

In October 1870 Clausius published his virial theorem, 'a very valuable contribution to molecular science' as Gibbs called it in 1889 [*19*, p. 265]. The theorem stated that the mean *vis viva* of a system is equal to its virial. In modern notation

$$\tfrac{1}{2}\langle m_i v_i^2\rangle = -\tfrac{1}{2}\langle \boldsymbol{f}(\boldsymbol{r}_i)\cdot \boldsymbol{r}_i\rangle \tag{3.4}$$

where m_i and v_i are the mass and speed of the ith molecule, and $\boldsymbol{f}(\boldsymbol{r}_i)$ is the force it experiences when at position $\boldsymbol{r}_i$. The theorem applies to stationary motions of a system, that is, motions in which both the positions and the speeds are bounded. The average is over the time of a complete cycle if the motion is periodic, or over a sufficiently long time for the averages to become steady if the motion is irregular. Clausius, who liked to coin resonant names, called the quantity on the right of eqn (3.4) the virial, from the Latin *vis*, a force.

Clausius took the simplest case in which the virial comprises only the terms arising first from a uniform external pressure at the boundary of the system which contributes $\frac{3}{2}PV$ and, secondly, a central force between each pair of 'particles a body consists of', which is a function of their separation $r_{ij} = |\mathbf{r}_i - \mathbf{r}_j|$. Attractive forces are reckoned to be positive, and repulsive forces negative. Clausius replaces the averaging over time by an average over the many different configurations found in a molecular system, a change made more formally later by Gibbs when he introduced the motion of an ensemble. Clausius continues, 'Denoting, further, by h the *vis viva* of the internal motions (which we call heat), we can form the following equation

$$h = \frac{1}{2} \sum_{i,j} r_{ij} f(r_{ij}) + \frac{3}{2}PV.\text{' [20]} \tag{3.5}$$

He arrived at this virial theorem in the course of his attempts at deducing the second law of thermodynamics from purely mechanical principles. In this he failed, and he never used the theorem in any of his later work [176].

Van der Waals, who no doubt kept a careful eye on each new paper by Clausius, saw at once that the theorem could be used to handle the attractive forces. In Chapter 2 of his thesis he repeats Clausius's derivation, replacing h on the left-hand side of eqn (3.5) by the sum of the kinetic energies,

$$\sum_i \frac{1}{2} m_i v_i^2 = \frac{1}{2} \sum_{i,j} r_{ij} f(r_{ij}) + \frac{3}{2}PV. \tag{3.6}$$

In a rarified gas the first term on the right may be neglected, and in a liquid the second is negligible compared with the first. For the gas

$$\sum_i \frac{1}{2} m_i v_i^2 = \frac{3}{2} P_1 V \tag{3.7}$$

and for the liquid

$$\sum_i \frac{1}{2} m_i v_i^2 = \frac{3}{2} P_2 V \tag{3.8}$$

where P_1 is the external pressure and P_2 is the molecular pressure that arises from the attractive force and which is balanced by 'the continual tendency to separation' resulting from the thermal motion. The last equation is arrived at by qualitative arguments that the treatment of liquids can be reduced to that of gases because the number of forces acting on any one molecule is large and so they are roughly in balance, an argument similar to that used by Laplace [183, p. 9,12]

Van der Waals uses two premises in his derivation, the first of which is that 'We may consider it as proved that molecular forces act at very small distances only. . . . Researches on the distance at which molecular forces become insensible have not so far yielded concordant results; they agree, however, in showing that this distance is very small' [p. 134]. On this point van der Waals and Bosscha were in full agreement; it is, indeed, a necessary consequence of the fact that the forces were negligible in gases. Van der Waals developed Laplace's notion that the attractions were 'insensible at sensible distances' [*21*] into the more precise, 'the forces themselves becoming insensible at distances greater than the radius of the sphere' (i.e. Clausius's sphere of action') [p. 135].

His second premise, 'fully proved' by experiment, is that a liquid at equilibrium has a uniform density at all points, except possibly in 'a very thin layer at the surface' whose thinness is a direct consequence of the short range of the molecular attractions. It is this uniformity of density that leads to there being no resultant force on a molecule in the bulk of a liquid, and so the uncompensated inward attraction on the outermost layer has the effect of an internal pressure, the P_2 of eqn (3.8), which can therefore be treated in the same way as, and added to, the external pressure P_1 of eqn (3.7). He summarizes his argument: 'We thus find a great similarity between the relations we have discovered for the particles of a liquid and for the particles of a gas. On the particles of a gas no forces act; on the particles within a liquid the forces neutralize each other. In both cases the motion will go on undisturbed so long as no collisions occur' [p. 135]. For intermediate states of density, in fact 'the only state which occurs in nature' [p.137], one can add the two terms

$$\sum_i \tfrac{1}{2} m_i v_i^2 = \tfrac{3}{2}(P_1 + P_2)V. \tag{3.9}$$

Since the left-hand side is a constant at a given temperature, he calls this equation the 'fundamental equation of the isothermals' [p. 138].

He notes that since in a liquid P_2 is much greater than P_1, we should not be surprised at the exceedingly small compressibility of liquids, but he is very cautious about identifying this pressure with Laplace's K, a step that he does not take until Chapter 9. He does, however, argue that it is likely that P_2 increases as the square of the density, rather than as the density itself. Such a variation would be consistent with the observed fact that for (most) gases the product of pressure and volume falls below that of an ideal gas as the pressure increases. He makes a brief reference to the fact that such a dependence on density would be consistent with the law of mass action, a law formulated in 1864 by Guldberg and Waage but not widely known until later [p. 137].

The 'fundamental equation of the isothermals' appears first at the end of Chapter 2; the next two chapters are variations on the theme of Laplace's theory of capillarity. He adopts provisionally Laplace's view of matter as a static continuum, and goes beyond his celebrated predecessor in one respect only. As van der Waals stressed many years later in his university lectures, Laplace (and his contemporaries) believed that liquids were incompressible, and so saw no need to introduce the densities into his calculations; these constants were simply irrelevant [91, p. 239]. Van der Waals restores the density to its proper place and so obtains, for Laplace's two integrals,

$$K = 2\pi D^2 \int_0^\infty \psi(x)\mathrm{d}x \tag{3.10}$$

and

$$H = 2\pi D^2 \int_0^\infty x\psi(x)\mathrm{d}x, \tag{3.11}$$

where D is the density that was omitted from Laplace's equation and $\psi(x)$ is the function whose derivative with respect to the separation x is the interparticle potential of the molecular attraction.

He opens Chapter 5 with the following statement: 'Hitherto we have treated molecules as points of mass, and have thus been led to a simplification of our problem which, however, does not in any way agree with the phenomena exhibited by matter. We must now, therefore, proceed to apply corrections to our theory in two different directions' [p. 160]. The first of these is to take account of the internal structure of the molecules, 'since these must be considered not as mere points of mass, but as aggregates built up of atoms' [p. 160]. The second correction, the subject of Chapter 6, is that arising from the fact that the molecules have size, or at 'definite extension in space'.

The kinetic energy on the left-hand side of eqn (3.4) will contain terms arising from translational motion of the molecule as a whole (its 'velocity of progressive motion', as Clausius called it), terms from the rotational motion of the molecule as a whole, and from the relative internal velocities of the atoms with respect to one another. On the right-hand side of the equation must be added the virial of the internal forces between these atoms. On the basis of this equation, and from Clausius's analysis of the ratio of the heat capacities of a gas at constant pressure and constant volume, van der Waals argues that we may use the virial equation for complex mclecules, provided that we ignore all kinetic energies on the left except for that of the 'progressive motion', and, on the right, ignore the virial of the internal forces. It seems, however, that he was far from being firmly convinced of the correctness of his reasoning in this delicate matter.

He turns next to the problem of extension and, although he might now be thought to be well equipped for the challenge of treating real molecules, he makes instead a sharp change of tactics. 'I have thought it well to adopt a method different from that hitherto pursued', he writes early in Chapter 6.

3.3.2 The volume correction

It would seem today to be quite natural to take into account the 'extension of the molecule' by introducing into the 'fundamental equation of the isothermals' the short-range repulsive forces that increase steeply as the molecules come close to each other. We have seen, however, that van der Waals would not accept the notion of repulsive forces and so he temporarily abandons eqn (3.9) for a simpler semi-quantitative argument.

Clearly the larger the molecules, the shorter the mean free path between collisions, other things being equal, and the shorter the mean free path the greater the pressure since we know that this is proportional to the number of collisions of the molecule with the walls. Clausius and Maxwell had calculated mean free paths but without allowance for the shortening due to the size of the molecules. It was, he said, as if they had given their molecules breadth but no thickness in the direction of their motion. They had over-estimated the length of the path, he said, just as we should over-estimate the distance we can throw a ball against a wall if we took this to be the distance of the centre of the ball from the wall, and not that distance less the radius of the ball. Clausius's expression for the mean free path

$$l_1 = V/\tfrac{4}{3}\pi s^2 N, \tag{3.12}$$

where V is the volume that contains N molecules, and s is their diameter or, more correctly, the diameter of their 'spheres of action', or, since thickness is neglected, 'discs of action' is, van der Waals suggests, the more appropriate expression. When the thickness is taken into account then we obtain instead, for direct collisions along the line of centres, a mean free path of $l_2 = l_1 - s$, or

$$l_2 = \frac{V - \tfrac{4}{3}\pi s^3 N}{\tfrac{4}{3}\pi s^2 N} \tag{3.13}$$

so that

$$\frac{l_2}{l_1} = \frac{V - 8b_1}{V} \tag{3.14}$$

where b_1 is the volume of the molecules. Direct collision along the line of centres is, however, only one form of impact, and averaging over all angles of collision (over all 'impact parameters' in modern parlance) reduces the factor of 8 by a half, so giving the final expression for the mean free path of

$$\frac{l_3}{l_1} = \frac{V - 4b_1}{V}. \tag{3.15}$$

He recognizes that this expression for the reduction of the length of the mean free path cannot 'be pushed to the extreme limit of condensation of matter' [p. 171]. It clearly breaks down if $V < 4b_1$, which is quite possible on physical grounds, but must become inaccurate long before this limit is reached; the multiplier of b_1 must be less than 4 at high densities. Moreover, the factor is correct at low densities only for spherical molecules since for other shapes the proportion of direct or central impacts is smaller than for spheres.

He is thus led to his correction for the extension of the molecules, which he writes b which will, 'up to a certain limit [of density] represent four times the volume of the molecules, but beyond this limit a smaller multiple' [p. 171]. The 'fundamental equation of the isothermals', eqn (3.9), therefore becomes

$$\sum \tfrac{1}{2} m_i v_i^2 = \tfrac{3}{2}(P_1 + P_2)(V - b). \tag{3.16}$$

3.3.3 *The equation of state*

Van der Waals now proceeds, in Chapter 7, to obtain an expression for P_2, the 'internal' or 'molecular pressure'. He does not, however, return to his earlier expression for Laplace's K, eqn (3.10). He had improved the original form by including the factor D^2, but he regarded the expression as still resting on a static and continuum picture of matter, and so he did not trust Laplace's method of obtaining K by an integration over each of the molecular attractions. In view of the short range of the attractive forces we should, for any one molecule, need to know 'the law of force to be able to estimate its attraction' [p. 173]. Forty years later he put the matter more clearly: 'Laplace regards his liquid actually as a continuum; at that time he was still unaware of molecules. And if we were dealing with stationary molecules it would have been inadmissible to reduce the attracting forces in the interior to a surface force' [103e, p. 256].

He assumed that it was the chaotic thermal motion of the molecules that saved him from this impasse: 'But if this molecule is in motion, and can occupy any part of the space indifferently, the above difficulty for the most part disappears.... Hence we see that we arrive at the same result by considering the molecules moving as by considering the matter continuous, as is done in integration' [p. 173]. It is hard to know how far he was able to convince himself by means of this reasoning but, after some apparently hard deliberations, he settled on a simpler argument that has found its way into every student textbook for the last hundred years. The number of molecules in a layer close to the surface is proportional to the density, and the attraction for them by the underlying fluid is also proportional to its density, so the internal pressure is proportional to the square of the density, or $P_2 = a/V^2$, where V is the volume per unit amount of matter and a, the constant of proportionality, he called the 'specific attraction'. Thus eqn (3.16) becomes

$$\left(P + \frac{a}{V^2}\right)(V - b) = \sum_i \tfrac{1}{3} m_i v_i^2 \tag{3.17}$$

And now, turning at last to Clausius's identification of the 'heat' in a body with the mean kinetic energy of its molecules, he comes, at the end of Chapter 7, to his famous equation of state:

$$\left(P + \frac{a}{V_2}\right)(V - b) = RT. \tag{3.18}$$

He realizes the audacity of equating the mean kinetic energy to $\tfrac{3}{2}RT$, a step which is tantamount to assuming that the energy of 'progressive motion' is independent of density and that, 'for instance, a molecule of water and a molecule of steam at 0°C have the same velocity of progressive motion' [p. 174]. He supports this assertion very briefly by the cunning assumption that the mean value of $\frac{1}{3}mv^2$ may be taken to be the definition of temperature.

3.3.4 Derivation or justification?

As Klein observes, 'The first five chapters of van der Waals's thesis have built up the expectation that he will now calculate the molecular pressure from the average virial of the molecular forces, taking into account both the attractions . . . and the repulsions that operate only during collisions, that is, the effect of the non-zero molecular volumes. . . . But we do not find the expected derivation from the virial theorem' [176, p. 39]. One might guess that van der Waals preferred an easier qualitative approach to a plunge into the complicated mathematics of the virial route which may not have been accessible to his readers. This is, however, probably not the reason, for there is considerable evidence that he had already obtained his equation before he knew of Clausius's remarkable theorem. This appeared in a paper in the issue of the *Annalen der Physik* that was 'closed' on 10 October 1870, and which was received by the Library of the Academy in January 1871, as is shown by the monthly lists of acquisitions. The fifth supplement (*Ergänzungsband*) of the *Annalen* came with it, and this contained the shortened version of Andrews's paper which proved to be so crucial to van der Waals. When he first mentions this paper in his thesis he writes: 'More recently Andrews published investigations on carbon dioxide – more suited to test our equation than any others' [p. 191]. This sentence followed the paragraph in which he tested his equation against the earlier results of Regnault (see below). It seems clear that he already had his equation when he first learnt of Andrews's and also of Clausius's results. Other hints scattered through the thesis support this reading of the order of events[6].

It may be that he even had both correction terms for some time before he combined them in his equation. He writes, '*At first* I considered that the difference between the external volume and the volume taken up by the molecules was the space within which the motion takes place' [p. 168] (our italics). Similarly in his Nobel Lecture he relates his astonishment at finding that the excluded volume was four times greater than this [103e, p. 255]. Kohnstamm once asked him about the far from rigorous route he had used in his thesis to calculate b, and received the reply: 'Oh, you see, when I was writing this I was well aware that it was not strict. But I knew the result was right and had no wish to search after a long roundabout way to get it' [206, p. 17].

We have already seen that the simple ideas that led him to the term a/V^2 included the law of mass action and the qualitative observation that the pressure of real gases was below that of ideal, that is, 'in imperfect gases the product of the pressure by the volume diminishes with increase of pressure' [p. 138]. Even his 'virial derivation' of the 'fundamental equation of the isothermals' appeals more to its agreement with experiment than to its logical rigour.

These arguments all support the view that the 'derivation' in the first seven chapters of the thesis is, rather, a justification of an equation that he had already arrived at on the basis of more elementary reasoning and some experimental information. Nevertheless his use of Clausius's virial theorem for the analysis of problems in kinetic theory was later acknowledged as a great contribution to molecular science.

3.3.5 Van der Waals's equation and Regnault's experimental results

To test his equation van der Waals used the vast set of experimental results on the deviation of gases from ideality published in the late 1840s and in the 1860s by the French physicist H.V. Regnault [*22*]. These results were intended mainly for calculating the corrections to gas-thermometers, and discussion of them was one of the most popular themes of the second half of the last century. Bosscha, for example, engaged in a heated argument with Regnault on questions of thermometry [*3,23*].

Van der Waals used three properties of gases that Regnault had measured:

(1) the so-called pressure coefficient, $T(\Delta P/\Delta T)_V/P$;

(2) the 'compressibility', or $(P_1 V_1/P_0 V_0)_T$, where V_1 is about half of V_0;

(3) the coefficient of expansion, or $T(\Delta V/\Delta T)_P/V$.

He deduced expressions for these three coefficients from his equation of state and compared the first two with Regnault's results for carbon dioxide, hydrogen, air and sulphur dioxide, so obtaining values of the constants a and b for each gas. He then tested the equation of state by comparing these values of b with those obtained from the third property, the coefficient of expansion.

The concordance found between the values of his constants, whether computed from measurements of the same type or from different types, is far from convincing. On the other hand, he knew that the accuracy of Regnault's results was itself not beyond reproach. Moreover there was no case where there was a clear discrepancy between the experimental evidence and the predictions of his equation. His confidence was strengthened by the fact that he had been able to show, at least qualitatively, how many kinds of deviations from ideal behaviour, previously quite unexplained, could be rationalized.

3.3.6 The equation of state and Andrews's isotherms

In his discussion of Regnault's results, van der Waals omitted any detailed analysis of carbon dioxide. It is virtually certain that this was because, while he was working on his thesis, he came to know, through its abstract in German, Andrews's paper, *On the continuity of the gaseous and liquid states of matter*, which is entirely devoted to carbon dioxide [*24*]. He did not realize then, however, that it had been published two years earlier, in 1869 [*25*].

Andrews presented, in graphs and tables, an extensive set of isotherms of carbon dioxide in the temperature range 13–48 °C, and the pressure range 1–110 atm. Van der Waals fitted these to his equation using the value of the two constants that he had already obtained from Regnault's results. 'There was some doubt,' he says, 'about using the constants for carbon dioxide determined from Regnault's investigations, in which the pressure was 27 atm. at most. The result [of the comparison with Andrews's results] was, however, so satisfactory that it seemed better to work with these numbers than to begin by seeking for the most probable values of a and b from Andrews's experiments themselves' [p. 191]. This last phrase is one more argument to justify the supposition made above that he had already obtained his equation before he knew of Andrews's work.

This satisfactory result to which he refers is the close agreement between the values of b already computed and those obtained directly from Andrews's results. Thus from Regnault he obtained a value of 51 $cm^3\,mol^{-1}$ (in modern units), and from the measurements of Andrews between 13.1 and 35.5 °C and for molar volumes from 110 to 356 $cm^3\,mol^{-1}$ he obtained values of b from 51.9 to 56.3 $cm^3\,mol^{-1}$.

Some random variations in the calculated values of b are to be expected but he found however, that 'at widely differing temperatures a smaller value of b always corresponds to a smaller volume, so that any possible influence of temperature disappears as far as b is concerned' [p. 193]. He describes this as 'a most important result', and observes that such a natural regularity was to be expected from his theory. We know, however, that at first he thought that b was nothing other than the sum of the molecular volumes, and on this supposition there is no reason to expect any change. It seems that it was the observed change that made him doubt his original intention and think more seriously about the volume correction, for there is otherwise no reason to worry about the factor relating b to the molecular volumes; he knew well enough that no analysis of the isotherms would tell him anything of this factor.

Andrews had, for the first time, unambiguously demonstrated the continuous nature of the transition between gas and liquid. His graph of the isotherms was reproduced by Maxwell in his *Theory of Heat* [*17*, p. 120] who turned it on its side and added a dotted line to mark the saturated liquid and gaseous states. It was Maxwell's diagram that van der Waals reproduced as Fig. 8 of his thesis, and which is shown here, Fig. 3.1.

In his analysis of these results van der Waals noted with satisfaction that the values of b from the liquid state at 21.5 °C agreed (within about 15 per cent in fact) with those for the gaseous state at 32.5 °C, and the same range of volumes. This agreement clearly established that his equation was valid for both states of matter or, as he put it in the Preface, 'both portions of the isotherm belong to *one* curve' [p. 125].

3.3.7 The critical parameters

As a teacher in the HBS van der Waals had, no doubt, had to explain to his pupils that a cubic equation has either one or three real roots, and he cannot

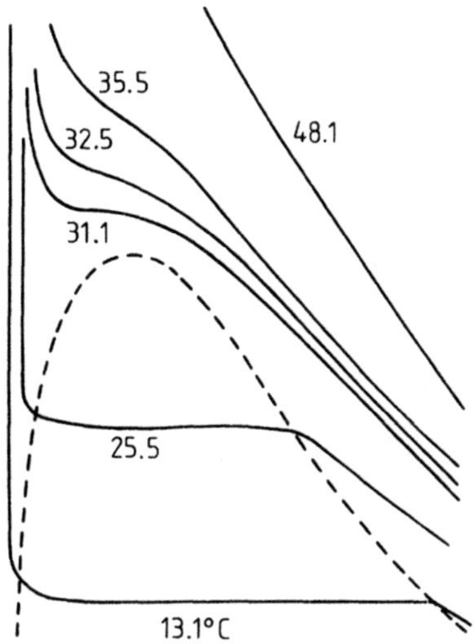

Fig. 3.1 The isotherms of carbon dioxide, from Andrews's measurements, as shown in Maxwell's *Theory of Heat* (1871), and reproduced as Fig. 8 of van der Waals's thesis.

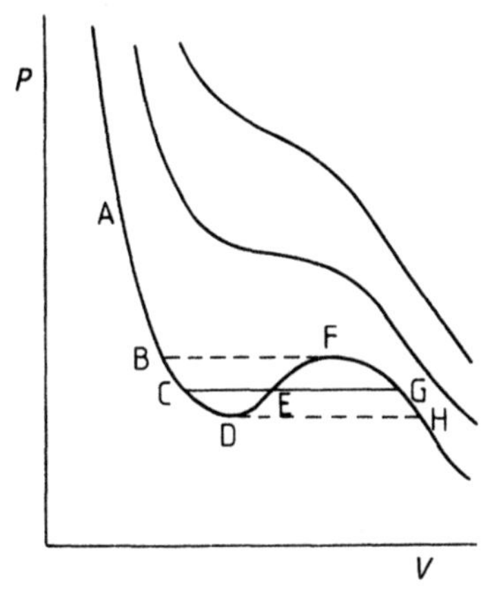

Fig. 3.2 Sketch of continuous isotherms in Maxwell's *Theory of Heat*, reproduced as Fig. 7 of van der Waals's thesis.

have failed to notice that his equation of state is a cubic in the volume V. He could also have found multiple solutions for V when comparing his equation with experiment.

On a graph of P as a function of V a straight line parallel to the V-axis cuts an isotherm in one or three points, as is shown in Figs 3.2 and 3.3. The first of these is Fig. 7 of the thesis and is taken from Maxwell's *Theory of Heat* of 1871 [*17*, p. 120]. It is purely qualitative. The second is Fig. 6 of the thesis and is van der Waals's own accurate representation of his equation.

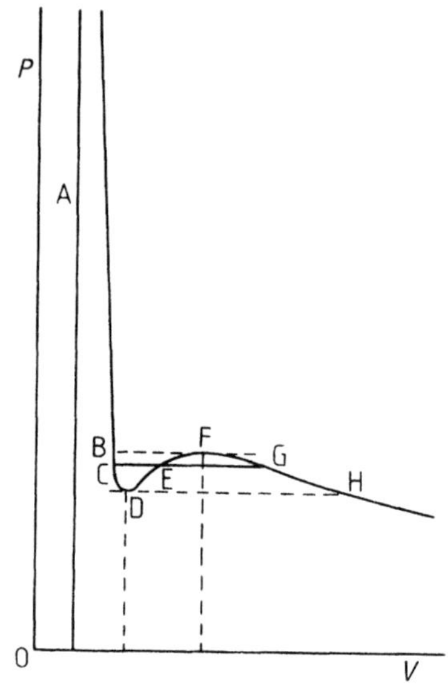

Fig. 3.3 An isotherm from van der Waals's equation of state, Fig. 6 of his thesis

Maxwell's textbook was one of a series 'adapted for the use of artisans and of students in public and science schools'. It must have appeared in Britain towards the end of 1871 since it contains a reference to a paper [*26*] of James Thomson (the elder brother of William) that was not published until November (although Maxwell may have known of it earlier), and it was available on the Continent soon afterwards since there is an anonymous review of it in the French weekly *Les Mondes* of 11 January 1872. It is almost certain that van der Waals had been working on Andrews's results since the middle of 1871, but these (Fig. 3.1) naturally did not show the S-shaped isotherm, BDFH of Fig. 3.2 – that was an idea for which he was almost certainly indebted to Maxwell's book, which led him to appreciate the unexpected bonus of the cubic curve that had been hiding in his equation. The idea was, in fact, not Maxwell's but James Thomson's [*26*] who had arrived at it by contemplating Andrews's results. Van der Waals's source for this idea may have accounted for his warm comment on Maxwell's book, describing it as 'the little book which is surely to be found in the hands of every physicist'. He followed Thomson and Maxwell in interpreting the outermost roots, e.g. C and G in Fig. 3.2, as the volumes of the liquid and gas respectively, and in dismissing the third root E as unreliable because $(\partial P/\partial V)_T$ is there positive, so that the compressibility would be negative and the system mechanically unstable.

As the temperature rises, the length of the horizontal line CG that corresponds to Andrews's 'mixture of gas and liquid' [*25*, p. 581] shrinks, becoming zero (Fig. 3.1) at a temperature that Andrews called the critical temperature. From his equation of state van der Waals calculated its value to be

$$RT^c = 8a/27b \tag{3.19}$$

and for the critical volume and pressure,

$$V^c = 3b, \quad P^c = a/27b^2. \tag{3.20}$$

His definition of V^c does not have the symmetry between gas and liquid that one might have expected: 'What I call the critical volume is the volume approached by the liquid under the pressure of its saturated vapour in the neighbourhood of the critical temperature. The critical pressure is the pressure at this limit' [p. 201].

One cannot but be impressed by the remarkable harmony between the physical meaning attached to T^c by van der Waals and the experimental result of Andrews that below this temperature 'a substance can exist in the so-called gaseous as well as the so-called liquid state, according to the external pressure. At certain pressures the substance may assume either state; but above the temperature in question it can exist only in one, whatever be the pressure. When the temperature is above the critical, it is useless to try to liquefy the substance' [p. 200]. He then acknowledges Andrews's contribution with the words: 'The honour of this remarkable discovery, which alters our views as to the so-called permanent gases, and the liquefaction of gases generally, belongs to Andrews' [p. 200–201], but he adds in a footnote that the same idea seems to have occurred also to Faraday. The agreement between the observed critical temperature, 30.9 °C, and the value of a and b that van der Waals deduced from Regnault's results is remarkable, indeed much closer than one has any right to expect for such an equation.

The only results available to van der Waals for determining the critical constants of other gases were those of Cagniard de la Tour in 1822 and 1823. Their accuracy left much to be desired but he noted that the results for ether and for carbon disulphide did, at least, not contradict his prediction for the value of the critical ratio,

$$P^cV^c/RT^c = 3/8. \tag{3.21}$$

In passing, he estimates the critical temperature of water to be 412 °C, a figure probably based on Cagniard de la Tour's results. (The modern value is 374 °C.)

Cailletet had published in 1870 results for the change with pressure of the product PV for air and for hydrogen at 15 °C. From these, van der Waals estimated the critical temperature of air to be −158 °C (modern value −141 °C) and of hydrogen, for which he found a to be too small to calculate, a critical temperature of −273 °C (modern value −240°C). These figures mark the first exploration of the domain of the very low temperatures needed for the liquefaction of the so-called permanent gases. His thesis is the conceptual starting point for the development of low-temperature physics which has become such an important branch of the study of condensed matter, and which later became Leiden's most celebrated contribution to physics. Van der Waals notes the change of perspective his equation has brought about; 'So the customary assertion that a gas becomes liquefied by pressure, and that cooling (which really diminishes pressure) promotes this liquefaction, must be reversed. A gas may become liquid by cooling without any remarkable pressure being required' [p. 201].

Summing up his analysis of the critical parameters he writes with obvious pride: 'I consider that I have thus explained the hitherto inexplicable fact – at least if it can be called an explanation of a phenomenon, when it is shown to be a consequence of a universal physical law. By the three given equations we can express very simply the critical temperature, critical volume and critical pressure, when the constants a and b for the substance are known' [p. 201].

He left open the difficult question of the loop BCDEFGH. This, as we have seen, was first sketched free-hand by James Thomson as a means of joining the two parts of Andrews's isotherm. Van der Waals learned of this 'happy idea' [p. 196] from Maxwell's book (he quotes Thomson's paper to the British Association in the autumn of 1871 but this cannot have been available to him until the summer of 1872). He accepts Thomson's and Maxwell's view that the outer portions of the loop, BD and HF, represent superheated liquid and supercooled vapour respectively, but has no sympathy with Maxwell's joining C to G by a straight line (which was simply what Andrews had found experimentally), saying that the idea 'was not a happy one, but calculated to throw one off the right track'[7] [p. 196].

To show that there is true and not merely theoretical continuity between the gaseous and the liquid states, he describes a practical way of continuously changing one to the other by means of a simple sequence of changes of temperature and pressure that takes the substance 'round' the critical point. He says that he has borrowed this idea from Maxwell, and overlooked the fact that Andrews had actually carried it out in the laboratory [*25*].

3.3.8 Molecular dimensions

With his equation of state, the relation $K = a/V^2$ and the computed values of the constant a, van der Waals seems to be well equipped to realize his original goal of deducing the molecular pressure from theoretical considerations. He is, however, apparently now less enthusiastic about what he has achieved in this respect. 'In a sense', writes Klein, 'his new theory made K a much less interesting quantity than he had expected because it was now possible to go further in determining magnitudes on a molecular scale' [176, p. 42]. So only the shortest chapter in the thesis, Chapter 9, is devoted to estimations of the values of K for different substances. With reservations about the poor accuracy of the experimental data (many from the old results of Cagniard de la Tour), he arrives, in Chapter 10, at the figures shown in the first column of Table 3.1 (x_1 and s are discussed below). These values of K are for the liquid at room temperature and pressure. He is thus able to show convincingly that K for a liquid is much larger than the external pressure of 1 atm.

From the early years of this century it has been possible to make an estimate

Table 3.1

	K(atm)	$x_1 \times 10^8$ (cm)	$s \times 10^8$(cm)
Ether	1300	2.9	4.0
Alcohol	2100	2.5	2.7
Carbon disulphide	2900	2.3	
Water	5000	3.1	
Mercury	22500	4.9	

of molecular size from the fact that $\frac{1}{4}b$ is the sum of the volumes of one mole of molecules, and so division by Avogadro's constant gives the volume of one molecule. This route was not open to van der Waals since there was no reliable value of this constant in 1871. He chose therefore different and less direct routes. He first took Laplace's two integrals, eqns (3.10) and (3.11), and evaluated each, K from the values in Table 3.1 and H, more easily, from the surface tensions. Their ratio is the mean value x_1 of the range over which $\psi(x)$, the integral of the intermolecular potential, is appreciable. (This was a calculation that, unknown to van der Waals, Young had made in 1816 [183, p. 9].) His values of this mean range of the forces are shown in the second column of Table 3.1. He then takes his corrected version of Clausius's expression for the mean-free path l to relate this quantity to the molecular diameter s,

$$l = (V - b)s/2b.$$

The only known value of l was that calculated by Maxwell [27] for air at 15 °C and 1 atm, which was 6.5×10^{-5} cm, which he 'rounds up' to 7×10^{-5} cm and so obtains $s = 2.8 \times 10^{-8}$ cm for air. From this he calculates the other values of s shown in the table, by noting that values of l should scale as $b^{-2/3}$.

He could at this point have used Avogadro's Law, which unlike the constant he should have known, in order to estimate s for each substance from its proportionality to b, but he missed the opportunity and took the more roundabout route via the mean-free path. He does, however, use his values of b to estimate what we now call Loschmidt's number, that is, the number of molecules in a cubic centimetre of gas at 0 °C and 1 atm pressure. His result, 5×10^{19}, is not far from the modern value of 2.7×10^{19}. His value is more accurate than Loschmidt's original estimate of 1865 (2.5×10^{18}) [28] or Stoney's estimate of 1868 (1×10^{21}) [29]. He notes that this last value provides some indirect support for his demonstration that b was four times the volume of the molecules, and not eight as Clausius, and Stoney after him, had supposed.

From this foray into molecular quantities he concludes with more confidence than ever that the range of the molecular attraction is about the same as the molecular diameter. He wrote; 'This affords us quite a new standpoint from which to consider the meaning of the encounters. Hitherto they have been regarded as only opposing a diminution of volume; now they appear as producing an opposite effect in that they cause the exertion of a sensible attraction' [p. 217–218]. Chapter 9 concludes with the bald statement: 'The absolute weight of a molecule of hydrogen is about 10^{-21} mg' [p. 219], a result obtained, presumably, from his estimate of Loschmidt's number and the density of the gas. (The modern value is 3.3×10^{-21} mg.)

The last two chapters are devoted, firstly, to a thermodynamic calculation of the Joule–Thomson coefficient and 'to the well-known problem of a gas expanding adiabatically' [p. 230] and, secondly, to a fruitless attempt to calculate the saturated vapour pressure at a given temperature. These chapters have not the same historical interest as those that have gone before, but it must be stressed that he was the first to calculate values of the Joule–Thomson coefficient, which plays a fundamental rôle in modern low-temperature technology. His discussion of the saturated vapour pressure is, however, without any concrete result since he lacked the essential condition that Maxwell, and later Clausius, discovered, namely the 'equal-areas' rule.

In a short peroration he modestly but firmly appraises his achievements, and concludes with a solemn speech by William Thomson to the British Association in Edinburgh in 1871 [*30*] in which he looks forward to a beautiful future for science when 'the motion of the planets and music of the spheres will be neglected for a while in admiration of the maze in which the tiny atoms run' [p. 239]. Van der Waals did not know that the words were not in fact Thomson's but were a quotation from an anonymous essay in the *North British Review*, whose author, it is now known, was Thomson's partner in many of his engineering enterprises, H. C. Fleeming Jenkin [183, p. 14].

It is not necessary to stress how much van der Waals owed to Andrews's experimental results, but it seems that he had less admiration for what he saw as the Irish scientist's refusal to be drawn into theoretical attempts to understand the observations, and for his generally rather old-fashioned views. The furthest Andrews went in his famous paper of 1869 was that there were 'some internal forces having an expansive or repulsive character' [*24*, p. 85]. Van der Waals could not accept that such a vague generality was all that could be obtained from the beautiful structure of the kinetic theory, and it may well have provoked the implied criticism of the opening sentence of his thesis: 'The doctrine according to which the molecules of a body in molecular equilibrium remain at rest, and according to which the invariability of the distance of the molecules from one another depends on a repulsive force, has generally been abandoned. Such a doctrine is in fact in direct opposition to certain consequences drawn from the principle of the conservation of energy, and is in consequence untenable' [p. 127]. Even his direct but unacknowledged borrowing[8] of the title of his thesis from that of Andrews's paper may have carried a critical sting, for there was a parallel instance a few years earlier which must have been known to van der Waals. In 1857 Clausius published his epoch-making paper *Ueber die Art der Bewegung, welche wir Wärme nennen*, and later the same year Buys Ballot raised the objection to it discussed above; his paper was entitled *Ueber die Art der Bewegung, welche wir Wärme und Elektricität nennen*.

Van der Waals was greatly excited by what he was achieving. His equation, found 'to some extent by inspiration', according to Boltzmann [*31*, p. 155], astonished him by the range of possibilities it disclosed: the calculation of the molecular pressure, explanations of the results of Regnault and of Andrews, a theory of critical phenomena, the calculation of the Joule–Thomson coefficient and, above all, estimates of molecular sizes from the deviation of gases from ideality. He had solved a whole range of problems that had been given up as hopeless by all other physicists. Many years later he was to say that he used to think at this time, 'Either I am mad, or I am doing a great thing' [206, p. 11].

In 1905 Boltzmann and Nabl were to describe him as 'the Newton of the theory of the deviations of gases from ideality' [*32*, p. 550]. But what kind of reception did his thesis receive in 1873?

3.4 The defence of the thesis

On Saturday 14 June 1873, at 3 p.m., van der Waals appeared before the Senate of Leiden University as an aspirant to the degree of Doctor of Mathematics and Physics. According to tradition, the defence of the thesis took place in the old Senate Hall whose walls were hung with the formal portraits of the professors who had taught there during the three centuries of the University's existence. The chair was taken by the Rector Magnificus who was that year M. de Vries, the founder of Dutch linguistics and the uncle of the botanist Hugo de Vries. Custom required that the candidate give his address in front of a long table, around which were seated sixteen of the professors from all disciplines – chemistry and botany, philosophy and history, philology and linguistics. Among them were naturally the physicist P.L. Rijke, the mathematicians D. Bierens de Haan and P. L. van Geer, and the astronomer H. G. van de Sande Bakhuyzen, and these four were the only ones who could have understood anything of what the candidate was saying. Of these, his *Promotor* Rijke should have been the best informed.

In a short obligatory introduction van der Waals made his acknowledgements; 'Now when I am invited to defend this thesis I am filled with several kinds of feelings. . . . It is gratitude that prevails, and, in particular, to you, most learned Rijke, the highly esteemed promotor, whose teaching aroused in my heart a love of physics. With sadness I recall those who, in the few years that have passed since I studied at the University, have been taken by death from the study of science, and to whom I owe so much' [1, p. vi]. Although he did not mention him by name it was Kaiser to whom he referred, the popular Leiden professor who had died the previous year. The thesis has no dedication, although one to the candidate's parents was the general rule. Perhaps this omission was because he received no help from them, unlike most students whose start in science was only possible because of many years' support by their families.

According to custom, which is still followed, the thesis ended with *stellingen* or propositions which were not directly related to the subject in hand but were chosen by the candidate to air his views on cognate matters which he thought to be important. There were nineteen of these and since they have never been reproduced since their appearance in the original Dutch version of 1873, it is not without interest to list some of them here.

1. There is no essential difference between gases, vapours and liquids. Not only are they but different forms of a substance in the same state of aggregation, but one cannot even draw the boundaries where one state stops and another starts.
5. The controversy as to whether the sun is a gas or a liquid is simply the to whether its substance is above or below its critical temperature.
7. Gases possess a so-called surface tension and so can exhibit capillary phenomena.
9. Thomson's definition of absolute temperature is surely the most rational, but the temperature so defined is nothing but that measured by an air thermometer.
10. [J. R.] Mayer's hypothesis [of 1868] that the persistence of the sun's heat is due to the impact of meteors is untenable.
12. Water is not an electrolyte.
15. When the forces causing excitations of any kind are increased in geometric progressions, the responses increase in arithmetic progression (Weber's Law[9]).
16. 'It is inconceivable that inanimate brute matter should, without the mediation of something else, which is not material, operate on, and affect other matter without mutual contact; as it must do, if gravitation, in the sense of Epicurus, be essential and inherent in it' (Newton[10]).
18. Poinsot exaggerates when he says 'La vraie méthode n'est que cet heureux mélange de l'analyse et de la synthèse, où le calcul n'est employé que comme un instrument; instrument précieux et nécessaire sans doute, parce qu'il assure et facilite notre marche, *mais n'a par lui-même aucune vertu propre*'[11].

The outcome of the defence was not seriously in doubt. Van der Waals deserved to be held in good esteem; he was learned in science, of high ability and with a reputation for scrupulous work in his teaching, and by now of mature years. Moreover there were still those in Leiden who remembered his triumphs in the Akte examinations. The fact that Rijke had agreed to be the promotor virtually guaranteed success, although it was unlikely that this rather old-fashioned experimentalist ever mastered the contents of the thesis. We can only guess that the defence was merely formal and that the Senate was not greatly impressed. No distinction of *cum laude* or *summa cum laude* was conferred, although such commendations were not infrequent. This was, moreover, a time when the requirements for a thesis were much lower than nowadays; it was quite sufficient to present a review of the papers on some scientific problem and one or two results of one's own, without making any real contribution to the subject. It is little wonder that with so difficult a topic the Leiden press took no interest in van der Waals's examination, although it generally covered University affairs at length [236].

'I still remember quite vividly', wrote Lorentz in 1910, 'the impression produced in 1873 by the thesis *On the continuity of the gaseous and liquid states.* Our elders seemed to be impressed most by the fact that some of the most difficult problems of theoretical physics were tackled and resolved by a man who had, for many years, been engaged in teaching and who was burdened with many duties. We, the younger generation, did not yet understand what difficulties such circumstances could produce for a research worker, but we were fully able to admire the audacity and the ingenuity of this attack on the fundamentals of matter and its forces. The novelty of the title itself was enough to amaze the neophyte; how could one discuss the continuity of states which, at first sight, seemed to be as completely different as gases and liquids' [195, p. 308]. This recollection of Lorentz is valuable since he was almost a witness; he completed his doctoral examinations [12] the very day of the defence [*33*]. He was, however, not present in the Senate Hall since no onlookers were allowed to attend. His comments relate not to the defence itself but to a slightly later period when van der Waals was beginning to acquire a reputation.

3.5 Stat sua cuique dies [13]

On 7 August the new doctor of mathematics and physics received his greatest gift when Anna Magdalena presented him with a son and heir, Johannes Diderik Jr. (it cannot have taken long to decide on the names) who was later to joke that the famous thesis was his twin sister [173, p. 263]. And, with these names, did the father dream that one day his son would succeed him as a professor of theoretical physics? The doctorate led to a definite advance in his career; from 1 January 1874 he became a deputy director of his HBS with a salary raised from fl. 2500 to fl. 2750 [234,238]. By then the first notice of his thesis had appeared.

At the very end of 1873 there was published a massive work in two volumes by J. A. F. Plateau, the blind Belgian experimental physicist, *Statique expérimentale et théorique des liquides soumis aux seules forces moléculaires.* This 'encyclopaedia of surface tension' contained an extensive bibliography of works on capillarity, and at the end of the second volume was a list of recent publications which included van der Waals's thesis as item N44 [*34*, v. 2 p. 473]. The same list included two of Bosscha's papers on capillarity published in the autumn of 1873. Plateau could hardly have included a thesis by an unknown author with a title that had no apparent relation to capillarity unless he had been prompted to do so. Almost certainly it was Bosscha who told him of his own two papers and of this new work by his colleague at The Hague; no one else was so *au fait* with van der Waals's affairs.

Plateau's work made an immediate impression in the world of physics and was well received. The best known review, entitled 'Plateau on soap-bubbles' was by Maxwell and appeared in *Nature* on 18 June 1874. In this fine essay he wrote, 'M. van der Waals, whose academic thesis *Over de Continuiteit van den Gas- en Vloeistoftoestand*, is a most valuable contribution to molecular physics, has attempted to calculate approximately the thickness of the stratum [i.e. the capillary layer] within which this continuous change of energy is accomplished, and finds it for water about 0.0000003 millimetre' (see Table 3.1, above). 'Whatever we may think of these calculations, it is at least manifest that the only path in which we may hope to arrive at a knowledge of the size of the molecules of ordinary matter is to be traced among those phenomena which come into prominence when the dimensions of bodies are greatly reduced, as in the superficial layer of a liquid' [*35*, p. 120].

How should we assess Maxwell's estimate of the thesis? There is both generous praise, 'a most valuable contribution to molecular physics', but also the rather ominous words, 'whatever we may think of these calculations'. He is clearly not enthusiastic about the calculations, but he has as yet only looked through the thesis. Moreover his assertion that it is only by study of small systems, like the surface layer, that we can determine molecular sizes, is just the opposite of van der Waals's view that these sizes can be estimated from such bulk properties as the deviations of gases from the ideal-gas laws.

Such a favourable reference by a prominent scientist in a widely read foreign weekly can hardly have passed unnoticed in the Netherlands, which had for a long time had little to boast of in science. It was, however, only a prelude; on 18 October 1874 Maxwell published, again in *Nature*, a substantial review of 'Van der Waals on the continuity of the gaseous and liquid states' [*36*]. At last the Dutchman comes on to the stage in his own right, and not merely as an 'extra' in a list of articles, or a person mentioned in passing in the review of another's book.

It is well known that Maxwell included a gift for languages among his many talents, but his biographer, who knew him well, wrote that, 'The only language he had any difficulty in mastering was Dutch' [*37*, p. 325]. Nevertheless he made a detailed study of the text, not missing for example the brief reference to the observations of Bosscha and van der Waals on capillary effects in tobacco smoke.

In its tone, Maxwell's review is extremely benevolent. 'M. van der Waals', he writes, 'in entering on this very difficult inquiry, has shewn his appreciation of its importance in the present state of science; many of his investigations are conducted in an extremely original and clear manner; and he is continually throwing out new and suggestive ideas; so that there can be no doubt that his name will soon be among the foremost in molecular science. . . . His treatment of "molecular pressure" . . . seems ingenious. . . . I would call attention to the able manner in which he expounds the theory of capillarity'. But for all that he is, at bottom, severely critical. He has several objections any one of which is serious enough to discredit van der Waals's deductions. Firstly, 'It appears to me that the equation [of state] does not agree with the theorem of Clausius on which it is founded. In that theorem p is the pressure of [on] the sides of the vessel, and V is the volume of the vessel. Neither of these quantities is subject to correction'. Secondly, the volume correction is not deduced from the virial theorem. Thirdly, the volume correction, when derived properly, has an entirely different form. Fourthly, the equating of the mean kinetic energy to $\frac{3}{2}RT$ has been proved to be correct only for a gas. 'M. van der Waals seems, therefore, to be somewhat too hasty' in extending this assumption to a liquid. The first two objections are criticisms of van der Waals's derivations and not of his results. The third and fourth criticisms are more serious but in these cases van der Waals was right and Maxwell wrong. In his review Maxwell gave, without derivation, his own calculation of the volume correction but it was, as has been observed, 'almost ludicrously wrong' [183, p. 15–16].

Maxwell continues: 'The results obtained by M. van der Waals, by a comparison of this equation with the determinations of Regnault and Andrews are very striking, and would almost persuade us that the equation represents the true state of the case. But though this agreement would be strong evidence in favour of the accuracy of an empirical formula devised to represent the experimental results, the equation of M. van der Waals, professing as it does to be derived from the dynamical theory, must be subjected to a much more severe criticism.' As we shall see below, Maxwell never changed his view that the equation was an empirical one but, in contrast with many others for whom his refutation meant that the equation was wrong, he himself accepted that it was successful and he used it repeatedly in unpublished calculations. He simply rejected van der Waals's derivation of it.

The only real theoretical contribution that Maxwell recognized was van der Waals's introduction into kinetic theory of the 'very remarkable theorem of Clausius. . . . In this country the importance of this theorem seems hardly to be appreciated. . . . I must not . . . forget to add that to him alone [i.e. van der Waals] are due the suggestions which led me to examine the theory of the virial more carefully.' Maxwell had read Clausius's paper in 1871 at the same time as van der Waals [*12*, p. 317] but he missed its importance. Perhaps this was because of his mildly ironical view of the 'Heat philosopher's' [*37*, p. 336] partiality for introducing new and resonant scientific terms.

Maxwell evidently considered the thesis to be a promising work by a newcomer to the field, which was evidence of his great scientific promise and ability, notwithstanding that, 'He does not, however, seem to be equally familiar, as yet, with all parts of the subject.' This review in *Nature* must have raised van der Waals's prestige among his compatriots; the criticisms would be seen as merely technical points, while the praise was clear and impressive. This was, however, not the end of the story.

On 18 February 1875, Maxwell gave a lecture to the Chemical Society in London on the principles of molecular dynamics. By means of purely qualitative arguments and a reference to the virial theorem, but without any mention of the Leiden thesis, he deduced the general course of van der Waals's isotherms, and then added to them his now well-known 'equal-areas' rule by means of which he could deduce the value of the saturated vapour pressure. Later, he does mention van der Waals: 'The molecular theory of the continuity of the liquid and gaseous states forms the subject of an exceedingly ingenious thesis by Mr. Johannes Diderick van der Waals, a graduate of Leyden. There are certain points in which I think he has fallen into mathematical errors, and his final result is certainly not a complete expression for the interaction of real molecules, but his attack on this difficult question is so able and so brave, that it cannot fail to give a notable impulse to molecular science.' And then Maxwell pronounced the famous phrase, known to any Dutch physicist, which has been quoted for over a century whenever van der Waals has been discussed. The 'attack' described in the thesis 'has certainly directed the attention of more than one inquirer to the study of the Low-Dutch[14] language in which it is written' [*38*, p. 359].

Maxwell again describes van der Waals as (merely) 'a graduate of Leyden', and his praise is accompanied by the deadly phrase, 'he has fallen into mathematical errors'. Such criticism cannot have impressed the Dutch so much as the fact that van der Waals's thesis had caused British scientists to learn their language.

This lecture was reported in *Nature* on 4 March 1875, and two months later a modest teacher at The Hague, with but one publication to his name, and that a mere dissertation, was elected an ordinary (*gewoon*) member of the Physico-Mathematical Division (PMD)[15] of the Royal Academy of Sciences. The sequence of events was as follows [235,240]. On 27 February at an extraordinary meeting of the PMD, a commission was set up to consider proposals for the election of new members of the Academy. It contained seven academicians, of whom only three were competent in physics, C. H. D. Buys Ballot, H. G. van de Sande Bakhuyzen, and the Delft chemist A. C. Oudemans. The commission met on 13 March and, at the extraordinary session of the PMD on 27 March, its chairman Buys Ballot proposed that six elections be made, two in the medical sciences, two in chemistry and one each in mathematics and physics. He announced a list of 21 candidates of whom, after a vote, there remained 12, two for each vacancy. After a long meeting of the PMD on 27 April the election took place, and the ballot for the vacancy in physics was won by van der Waals. On 4 May the new elections were confirmed by Royal decree. It is likely that Maxwell's compliments swayed the electors since van der Waals had had a serious rival. He was D.J. Steyn Parvé, inspector of secondary education in the province of South Holland (and so van der Waals's superior in the education service), a member of the HMW and of the *Diligentia*, a man known throughout the Netherlands as the architect of the HBS scheme, and Thorbecke's right-hand man in putting this scheme into practice. He had studied at the Deventer Athenaeum, graduated at Groningen in 1847, was for ten years a professor of mathematics in the Athanaeum at Maastricht, and the author of a textbook of physics for the HBS which ran to two editions. He had done no research in science but, according to the custom at that time in the Netherlands, had every chance of being elected to the Academy since cultural and social contributions were taken to be on a par with original research [235]:

'I should point out', wrote Brush, 'that van der Waals's own work was published in Dutch and would not have become so rapidly known in England and other countries except for the fact that the Scottish physicist James Clerk Maxwell published an enthusiastic review of it in *Nature* in 1874' [173, p. 263]. This statement needs qualification. We have seen that Maxwell's review was far from wholly enthusiastic, and Maxwell was not the only one to 'discover' van der Waals. He may have opened the eyes of van der Waals's countrymen, among whom only Bosscha perceived from the very beginning the importance of the thesis. In English-speaking countries Maxwell's compliments to 'a graduate of Leyden' impressed few, while his apparently authoritative criticisms persuaded others that the thesis was factually erroneous, and so not worth the effort of struggling with the Dutch. Thus in 1891 Tait wrote to Lord Rayleigh: 'Partly on account of its unfamiliar language, but more especially on account of a very definite unfavourable opinion expressed by Clerk-Maxwell, I did not attempt to read the pamphlet when it appeared; and it was not until 1888 that, in consequence of some hints from Dr H. Du Bois[16], I hastily perused it in its German form' [*39*, p. 546].

Maxwell himself had said that he was not the only British scientist to tackle the thesis ('has certainly directed the attention of more than one inquirer . . .'). Who were the others who had been forced to study the 'Low-Dutch language' in order to read this obscure Leiden pamphlet? It is not surprising that one was Andrews. On 5 February 1874, long before Maxwell's first public reference to van der Waals, he wrote to his former colleague James Thomson, now in Glasgow, 'Bessy [one of his three daughters] and I are hard at work on Diderik van der Waals' [*40*, p. lxxi]. This sentence would have made no sense if Thomson also did not already know who van der Waals was, and what he had done. In April he again wrote to Thomson to say, 'I am deep in the Dutch paper, and also fully occupied with my own results, which I believe will turn out more important than I supposed' [*40*, p. lxxi]. Andrews did not generally underestimate the importance of his own work; as early as 1865 he had written to his wife, 'I really begin to think that Dame Nature has at last been kind to me, and rewarded me with a discovery of a higher order than I ever expected to make' [*41*, p. xxxi]. Nevertheless in 1874 his opinion was enhanced, maybe as a result of the use made of his results in van der Waals's thesis.

At this time, the spring of 1874, Maxwell had not yet gone deeply into the thesis, as may be seen by the casual nature of the comment in his review of Plateau's book in June, and by his letter to Tait, apparently dated April[17]; 'Van der Waals makes the only effect of the extension of the molecules, or the diminution of the effective volume wherein the gas moves, by 4 times the volume of the molecules. I have not lynxed his proof enough' [*42*].

In his numerous papers of 1875–76 on the question of continuity, Andrews never mentions van der Waals's thesis, in spite of the effort he had put into studying it. By then Maxwell's review had appeared and Andrews, like Tait, may have been convinced, in view of Maxwell's great authority as a theoretical physicist, that the thesis was flawed. A second reason may have been that Andrews's later results seemed not to agree with van der Waals's equation.

3.6 The Royal Academy of Sciences in Amsterdam

For more than forty years after his election in May 1875, van der Waals was closely involved with the affairs of the Academy of Sciences. It is appropriate, therefore, to describe briefly this institution, the highest in the Dutch scientific hierarchy.

It arose from the Amsterdam Royal Institute of Sciences, Letters and Arts, established in 1808 by a decree of King Louis Napoleon, the next-to-youngest brother of Napoleon Bonaparte. The Netherlands was later than most European countries in establishing a national scientific body, but this highly decentralized country had not felt the lack of such an institution in pre-Napoleonic times. The decision to establish this body in Amsterdam arose from a disagreement of Louis Napoleon (Lodewijk, in Dutch) with the HMW, which was the most influential society and could have expected to become the national body. Lodewijk nominated himself *Président perpetuel* of the HMW and originally planned to transform it into a highly centralized body along the lines of the recently established Institut de France in Paris. The Board of Directors of the HMW, in spite of their dislike of this plan, were about to yield to all His Majesty's demands, but finally dug their heels in when he insisted that no new members should be elected to the Board. This body was composed of representatives of the aristocracy, wealthy merchants and politicians each of whom was elected for life so, under Lodewijk's plan, the body would have slowly disappeared. Faced with this intransigence, he abandoned the HMW and opted for a new institution in Amsterdam which he could mould to his design.

After the fall of Napoleon and the restoration of Dutch independence, this Royal Institute was preserved and functioned in a desultory way for several decades. In 1851 it was decided that it was too expensive and a curb on the freedom of scientific research, and it was abolished to be replaced by the Royal Academy of Sciences which was designed to assist the progress of Dutch physics and mathematics; Thorbecke was again the instigator of reform. Four years later a Division of Letters was added to cater for work in history and philosophy. The Academy now covered all branches of knowledge and culture, since medicine was included among the natural sciences, and economics, politics and the arts were covered by the new Division. It has preserved the essential features of this structure to the present day.

It is still housed in a building part of which had been given to the Royal Institute when the French ruled the country. This building is a magnificent mansion on the Kloveniersburgwal canal, built in 1662 by a famous architect, J. Vingboons, for the brothers Trip, who were rich Amsterdam merchants; it is now called the Trippenhuis. It was built for its original owners in two non-communicating halves, one of which was put at the disposal of the Academy in 1812, and the other of which accommodated the Rijksmuseum from 1816 until 1885, when it moved to its present building[18], and the Academy was left in sole possession of the Trippenhuis.

According to its statutes the Royal Academy advises the Government on matters of science, scholarship and research. It plays the leading rôle in national and international cooperation in these fields. The word 'Royal' in its title is not merely decorative since formally the Academy is accountable only to the monarch, to whom it presents its annual reports and whose consent is required for all important decisions and for the appointment of its officers. It comprises two Divisions, the PMD (or Section of Sciences) and the Division of Letters, each with a Chairman, Vice-Chairman and Secretary, the first two of whom are elected annually and the last nominally for five years although, since re-election is usually a formality, they are generally in office for much longer. It is the Secretary who is, in practice, in charge of the affairs of the Division. The Chairmen alternate each year as President of the Academy, but it is the PMD Secretary who is, as a rule, appointed to be the General (or Permanent) Secretary of the Academy. Each Division meets once a month, except during July and August, usually on the last Saturday of the month. The President, at the time of van der Waals's election, was F. C. Donders, an ophthalmologist and the Dutch scientist with the widest European reputation.

3.7 From The Hague to Amsterdam

On 29 May 1875, van der Waals and the other four newly elected members came for the first time to the Trippenhuis to be welcomed by the President. Among those who were present at that session of the PMD were his two Leiden professors, Rijke and Bierens de Haan, Bosscha and his teacher V. S. M. van der Willigen, and van der Waals's predecessor at The Hague, van de Sande Bakhuyzen who had also been his colleague in Kaiser's courses and who had been elected to the Academy two years earlier. There were also Buys Ballot from Utrecht, the Secretary of the HMW, the chemist E. H. von Baumhauer, J. M. van Bemmelen, one of the founders of colloid chemistry, G. van Diesen, a civil engineer, and the mathematical genius T. J. Stieltjes who after many unsuccessful attempts to get a position in a Dutch university (he had no university diploma) had to emigrate to France. Van der Waals may have already met some of the members from The Hague at meetings of the *Diligentia*.

This first session in the Trippenhuis was followed by others — a new phase of his life had begun. He was scrupulous in his attendance, never missing a session, but at first he contributed nothing. It is remarkable that, although now a member of the Academy and Deputy Director of the HBS, he did not lighten his teaching load, merely rescheduling his lessons so that he had Saturdays free to attend the sessions [234]; the journey there and back took him about three hours. He resumed his scientific work at the session of 18

December 1875, when he presented a paper for publication in the Academy's 'Reports and Communications' on the influence of the size of molecules on their collisions [4]. This paper and its French translation [4f] are both dated 17 December, and it is curious that he was in such a hurry to publish since two and a half years had passed since he had submitted his thesis. Why so long a wait? It is certain that his thesis was not written merely to advance his career and that he had then lost all interest in science. It may well be that he was disappointed by the indifference with which most of the scientific community had received what he had put so much effort into. Then had come the criticisms by Maxwell, a scientist he ranked so highly. Whatever other Dutch scientists may have thought, he at least would have known how to evaluate these and he would not have been mollified by Maxwell's praise of him as a promising novice. We remember his outburst, 'Either I am mad or I am doing a great thing' [206, p. 11]. Perhaps the reactions of his fellow physicists had made him wonder whether the question were more than rhetorical.

'Have you seen Lorentz's (of Leyden) *Over de Terugkaatsing &c van het Licht*[19]' wrote Maxwell to Tait on 14 February 1876 [*43*]. He was referring to Lorentz's thesis presented at Leiden on 11 December 1875. The published version of the thesis did not appear until 1877 so Maxwell had clearly received one of the original copies [*43*, p. 510], much in the same way that he and Andrews (and possibly James Thomson) had received van der Waals's thesis two years earlier. Rijke was the promotor of both theses, so could well have been the sender [231]. If so it would be surprising if a copy of *Continuity* had not been sent also to Clausius. This inference can be drawn from van der Waals's second paper [4] in which he writes, 'Quite recently Clausius has published a paper in which he, perhaps not knowing that it had already been done, investigates the influence of the thicknes of the molecules on the collision number' [4f, p. 211]. This had been done, however, only in his thesis, and so the word 'perhaps' may imply that the thesis had been sent to Clausius in Bonn, but that he had not read it[20]. Clausius first mentions van der Waals in September 1879, in a report [*44*] to the Annual Meeting of the German Naturalists and Physicians, when he quotes from the original Dutch text and not from the long German abstract of 1877 [$1c_2$]. Apparently the thesis had been gathering dust on his shelves for five years.

The paper of Clausius to which van der Waals referred was published in the *Ergänzungsband* of the *Annalen der Physik* of 1876, and so it is surprising that van der Waals's paper is dated December 1875. He may have read an earlier version of it, a report delivered on 9 November 1874 and of which an English translation appeared in 1875 [*45*]. Clausius had calculated the effective volume available for the thermal motion of the molecules and had found it to be the total volume less eight times the volume of the molecules. The aim of van der Waals's paper was to correct this error and re-derive his factor of four in a more conclusive way than he had done in his thesis.

Apart from the provocation of Clausius's paper, this re-derivation may have been another reason for van der Waals to have returned to one of the principal topics of his thesis. On 18 December, at the same meeting of the PMD at which he presented his paper, it was announced that a paper [*46*] had been received from D. J. Korteweg, a HBS teacher of mathematics at Breda, in which van der Waals's value for the volume correction was confirmed by an alternative and mathematically more rigorous derivation. Korteweg's paper was dated 11 November and van der Waals referred to it; 'Although the way I then presented the result is open to criticism, it is, nevertheless, perfectly accurate, as is shown by Mr Korteweg' [4f, p. 211].

At the next session, on 29 January 1876, van der Waals and Korteweg each presented a paper in which, in different and more rigorous ways, they justified anew the calculation of the volume correction presented in *Continuity*. We do not know which of these papers is the earlier since Korteweg's [*47*] has no date; van der Waals's carries the subscription 'The Hague, 29 December, 1875' [5]. After this début before the Academy, van der Waals embarks on regular scientific activity, in so far as this was allowed by his teaching duties. At first this was solely experimental. In the physics room of the HBS he studied the line spectrum of carbon [3], and investigated how the temperature of the well-known maximum density of water changes with pressure [7].

Korteweg will be mentioned repeatedly in what follows, so it is worth saying a few words about him. He was born in 1848 of an ancient Brabant family; he proved to be a gifted mathematician who passed, with some difficulty, the Akte examination for entry into the recently established Delft Polytechnicum. It soon became clear that he had no taste for engineering and after five years of study he emerged, not with the Diploma of Engineering, but only with the certificate needed to be a mathematics school teacher. His first employment was in the HBS at Tilburg, and then, from 1873, in the HBS at Breda. Meanwhile he acquired in Dutch scientific circles the reputation of a mathematical virtuoso, winning a number of competitions. In 1873 he had a dispute with a Leiden mathematician and, at the end of that year, won a further prize for a crushing refutation of a German treatise in which the author stated that he had at last put 'molecular mechanics' on a firm mathematical foundation. He had assumed that atoms were rotating discs and never mentioned kinetic theory. This treatise was mentioned by A.P.N. Franchimont in May 1874 in his inaugural lecture as Professor of Chemistry at Leiden.

It seems likely that some of those at Leiden had asked Korteweg, as an expert in mathematics, to clarify the correctness of van der Waals's results which had aroused such critical comment from Britain. It is noteworthy that Korteweg had never previously worked on the kinetic theory of gases; he makes no mention of Maxwell's review or Clausius's paper of 1875, and one is left with the impression that his only aim was to check the correctness of van der Waals's results. The date on the first of his papers, 11 November 1875, is consistent with this conjecture, since it is known that Andrews, who had by then already discarded van der Waals's theory, visited the Netherlands early in the autumn of that year [*41*, p. xxxii]. It was agreed then to carry out in

Leiden experiments similar to those already made in Belfast, but with nitrous oxide as the gas. Rijke charged his student W. J. Janssen with this task and sent him at once to Belfast with the aim of learning Andrews's technique and of bringing back to Leiden 'one of Dr. Andrews' pressure apparatus', the necessary piece of equipment [*41*, p. xxxi].

In April 1876 Andrews gave his second Bakerian Lecture [*48*] in which he summarized his new work on the isotherms of carbon dioxide which seemed not to agree with van der Waals's equation, although van der Waals himself later disputed this conclusion [6]. Andrews did not even mention van der Waals although one can easily see a rejection of his equation in the following sentences: 'It would not be difficult from the experimental results to obtain an empirical formula.... But I have refrained from doing so ...; and for scientific purposes, as I have had occasion in the course of this inquiry more than once to observe, the results of empirical formulae are apt to mislead the investigator and rarely aid research' [*48*, p. 434].

When Andrews presented his paper in August 1876 for publication in the *Philosophical Transactions of the Royal Society*, it was Maxwell who was asked to be the referee. He was upset by Andrews's criticisms of empirical formulae and he had no doubt that it was van der Waals's equation that was meant. He was not about to change his opinion of it, but he could not equate 'empirical' with 'useless' or, even worse, 'harmful'. He wrote: 'The dynamical reasoning in support of this formula is by no means sound, but it expresses with considerable felicity the march of the phenomena of the compression of carbonic acid' [*49*]. He meditated further on what is empirical and what is theoretical: 'The degree of accuracy of an empirical formula is a test of its merit, but not of whether it is empirical or not. Ohm's law of the resistance of conductors appears to be accurate to a very high degree.... But it is a purely empirical law.... Boyle's law of gases on the other hand was an empirical law till it was shown to be a result of a particular hypothesis about the molecules of gases.... If this hypothesis were correct Boyle's law would be strictly true, and it would not be an empirical formula, but a law of nature. But Boyle's law is not strictly true. What then is its proper designation?' [*49*].

By the autumn of 1876 Janssen had completed his measurements in Leiden of the isotherms of nitrous oxide. Andrews had continued to help him from Belfast with letters that 'contain valuable practical details (sometimes of the minutest character)' [*41*, p. xxxi]. In September Janssen was invited to Glasgow to present his results to the Annual Meeting of the British Association; Andrews was President that year. The young Dutchman had a good reception, he was elected a Corresponding Member and, on a resolution of the steering committee, his report, with only a few others, was published *in extenso* in the transactions. His results were similar to those of Andrews on carbon dioxide, and he followed his Irish mentor in making no mention of van der Waals.

Van der Waals's name was, however, mentioned by Andrews in his Presidential Address. As well as reviewing recent outstanding scientific work he touched on some issues in higher education which was always of great interest to him. He pointed out that the scientific standards of British universities were considerably lower than of those on the Continent. He illustrated his argument thus: 'Two proof essays recently written for the degree of Doctor of Science in Leyden, one by van der Waals, the other by Lorenz (*sic*), are works of unusual merit, and another pupil of Professor Rijke is now engaged in an elaborate experimental research as a qualification for the same degree' [*50*, p. 411]. The reference to Lorentz is surprising, for it is most unlikely that a pure experimentalist of Andrews's generation could have made much of Lorentz's mathematical physics, most of which was applied to a discussion of Maxwell's electrodynamical equations, then far from universally accepted. Maxwell himself gave a paper at the meeting in which he mentioned Lorentz's thesis [*51*], and it was presumably he who prompted Andrews's praise of it. And the inclusion of van der Waals's thesis among the 'proof essays' that deserved the award of a doctorate is no more than Maxwell's view in his article in *Nature*.

Janssen was examined by the Senate of Leiden University on 30 June 1877. He seems to have been received with more enthusiasm than was shown to van der Waals, and was at once appointed Rijke's assistant, a great step for any young Dutch physicist. There is something odd in all this. The defence took place in the same hall where van der Waals had defended his thesis four years earlier; Rijke was again the promotor; the Senate was essentially unchanged, Bierens de Haan, van de Sande Bakhuyzen and de Geer were all present, but van der Waals was not mentioned in Janssen's thesis – not in the preface, nor in the main text, nor even in the *stellingen* [*52*]. So there was no word of the only theory in terms of which one could discuss his results. It is true that the dissertation was purely experimental and at that time it was not thought obligatory to compare the results with the predictions of theoretical schemes, but it is still hard to explain the complete ignoring of van der Waals's work. It becomes even more puzzling when one realizes that by 1877 van der Waals was acquiring some standing in the Netherlands. He had been a member of the Academy for two years and was being quoted more frequently by other Dutch scientists. Moreover he received real international recognition in 1877. Janssen must have had a strong reason for not mentioning van der Waals, and there can be little doubt that this was Andrews's rejection of *Continuity*; Janssen acknowledges Andrews's help 'in word and deed' in the preface to his thesis [*52*, p. ii]. Rijke retired in 1882, to be succeeded by Kamerlingh Onnes, and two years later Janssen left Leiden to live in Switzerland[21]. (Some thirty years later, van Laar left the Netherlands for Switzerland after a period of strained relations with van der Waals [*53*].)

The international recognition of van der Waals in 1877 came when a new monthly series of abstracts, the *Beiblätter*, was added to the *Annalen der Physik*. The idea was that of J. C. Poggendorff, a physicist and an historian of physics, but better remembered as the editor of the *Annalen*. It was a reaction to the 'information explosion' of the day; the editor had had, for example, to devote 700 pages to the translation of Regnault's works. The eighty-year-old

Poggendorff died, however, in January 1877, and the editorship passed to Gustav Wiedemann, Professor of Physical Chemistry at Leipzig, and his son Eilhard[22], a twenty-five-year-old *Privat Dozent* who had taken his *Habilitation* the year before with an experimental study of the heat capacity of gases. Eilhard Wiedemann, who was the driving force behind the new journal, became well known as a physicist and as an historian of Arabian science. His linguistic abilities were put to good use in abstracting non-German papers for the *Beiblätter*, which quickly won a wide popularity.

The first issue of the *Beiblätter*, which still had Poggendorff's name on the title-page, appeared early in 1877. Among the many abstracts written by Eilhard Wiedemann of papers in English, French, Italian, Swedish and Dutch, was a long and clear account of *Continuity* prepared 'in collaboration with the author'. In a footnote to the title it said the 'Dutch original is unfortunately little known' [$1c_2$, p. 10][23]. With this issue of the *Beiblätter* van der Waals's work at last received international recognition, and he acknowledged his debt to Wiedemann in his Nobel Lecture [103e, p. 256].

At the session of the Academy of Science in Vienna of 11 May 1877, Boltzmann described the relation between surface tension and the sizes of molecules [*54*]. He referred both to the abstract in the *Beiblätter* and to the Dutch original of *Continuity*, but it seems that he read only the former since he misquotes the Dutch title (it is not printed in the abstract); perhaps he saw only a brochure sent from Leiden to his colleague and former teacher, Josef Stefan.

Finally, and also in 1877, O. E. Meyer, the brother of the chemist Lothar Meyer, published an important monograph on the kinetic theory of gases [*18*] which was soon accepted as the fullest and most up-to-date text on the subject. He referred both to Wiedemann's abstract and to the original Dutch version of *Continuity* and gave the work an important place in his treatise.

In 1881 an Amsterdam bookseller wrote to van der Waals: 'From England, Germany and America we receive continual requests for your book', which he was obliged to turn down because there were no more copies to be had [231, p. 149].

Meanwhile life was following its accustomed course at The Hague, with its lessons, examinations and then, in the spring of 1877, an administrative change when the mathematician A. T. van Aken resigned as Director [234,238]. There could be little doubt about who was best qualified for the post and the Commission for the supervision of secondary education asked the Hague Magistrate to seek Inspector Steyn Parvé's approval of van der Waals's appointment. He agreed but expressed some doubt as to whether van der Waals would then be able to continue with all his physics lessons. The Inspector thought, however, that van der Waals would not wish to give up any of them. On 4 May the Magistrate wrote in confidence to van der Waals to clear up this point. His answer is not known, but everyone seemed to be satisfied since three days later the Magistrate wrote to the municipal authorities to recommend that van der Waals be appointed without any other candidate being considered. His reasons were, first, that van der Waals had already had to perform the duties of the director during van Aken's absences, and second, that 'he has within himself all the qualities needed for the post', and 'according to the general view, he is particularly qualified for it'. Throughout all this correspondence there is no mention of van der Waals's membership of the Academy, perhaps out of deference to the feelings of Steyn Parvé. On 15 May the appointment was approved by the municipality and the next day the Magistrate confirmed it and fixed his salary at fl. 3500 per annum.

He was, however, to occupy the Director's office for less than six months, for at the start of the next school year he accepted the post of Professor of Physics at the University of Amsterdam; he applied to the municipality for leave to resign on 20 September. In the annual report of the municipality his resignation is acknowledged with the words, 'We believe that it is irrelevant here to describe his merits, but one cannot overlook how much he did for the well-being of the school. And parting from him with regret, we may rejoice, however, in the honourable position he was destined to be given, namely the post of a University Professor.' His successor was the chemist W. F. Koppeschaar[24] and the same report expressed the hope that he 'would follow in his predecessor's footsteps'.

On 23 October 1877, van der Waals and his family left The Hague for Amsterdam.

Notes

1. J. C. Maxwell on van der Waals's dissertation [*38*, p. 359].
2. Kamerlingh Onnes's emphasis.
3. In this chapter, references of the type [p. *n*] stand for [$1e_2$, p. *n*].
4. Van der Waals means the conception of molecules as point-like masses.
5. See also [*10*, p. 48].
6. Consider, for example, his account of his attempts at making sense of the jungle of experimental results: 'I therefore considered every circumstance . . . but without any satisfactory result' [p. 177], and later, 'At one time I went into this subject in detail' [p. 180].
7. This remark was omitted from the early translations of *Continuity*.
8. Van der Waals uses the singular *toestand* rather than Andrews's plural *states*; in Dutch the plural, *toestanden*, would also be the more natural word.
9. The law of psycho-physics of W.E. Weber and G. Fechner (1858).
10. Quoted in English from Newton's letter of 1693 to the Rev. R. Bentley [*55*, p. 302–303].
11. The emphasis is by van der Waals. This criticism of Poinsot [*56*, p. 507] may be related to the great impression produced on him by an equation of state which yielded quite unforeseen results; or it may be that this and some of the other *stellingen* were designed more to impress the Senate with his erudition and knowledge of foreign languages.
12. Lorentz took his doctorate, *summa cum laude*, in 1875.
13. 'To each man comes his appointed day', Virgil, *Aeneid*, x, 467.
14. It is often said that *Low-Dutch* was a slip of the tongue or even a mistake of Maxwell's. Thus in the 1967 commentary on van der Waals's Nobel Lecture we read:

'Maxwell probably meant to say 'Low-German', which would also be incorrect, since Dutch is a language in its own right' [103e, p. 216]. Klein remarks that this expression 'must give rise to a pained smile in Holland' [176, p. 42]. However in Britain until the eighteenth century (see e.g. [*57*]), and in America until the nineteenth, the word *Dutch* was used in the contemporary sense of *German* (*Deutsch*), for both language and people. Thus the language of the Netherlands was *low Dutch*, in contrast with the *high Dutch* of Germany. Maxwell's language may have been old-fashioned but was linguistically defensible. In a letter of 1862 he speculates whether the Scots word *wintershines* (or more commonly *widdershins*) 'is equivalent to *wider die Sonne* in High Dutch' [*37*, p. 383].

15. AW-NW, Afdeeling der Wis – en Natuurkundige Wetenschappen. Two years later Maxwell became a foreign member of the PMD, the Section of Sciences.
16. A Dutch physicist who worked for many years in Britain and Germany.
17. The anonymous dating of this letter to April is dubious since some of it corresponds closely with Maxwell's review of the thesis on 18 October. It contains an erroneous expression for the volume correction, with a factor of 16 instead of van der Waals's 4; 'The form of the result is quite different from the Dutchman's.' In 1874–75 Maxwell returned several times to the problems raised by the Leiden thesis; some important calculations are to be found in his referee's report on Andrews's paper [*49*] (see below, and also Rowlinson's discussion of this report [183, p. 16–17]), and further comments are in unpublished notes in Maxwell's papers at Cambridge. In one of these he uses van der Waals's equation and Regnault's experimental results to estimate the critical temperature of water. His value, 434 °C, is worse than van der Waals's result, 412 °C. Maxwell's critical parameters were published by his former student, W. Garnett, in the article on 'Evaporation' in the ninth edition of *Encyclopaedia Britannica* (1879), in which there is a brief mention of van der Waals. Maxwell revised his opinion of the importance of the Joule–Thomson effect after reading *Continuity*. In an article he wrote for *Encyclopeadia Britannica* in 1874 [*37*, p. 390], he wrote that it implied that 'the force between the molecules is small but, on the whole, attractive' [*58*, p. 185].
18. From 1870 to 1874 the Director of the Rijksmuseum was the engraver J. W. Kaiser, the younger brother of F. Kaiser, the astronomer, and Lorentz's father-in-law.
19. *Over de theorie der terugkaatsing en breking van het licht* (On the theory of the reflection and refraction of light).
20. It is possible that by then Clausius had read Maxwell's review of 18 October, and had been persuaded by it that he need not trouble to tackle the Dutch thesis.
21. Janssen's moves can be followed from the list of addresses of corresponding members of the British Association published in its Annual Reports.
22. His maternal grandfather was the chemist, Eilhard Mitscherlich.
23. The situation was similar in Britain. The physicist R. E. Baynes, an editor of *Encyclopaedia Britannica*, who was by then familiar with van der Waals's work, wrote in 1880 that it 'is not sufficiently known in England', and that his theory 'has not hitherto appeared in an English dress' [*59*].
24. J. J. van Laar wrote that by the age of fourteen (i.e. 1874) he knew by heart the three volumes of Koppeschaar's HBS textbook of chemistry [*53*, p. 22].

References

1. Hertz, H., Einleitende Übersicht. In *Gesammelte Werke*, Bd 2, *Untersuchungen über die Ausbreitung der Elektrische Kraft*, 3. Aufl., Leipzig, 1914, S. 1–31.

2. Clausius, R., Ueber die Art der Bewegung, welche wir Wärme nennen, *A. Ph.*, 1857, Bd 100, S. 353–380.
3. Bosscha, J., Sur la dilatation absolue du mercure, d'après les experiénces de M. Regnault, *A.N.*, 1869, t. 4, p. 167–196; *A. Ph.*, Ergänzbd 5, S. 276–304.
4. Clausius, R., Ueber die bewegende Kraft der Wärme und die Gesetze, welche sich daraus für die Wärmelehre selbst ableiten lassen, *A. Ph.*, 1850, Bd 79, S. 368–397, 500–524.
5. Bierens de Haan, J. A., *De Hollandsche Maatschappij der Wetenschappen 1752–1952*, Haarlem, 1952.
6. Brush, S. G., *The kind of motion we call heat — A history of the kinetic theory of gases in the nineteenth century*, Amsterdam, 1976.
7. Buijs-Ballot (Buys Ballot), C.H.D., Ueber die Art von Bewegung, welche wir Wärme und Elektricität nennen, *A. Ph.*, 1858, Bd 103, S. 240–259.
8. Snelders, H. A. M. and Schuurmans, C. J. E., C. H. D. Buys Ballot. In *Van Stevin tot Lorentz. Portretten van Nederlandse natuurwetenschappers*, Red. A. J. Kox, Amsterdam, 1980, b. 123–134.
9. Clausius, R., Ueber die mittlere Länge der Wege, welche bei der Molecularbewegung gasförmiger Körper von den einzelnen Molecülen zurückgelegt werden; nebst einigen anderen Bemerkungen über die mechanische Wärmetheorie, *A. Ph.*, 1858, Bd 105, S. 239–258.
10. Brush, S. G., *Statistical physics and the atomic theory of matter from Boyle and Newton to Landau and Onsager*, Princeton, New Jersey, 1983.
11. Maxwell, J. C., Molecules, *Nature*, 1873, v. 8, p. 437–441.
12. Garber, E., Clausius' and Maxwell's kinetic of theory gases, *Hist. Stud. Phys. Sci.*, 1970, v. 2, p. 299–319.
13. Garber, E., Molecular science in late-nineteenth-century Britain, *Hist. Stud. Phys. Sci.*, 1978, v. 9, p. 265–297.
14. Maxwell, J. C., On the dynamical theory of gases, *Philos. Trans. R. Soc. London*, 1867, v. 157, p. 49–88.
15. Brush, S. G., Interatomic forces and gas theory from Newton to Lennard-Jones, *Arch. Rat. Mech. Anal.*, 1970, v. 39, p. 1–29.
16. Garber, E., Brush, S. G. and Everitt, C. W. F., Kinetic theory and the properties of gases: Maxwell's work in its nineteenth-century context. In *Maxwell on molecules and gases*, ed. E. Garber, S. G. Brush and C. W. F. Everitt, Cambridge, Massachusetts, 1986, p. 1–63.
17. Maxwell, J. C., *Theory of heat*, London, 1871.
18. Meyer, O. E. *Kinetische Theorie der Gase in elementarer Darstellung mit mathematischen Zusätzen*, Breslau, 1877.
19. Gibbs, J. W., Rudolf Julius Emanuel Clausius. In *Collected Works*, New York, 1928, v. 2, p. 261–267.
20. Clausius, R., Ueber einem auf die Wärme anwendbaren mechanischen Satz, *A. Ph.*, 1870, Bd 141, S. 124–130.
21. Laplace, P. S., *Traité de méchanique céleste; Supplément au dixième livre, Sur l'action capillaire*, Paris [1806].
22. Regnault, H. V., Relation des expériences entreprises par l'ordre de M. le Ministre des Travaux publics et sur la proposition de la Commission centrale de machines à vapeur, pour déterminer les principales lois et les données numériques qui entrent dans le calcul des machines à vapeur, *Mém. Acad. Sci. Inst. Fr.* 1847, t. 21; 1862, t. 26.
23. Bosscha, J., Sur la dilation apparente du mercure . . . d'après les expériences de M. Regnault, *A. N.*, 1869, t. 4, p. 197–208; *A. Ph.*, 1871, Ergänzbd 5, S. 444–469.

24. Andrews, T., Ueber die Continuität der gasigen und flüssigen Zustände der Materie, *A.Ph.*, 1871, Ergänzbd 5, S. 64–87.

25. Andrews, T., On the continuity of the gaseous and liquid states of matter, *Philos. Trans. R. Soc. London*, 1869, v. 159, p. 575–590.

26. Thomson, J., Considerations on the abrupt change at boiling or condensation in reference to the continuity of the fluid state of matter, *Proc. R. Soc.*, 1871, v. 20, p. 1–8.

27. Maxwell, J. C., Illustrations of the dynamical theory of gases. Part II. On the process of diffusion of two or more kinds of moving particles among one another, *Philos. Mag.*, 1860, v. 20, p. 21–37.

28. Loschmidt, J., Zur Grösse der Luftmolecüle, *Sitzungsber. Kais. Akad. Wiss, Wien Math — Naturwiss. Kl.*, 1865, Sekt. II, Bd 52, S. 395–413.

29. Stoney, G. J., The internal motions of gases compared with the motions of waves of light, *Philos. Mag.*, 1868, v. 36, p. 132–141.

30. Thomson, W., Presidential Address to the British Association meeting at Edinburgh, *Rep. BAAS*, 1871, v. 41, p. lxxxiv–cv.

31. Boltzmann, L., *Vorlesungen über Gastheorie*, Bd 2, Leipzig, 1898; *Lectures on Gas Theory*, trans. S. G. Brush, Berkeley, California, 1964.

32. Boltzmann, L. and Nabl, J., Kinetische Theorie der Materie. In *Encykl. d. Math. Wiss.*, 1905, Bd 5, T. 1, H. 4, S. 493–557.

33. Fokker, A. D., H. A. Lorentz. In Sevensma, T. P., *Nederlandsche helden der wetenschappen*, Amsterdam, 1946, b. 53–93.

34. Plateau, J. *Statique experimentale et théorique des liquides soumis aux seules forces moléculaires*, 2 vols, Paris, 1873.

35. Maxwell, J. C., Plateau on soap-bubbles, *Nature*, 1874, v. 10, p. 119–121.

36. Maxwell, J. C., Van der Waals on the continuity of the gaseous and liquid states, *Nature*, 1874, v. 10, p. 477–480.

37. Campbell, L. and Garnett, W., *The life of James Clerk Maxwell*, London, 1882.

38. Maxwell, J. C., On the dynamical evidence of the molecular constitution of bodies, *Nature*, 1875, v. 11, p. 357–359, 374–377.

39. Tait, P. G. On van der Waals's treatment of Laplace's pressure in the virial equation: in answer to Lord Rayleigh, *Nature* 1891, v. 44, p. 546–547.

40. Thomson, J., *Collected papers in Physics and Engineering*, Cambridge, 1912.

41. Tait, P. G. and Crum Brown, A., Memoir. In Andrews, T., *The Scientific Papers*, London, 1889, p. ix–lxii.

42. Maxwell, J. C. to Tait, P. G., Letter anonymously dated April, 1874, Maxwell Collection, Cambridge University.

43. Maxwell, J. C. to Tait, P. G., Letter of 14 February 1876, In *Maxwell on molecules and gases*, ed. E. Garber, S. G. Brush and C. W. F. Everitt, Cambridge, Massachusetts, 1986, p. 508–510.

44. Clausius, R., Ueber das Verhalten der Kohlensäure in Bezug auf Druck, Volumen und Temperatur, *A. Ph.*, 1880, Bd 9, S. 337–357.

45. Clausius, R., Ueber den Satz vom mittleren Ergal und seine Anwendung auf die Molecularbewegungen der Gase, *A. Ph.*, 1876, Ergänzbd 7, S. 215–280; *Sitzungsber. Niederrein, Ges. Bonn*, 1874, S. 183–231; *Philos. Mag.*, 1875, v. 50, p. 26–46, 101–117, 191–200.

46. Korteweg, D. J., Over de berekening van den gemiddelden botsingsafstand der gasmoleculen, met in achtneming van al hunne afmetingen, *V. M.*, 1876, d. 10, b. 349–362; *A. N.*, 1877, t. 12, p. 241–253.

47. Korteweg, D. J., Over de berekening van de vermeerdering welke de spanning van een gas tengevolge van de botsingen der moleculen ondergaat, *V. M.*, 1876, d. 10, b. 363–370; *A. N.*, 1877, t. 12, p. 254–261.

48. Andrews, T., On the gaseous state of matter, *Philos. Trans. R. Soc. London*, 1876, v. 166, p. 421–449.
49. Maxwell, J. C., Report on Dr. Andrews' paper 'On the gaseous state of matter', A collection of scientific papers and correspondence by and addressed to Thomas Andrews, item MS2/16–1, Library of the Queen's University, Belfast.
50. Andrews, T., Presidential address. In *The Scientific Papers*, London, 1889, p. 393–417; *BAAS Rep.*, 1876, v. 46, p. lxviii–lxxxvi.
51. Maxwell, J. C., in Report of the Committee . . . for testing experimentally Ohm's law, *BAAS Rep.*, 1876, v. 46, p. 49–52.
52. Janssen, W. J., *Stickstofoxydule in den vloeibaren en gasvormigen toestand*, Leiden, 1877.
53. van Emmerik, E. P., *J. J. van Laar (1860–1938). A mathematical chemist*, Thesis, Delft University of Technology, 1991.
54. Boltzmann, L., Ueber eine neue Bestimmung einer auf die Messung der Moleküle Bezug habenden Grösse aus der Theorie der Capillarität, *Sitzungsber. Kais. Akad. Wiss. Wien, Math.-Naturwiss. Kl.*, 1877 Sekt. II Bd 75 S. 801–813.
55. Newton, I., *The correspondence*, ed. H. W. Turnbull, v. 3, p. 253–254, Cambridge, 1961.
56. Poinsot, L., *Eléments de statique*, Paris, 1848.
57. Ogilvie, J., *The Imperial Dictionary of the English Language*, v. 2, London, 1883.
58. Maxwell, J. C., Atom, In *Maxwell on molecules and gases*, ed. E. Garber, S. G. Brush and C. W. F. Everitt, Cambridge, Massachusetts, 1986, p. 176–215.
59. Baynes, R. E., Critical temperature of ethylene, *Nature*, 1880, v. 23, p. 186.

CHAPTER 4

The University of Amsterdam

On 23 October 1877 van der Waals and his family left The Hague for Amsterdam, which was by then a city of more than 300 000 inhabitants and had been for a long time the biggest city in the Netherlands. This busy commercial and industrial metropolis must have been a contrast to the quietness of Leiden. The French artist and writer Eugène Fromentin had described it in 1875 as an 'oldish-looking, middle-class, stuffy, busy, swarming' city [*1*, p. 233]. It must be remembered, however, that the population of London was by then ten times that of Amsterdam and to some European travellers the Dutch city seemed to be a quiet green town, particularly in the new suburbs. It was in one of these fashionable districts, near the most beautiful of the city parks, named after Joost van den Vondel, that van der Waals chose the family's new home. It was *N57 P. C. Hooftstraat*, named after Pieter Corneliszoon Hooft, like van den Vondel a writer of the seventeenth century. In 1896 the number was changed to *177* on the incorporation of the suburbs into the city [238]; van der Waals lived there until his death. The district is no longer as fashionable as it was and the house has lost its front door on its incorporation into *175–179*.

His departure for Amsterdam coincided with a peak in the rate of development of the Netherlands. It was the period when, as they said 'The flood begins to flow faster' [*2*, p. 25], when the country which had for many years been in the rearguard of European development was at last beginning to take its place as one of the advanced states of the world.

A strong impetus to Dutch trade followed the opening of the Suez Canal in 1869, thus reducing the distance between the Dutch colonies and the centre of the empire. Money poured into the country, and was wisely used to build many more Dutch canals. In 1872 the *Nieuwe Waterweg* (New Waterway) was completed. This bold project by the gifted young engineer Pieter Caland connected Rotterdam with the North Sea and allowed ships to sail into the centre of the city. Rotterdam soon became the leading seaport of Europe. A second sea canal, which came into use in 1876, connected Amsterdam with the North Sea. Other industrial development went ahead equally rapidly; let it suffice to mention here that the number of industrial steam engines doubled between 1870 and 1880.

The cultural and scientific life of the country blossomed also. As so often, the first signs of the national revival became clear in the field of literature. In the middle of the nineteenth century there appeared several outstanding works of contemporary realism. A pastor Nicolaas Beets (1814–1903, writing under the pen-name Hildebrand) published from 1839 to 1851 a collection of short stories entitled *Camera Obscura*, which came to be generally recognized as realistic vignettes of Dutch life, and Eduard Douwes Dekker (1820–87, under the pen-name Multatuli or 'I have suffered much') wrote the famous anti-colonial novel *Max Havelaar* (1860) which marked a turning-point in Dutch literature. A group of progressive writers and critics calling themselves 'The Young Netherlands' led the literary revival. In 1884 they launched a magazine *De Nieuwe Gids* (The New Guide) which challenged the established journal *De Gids* and became the mouthpiece of those who wanted to overcome national complacency and provincialism.

Dutch architecture, after a long period of decline, was regenerated by men such as Pieter J. H. Cuypers (1827–1921), who combined modern ideas with the traditional national styles of gothic and baroque. Among his best-known works are the Central Railway Station and the Rijksmuseum (1877–85) in Amsterdam. The revival of Dutch painting was associated with the 'Hague School', which gathered around the old artist Josef Israels and the enthusiastic picture-dealer G. Tersteeg. It was from this school that Vincent van Gogh first emerged.

The 1870s saw also a scientific renaissance in the Netherlands; the benefits of Thorbecke's law of 1863 were beginning to be felt. New reinforcements poured into the universities, the only centres in which science was then pursued. Many hundreds of former students at the HBS passed on to the universities through the state examination in Latin and Greek. Bosscha, in a paper to the International Conference on Education of 1884, wrote, 'For in 1863 the stipulation was still in force that young men who had not studied at a 'gymnasium' could be admitted to the University, only if they could pass an examination before a committee of examiners from the literary faculty at each university. As may easily be understood, this examination had, through the effects of different influences, in reality become so easy, that a cursory study of a few months was sufficient to acquire as many Latin and Greek words as were necessary to be admitted to the University' [*3*, p. 197]. Moreover the setting-up of the HBS created many jobs for university graduates. A teacher at a HBS had a well-paid job that carried considerable prestige, and the holding of this post was effective training for a future university teacher. The rôle of the HBS in the rapid advance of Dutch science from the 1870s is generally recognized [*4*] and can be demonstrated by the example of three future Nobel prize winners. In 1873, van der Waals, a teacher in a HBS at The Hague, gained his doctor's degree for his thesis on *Continuity*. In 1874, van 't Hoff, a former student of a HBS in Rotterdam, laid down the foundations of stereochemistry. In 1875, Lorentz, a former student and then teacher at a HBS at Arnhem, defended a thesis that contained significant advances in electrodynamics.

Perhaps the most striking and significant sign of the national renaissance was the founding of a fourth university in the Netherlands, the University of Amsterdam. This event was preceded by some important changes in the academic life of the country.

In the summer of 1874 the Liberal government of de Vries fell and was replaced by a Conservative ministry led by J. Heemskerk (1818–97), which immediately introduced a bill for the reform of higher education. The time was ripe, for modernization of the universities was badly needed, but the only change actually proposed in the bill was the abolition of the theological faculties. Heemskerk was no atheist, but he believed that it was the church that should pay for the training of its pastors, not the state which financed the universities. The Liberals predictably stood up for the preservation of the theological faculties. Two years were spent in hot debate in the Second House of the States General before the bill was passed, but only after there had been about 160 amendments which had transformed the original proposals beyond recognition.

The new Higher Education Law of 1876 contained two provisions that affected the career of van der Waals. The first provided for additional university posts. The theological faculties were preserved, but the number of places in them was limited. At Leiden a second chair of physics was established through the initiative of Rijke. He himself retained the chair of experimental physics and the new post was in theoretical physics; there were similar moves at Utrecht and Groningen.

A second article of the new law concerned the *Athenaeum Illustre* of Amsterdam. This had been founded in 1632 when the newly rich merchants of Amsterdam decided to give their sons the opportunity of being educated in their native city, with its new independence, wealth and importance. The city fathers failed, however, to secure the status of a university for their Athenaeum. They were strongly opposed by Leiden, then the only university, which was afraid of losing its privileged position. For two and a half centuries the Athenaeum of Amsterdam pursued its sound but unexciting way. It justified the hopes of its founders but it did not achieve the same scientific and cultural success as Leiden. The situation started to change in the 1870s, when the Athenaeum, like the city itself, began to grow rapidly. The old custom of classes in the professors' homes was abolished, and normal lectures were instituted. The numbers of students increased, particularly in the medical faculty, and the standard of teaching rose also. However, the young men of Amsterdam still had to betake themselves to another town, usually to Leiden, to get a university degree. The move to give the Athenaeum the status of a university came from its students and professors and even from the man-in-the-street, until finally it obtained the support of the city authorities. The Conservatives were against this, including Heemskerk, although he was a curator of the Athenaeum, which was supported financially by the municipal authority. The Conservatives did not oppose modernization and expansion of this ancient institution, but they did oppose the plan to change it into a municipal university. They held that the awarding of degrees in science should be under the control of the state, and they were against the extra cost of an additional state university. Heemskerk was defeated, however, and the new Higher Education Law required that the Athenaeum be reorganized as the *Gemeente Universiteit van Amsterdam* (Municipal University of Amsterdam). Leiden had, for the last time, tried to prevent the appearance of a rival and, after the bill had been passed by the Second House, sent a petition to the First claiming that the intellectual resources of the country were limited and insufficient to support four universities; if a University of Amsterdam were to be established then the Dutch system of higher education would be damaged. The petition failed, however, and on 15 October 1877 there was a solemn inauguration of the new University of Amsterdam.

At first it was housed in the former Athenaeum on the *Singelgracht*, the innermost of the concentric canals that enclose the city on the south and west sides. This building is now the University Library. In 1880 it obtained further accommodation in the centre of the city at the *Oudemannenhuis*, a former castle going back to the fourteenth century, later a nunnery, and then an almshouse. This remains the main building of the University to this day.

The transformation of the Athenaeum into the University led to many changes, including the establishment of a faculty of mathematics and physics, and the creation of a new chair in physics. For both posts, at Leiden and at Amsterdam, van der Waals was the obvious candidate as the rising star in Dutch physics. No doubt he would have chosen the famous University of his birthplace, where he first acquired his taste for science, and which he 'always considered to be a sanctuary of science' [157(a)]. Nevertheless he accepted the chair at Amsterdam. Why was this? The reason was probably quite prosaic – the Municipal University was responsible only to the Burgomaster of Amsterdam and could therefore follow the bureaucratic path required by the Higher Education Act for the appointment of new professors more quickly than the State University of Leiden which had to obtain a Royal edict for each appointment. Thus in the summer of 1877 van der Waals received a definite offer from Amsterdam while Leiden was still dithering. On 31 August Rijke wrote to van der Waals to say that he had had an audience with Heemskerk and asked van der Waals not to accept yet the offer from Amsterdam since he wanted him to become a professor in a state university. This letter only confirmed van der Waals's fear that a firm offer from Leiden was still not certain, and so in September he told the Amsterdam authorities that he accepted the post of head of the sub-faculty of physics in the new university [234]. Perhaps he was too hasty. The Royal edicts for the Leiden posts were signed in October, and the chair of theoretical physics that had been intended for van der Waals was occupied on 28 January 1878 by the twenty-five-year old Hendrik Antoon Lorentz.

This is a convenient place for a description of the university where van der Waals was to work for more than thirty years. Its main difference from the old universities of Leiden, Groningen and Utrecht was that it was governed by the city authorities, but in other respects its regulations were modelled on those of the state universities. There were five faculties – theology, law, medicine, physics and mathematics, and philology and philosophy. The supreme body in the University was the Senate, that is the Council of Professors with the Rector Magnificus at its head. The Rector was nominated each year by the

Burgomaster and Aldermen from a list of three Ordinary Professors proposed by the Senate; the retiring Rector was not eligible for re-election. The Rector and four Assistants or 'Assessors', one from each of the faculties other than that from which the Rector was chosen, together with the Secretary of the Senate, formed a Board for the management of the affairs of the University. The academic year began on the third Tuesday of September and ended on the Friday before the second Saturday of July, with recesses at Christmas and Easter.

Beside these general regulations there was a mass of detailed rules for the staff and students. For example, 'Every Ordinary Professor is required to attend the meetings of the Senate from start to finish. Non-attendance without a valid reason incurs a fine of one guilder' [*5*, p. 18]. Missing a session of the Faculty came more cheaply, the fine was only half a guilder. For comparison, we may note that the tuition fee was then 200 guilders per annum, and the annual salary for professors was 4500 guilders for the first five years, 5000 for the next five years, and 5500 thereafter.

A university is, in essence, its staff, and at first most of these came from the Athenaeum. Outstanding among these was T. M. R. Asser (1838–1913), a specialist in international law. A graduate of Leiden, he combined the work of professor of commercial and international law with a successful practice as an advocate. For some years he was a member of the government. He is remembered best for efforts in organizing cooperation in international law. He was one of the founders of the Institute of International Law (1873), and was an organizer of the Hague Conference on International Law in 1890, and of the Hague Peace Conferences of 1899 and 1907, at the first of which he brought about the establishment of the International Court at The Hague, which is still in being. In 1911 he was awarded the Nobel Prize for Peace.

The second important figure was B. J. Stokvis (1834–1902), Professor of Medicine in the Athenaeum from 1874. At the time of the transformation of the Athenaeum this was much the largest faculty with 310 of the 427 students. Stokvis was probably the best known medical man in the Netherlands; 'his card with a few words written in his own hand, was the best introduction in London, Paris or Berlin' [*6*]. Popular with his colleagues and students alike, he was also active in public affairs. He was the founder in 1887 of the Congress of Dutch Natural Scientists and Physicians. This met every two years, and the first meeting was opened with an interesting address by Stokvis on 'Nationality and Science'.

The oldest of the professors in the physics and mathematics faculty was C. J. Matthes (1811–82) who had held the chair of mathematics in the Athenaeum since 1847. A doctor of Leiden, and a founder member (1851) of the Academy, Matthes was a professor of the old school, whose earliest publications were in Latin. He wrote several textbooks and some addresses and other papers on public education, on the popularization of science, and biographies of mathematicians. His scientific work was in geometry and mechanics. He was the General Secretary of the Academy from 1864 to 1878. As a teacher he was probably not very popular, for after the transformation of the Athenaeum into the University its physics and mathematics faculty had only eight students. It was his chair that was then divided; he retained mathematics, mechanics and astronomy, and the new chair was in physics.

Two other professors at the Athenaeum who became colleagues of van der Waals in the University, and who were to become the last surviving links between the two institutions by the time of their retirement in 1896, were the botanist C. A. J. A. Oudemans (1825–1906) and the chemist J. W. Gunning (1827–1901). They were also his colleagues in the Section of Sciences of the Academy. Oudemans, who was the first Rector Magnificus, was the eldest of three brothers who were all academics. The second, J. A. C. Oudemans (1827–1906), was an astronomer at Utrecht, and the youngest, A. C. Oudemans (1831–95), a chemist at the Polytechnic at Delft, who succeeded Bosscha as its Director in 1885. He helped van der Waals by supplying him with organic compounds and determining their densities. For these services he was thanked in the first German translation of *Continuity* [1d, p. 162–164].

The eldest Oudemans was a doctor of medicine who then turned to botany and worked on the classification of plants. Physiology he did not like, and he left it to Hugo de Vries whom he had invited to assist him in 1877. 'Oudemans was an eloquent and clear speaker, a good teacher, and an earnest publicist' [7, p. 114]. In the Academy he was Vice-President of the Section of Sciences for several years before succeeding Matthes as General Secretary of the Academy in 1878.

Gunning was also trained in medicine, but then became a chemist; his doctor's thesis is on the chemistry of natural waters. As a professor in the Athenaeum from 1865 he supported the supremacy of medicine, treating chemistry and physics as auxiliary disciplines needed for its support. His great merit was his struggle for the independent status of the natural sciences in Amsterdam. He helped to draw up the bill on higher education under which the Faculty of Physics and Mathematics was established at Amsterdam, and saw that an independent Section of Physics was established in 'The Society of Medicine and Surgery' in 1874, so that its name was changed to 'The Society of Medical and Natural Sciences'. (Van der Waals became a member in 1887.) Gunning, like van der Waals, was elected to the Academy in 1875. On the establishment of the Faculty of Physics and Mathematics, Gunning retained for himself the courses in pharmacy and chemistry for the medical students; other courses were taken by his colleague J. H. van 't Hoff, whose stereochemical views he did not accept.

The two young scientists just mentioned, de Vries and van 't Hoff, are too well known for their achievements to be recorded here. Both had begun as *lectors*, but were promoted to professorships after one year. They were responsible for many different classes; thus van 't Hoff was entrusted with general chemistry, crystallography, geology and mineralogy.

Finally, there were two other professors in the Faculty whose names are not so well known. A. J. van Pesch was appointed in 1878 to help Matthes in mathematics and mechanics. An engineer by training, he worked first in Deventer and then in the Polytechnic at Delft where his research was on the application of the calculus of probabilities to social problems. His mortality tables won him wide recognition in the Netherlands, and he became a regular member of State Commissions on pensions and on statistics. Leiden awarded him an honorary doctorate in 1890. In the University he lectured mainly on the calculus, probably with some skill since most theses in physics and chemistry contained thanks to him.

In 1881 Matthes retired and was succeeded by Diederik Johannes Korteweg (1848–1941), whom we have already encountered as an advocate of *Continuity* in 1875. This gifted mathematician now became a colleague of van der Waals and thus began a friendship that lasted many years. A student at the University of Amsterdam, he passed the *candidaats* examination after one year, and then the doctor's examination. On 18 July 1878, on the last day of the academic year, he defended (cum laude) his doctor's thesis *On the velocity of wave propagation in elastic tubes*. This was the first thesis in the new University and van der Waals was the Promotor, but it is probable that his rôle was purely formal, as Rijke's had been in his case, since there was no acknowledgement in the published text of Korteweg's thesis. One of the 'Propositions' at the end, however, reproduced Korteweg's earlier calculation of the value of the parameter b in van der Waals's equation; 'If one assumes that gas molecules are solid spheres which neither attract nor repel each other, then, for a constant temperature, the approximate formula $p(v - 4a) = C$ is valid, where a is the total volume of the spheres' [*8*, p. 165].

Another of Korteweg's 'Propositions' is evidence of his confidence in the existence of atoms and his profound understanding of the atomic hypothesis; 'It is a pity that Stas sought to explain the approximate coincidence of most of the smaller atomic weights with integers as pure chance. On the contrary, this fact may be the key to important conclusions concerning the nature of atoms' [*8*, p. 166].

The theme of his inaugural lecture on 10 October 1881, 'Mathematics as an auxiliary science' was not chosen casually; it reflects both his early training at the Polytechnic and his personal contact with van der Waals. His many scientific papers were, for a long time, known only to a narrow circle of specialists, but his name is now remembered by physicists in connection with the concept of solitons, which appeared first in the thesis of 1894 of his student Gustav de Vries (1860–?, no relation of Hugo), and then in the paper of Korteweg and de Vries of 1895 [*9*]. Many of his papers were on topics studied by van der Waals. He wrote several papers on the equation of state and an important series that underpins van der Waals's work on binary mixtures. These papers were entitled *On plaitpoints, A general theory of plaitpoints and of the ψ-surface of van der Waals*, and *Plaitpoints and corresponding plaits in the neighbourhood of the sides of the ψ-surface* [*10–12*]. Van der Waals, in his turn, derived some of his ideas from Korteweg's theory of ponderomotive forces [14]. Clearly they had close professional interactions, but these were only at the level of discussions; there are no joint papers, perhaps both were too independent and distinctive in their ideas.

At the University, Korteweg delivered courses on many branches of mathematics, and also on astronomy and mechanics. He lectured not only to intending mathematicians but also to physicists and chemists. Ernst Cohen (1869–1944) recalled that Korteweg's lectures often resembled vivid talks, but anyone who tried to avoid work ran the risk of being asked 'But, mijnheer, why did you not become a grocer? Then you would not have had to bother with analytic geometry!' [*13*, p. 499].

Korteweg was by nature an active man who took part in many public aspects of science. For some years he ran the Mathematical Society of Amsterdam and edited the *Nieuw Arch's voor Wiskunde*, a mathematical magazine that attracted a group of gifted young contributors. He edited ten volumes of Huygens's correspondence and for that received, in 1906, the Huygens Medal of the Holland Society of Science (HMW). He was the Dutch representative in the compilation of the international bibliography *The Catalogue of Scientific Papers*. A member of the Academy from 1881 he was Vice-President of the Section of Sciences from 1902–19. He served on many committees of both the Academy and of the University, of which he was Rector in the academic year 1893–94. His Rectorial address on 8 January (the day of the founding of the Athenaeum in 1632) was entitled 'The time of the flowering of mathematical science in the Netherlands', and was devoted particularly to the work of Huygens[1].

To complete this selection from the portrait gallery of the new University we must say something of two of the professors of the humanities. The first of these was C. Bellaar Spruyt, a brilliant and formidable man, who had graduated in natural philosophy at Utrecht when he defended his doctoral thesis in 1867; it was entitled *The electromotive force of a Daniell cell at different temperatures*. He was an assistant teacher of physics and mathematics in the HBS at the Hague, then teacher of chemistry in the HBS at Utrecht. From 1871 onwards he published papers in which he opposed the positivist philosophy of C. W. Opzoomer (1821–92), and so opened a new chapter in the history of Dutch philosophy. His inaugural lecture at Amsterdam was on the relation between philosophy and specialized disciplines, a subject that then became the leading topic in his seminar classes. Bellaar Spruyt lectured on the history of philosophy and on logic and psychology. He had a great influence on Dutch intellectuals, including the many from the physics and mathematics faculty who came to his lectures. These included J. D. van der Waals Jr. (1873–1971) and Ph. A. Kohnstamm (1875–1951), who were much drawn to his philosophy all their lives, and who edited his lectures for posthumous publication.

Finally, we consider a professor in the theological faculty, P. D. Chantepie de la Saussaye (1848–1920), who was a very close friend of van der Waals and his family. A Protestant pastor, and the son of a pastor, he had had a thorough theological training at Utrecht, Bonn and Tübingen, and was appointed to

Amsterdam a year after van der Waals. His international reputation rested on his classic *History of Religion* (1st edn, 1887–89) but the unorthodox views, or even 'anti-orthodoxy' [7, p. 33] of this learned theologian had been displayed earlier in his doctoral thesis *On the methodology of studies on the origin of religion* of 1871. In 1893 he gave a course of public lectures on the importance of doubt ('Confidence and doubt') that was a resounding success. He wrote also on Confucius, Lao Tzu, Zoroaster and Buddha, and on the history of many religions. His papers of literary criticism included a biography of Nicolaas Beets, the author of *Camera Obscura*. He was the dominant figure on the editorial board of *Onze eeuw* (Our age), the favourite magazine of Dutch intellectuals. A man of great culture and of wide interests, he considered morals to be the foundation of his attitude and behaviour – 'His power was in his ethics' [7, p. 33]. Perhaps it was this belief that brought him close to van der Waals, who often expressed the wish that Chantepie would give the oration at his graveside, but this was not to be[2].

Van der Waals had the honour of opening the first academic year of the new University. On 15 October 1877 he read his inaugural lecture, 'A short outline of the purposes of physical investigations' (unfortunately only the title survives) [5, p. 43]. After the festive inauguration came the humdrum working life. He had to begin from nothing, not even an empty space; all he had at his disposal was a collection of mainly eighteenth century apparatus. He did not obtain his own room until three years had passed, when the University moved into the Oudemannenhuis. The only study had, however, been allocated to physics and he had to adapt it for lectures, for laboratory classes and for his own research. It is easy to imagine how hard he struggled to expand his department, struggles that were successful at last when a new physical laboratory was built in 1882 on the *Plantage Muidergracht*, near de Vries's botanical gardens. A detailed description of the Physical Laboratory has recently been published [250]. It can be seen from the plans that it had separate rooms for research and practical classes in magnetism on the ground floor, and for heat, light, sound and galvanic and static electricity on the first floor. The rooms for magnetism and galvanic electricity were particularly spacious, about $70\,m^2$ each. The professor's study was $4\,m \times 7\,m$. The laboratory had accommodation for a concierge, a few additional rooms and a large lecture theatre lit by windows in the ceiling. This could seat 140, and was often packed, a fact that caused Professor Kuhn, the wit of the University Council, to exclaim 'I should like to be as famous as our colleague van der Waals' [5, p. 84]. He had been trying for several years to get space for morbid anatomy and envied van der Waals not so much his crowded lecture theatre as his ability to produce results. The University could not keep pace with the professors' needs and it was 1891 before van 't Hoff obtained a new chemical laboratory.

Here an explanation is needed for the overcrowding of so large a hall when there were only fifty or so students in the physical and mathematical faculty[3]. It is that the course of physics was given also to students of medicine, pharmacy, biology and philosophy[4]. At this time such lectures were not left to the lectors or 'extraordinary' professors, but fell to van der Waals himself. He gave both an 'experimental' or general course of physics and a 'mathematical' or theoretical course; both were required for the physicists and chemists, but the rest took only the first.

A former student, probably G. C. Gerrits, described the first lecture they had from van der Waals in 1896; it was called 'On polarization'. When the Professor opened the autumn session with the lecture he was greeted with loud hand-clapping from the new students. 'The little "great man" in a frock coat, with turn-down collar and black necktie, with hair already grey and with a short white beard, came in swiftly, walked rapidly to his place and looked around impatiently from the rostrum, knitting his brows, causing deep furrows in his sallow forehead and from his cheeks to his eyes. Looking around impatiently, he rubbed his hands briskly, waiting for an end to this spontaneous ovation. . . . Applause at the start of lectures was not the custom. . . . He began to lecture, looking around angrily and sharply clipping his words; "We have two kinds of phenomena; phenomena that belong to ponderable matter and phenomena that belong to imponderable matter."' Van der Waals's lectures were 'crystal clear, intimidatingly clear!' [194].

Nevertheless these lectures on general physics to a class of 150–200 students did not come easily to van der Waals, as the following passage shows, 'Every time he entered the lecture hall he had to accustom himself to speaking before so many unknown men. Every time the opening of the lecture cost him much effort until here and there he found a few familiar faces. . . . Then everything went smoothly, and henceforth he addressed them only. But always, even after 15 or 20 years, he was uneasy before the start of such a lecture' [187, p. 57]. It was quite the opposite with the lectures on theoretical physics; then there was an audience of only 30–40, he knew each one personally and the lecture became a dialogue.

The difference between the teaching of biological and medical students and that of students of his own department was most apparent during the examinations. F. A. F. C. Went, a well-known biologist who studied at the University from 1880–85 recalled, many years later, that van der Waals 'demanded much, he was not an easy man. Many of the students were afraid of him and went into the examinations in fear and trembling. This was not the case for those who took science seriously, for van der Waals treated all of them with great kindness. But he could not stand any trifling with science, which aroused his displeasure in hard words, spoken in a shrill rasping voice. It is, therefore, easily understood why those who did not know him well were afraid of him' [202, p. 216–17].

There were probably many insufficiently earnest students, for the examination results show vividly that many had good reasons for their fears. In most subjects the rate of failure was about 10%, but in physics it was much higher. Extracts from the *Jaarboeken* for the First Physical Examination show the numbers entering and the numbers passing in each year.

1881–82	40 and 32		1889–90	55 and 43	
1882–83	42	34	1890–91	67	47
1883–84			1891–92	63	50
1884–85	56	31	1892–93		
1885–86	43	21	1893–94	53	37
1886–87	55	38	1894–95		
1887–88	55	32	1895–96	56	34
1888–89	43	25			

Those who failed had to sit the examination a second or even third time; van der Waals had patiently to receive them back into his class again. These figures are for the physical and mathematical faculty, but those for the medical were similar. One of the medical students confessed afterwards that 'the first year at the university was the hardest of them all' [190 p. 328].

The physics students had an advantage which became clear later. Van der Waals knew each one well and was usually able to advise him if he was ready for the examination, so giving him confidence and reducing the stress. Nevertheless the examination was no mere formality; van der Waals's standards were high and particular. He did not want an encyclopaedic knowledge but a deep understanding of the essence of each problem. He knew that it was not the extent of a student's knowledge that was important but his understanding and his ability to apply his knowledge. Often in the *candidaatsexamen* van der Waals allowed a student to choose his own subject for examination, but then he expected a thorough command of it.

What did he teach during the years of his professorship? For the years 1877–84 there is little information; we know that he gave the two courses mentioned above and that he held classes for the more advanced students, particularly on 'molecular theories'. For later years there is more detailed information in the *Jaarboeken* of the University. The list below gives an outline of the contents of his two courses; the figures in brackets show the number of hours per week.

Year	*Experimental physics*	*Mathematical physics*
1884–85	(4) Light and electricity	(5) The Faraday and Maxwell theory of electricity, Theory of absorption by gases
1885–86	(4) Constitution of bodies, Molecular forces, Heat, Sound	(4) Mechanical theory of heat and its applications (2) Repetition of the treatment of the preceding course (electricity and magnetism)

1886–87	(4) Light, magnetism and electricity	(4) Kinetic theory of gases
1887–88	(4) Introduction, Mechanics, Hydrostatics, Heat, Sound, Molecular forces (1) Colloquium	(2) Theories of Ampère, Grassmann, Weber and Helmholtz, (2) The thermodynamic equilibrium of a single substance
1888–89	(4) Light, Magnetism and electricity (1) Colloquium	(2) Sound (2) Molecular theory of mixtures
1889–90	(4) Mechanics, Heat, Molecular forces, Sound (1) Colloquium	(4) Theory of the potential, Static electricity
1890–91	(4) Light, Electricity and magnetism	(2) Mechanical theory of heat (2) Static electricity, Spherical harmonics, Magnetism, Dynamic electricity, Ampère's theory, The potential of closed circuits
1891–92	(4) General properties of matter, Motion and laws of motion, Force of gravity, Heat, Molecular forces, Sound	(2) Electricity and magnetism (2) Equilibrium laws of mixtures
1892–93	(4) Light, Magnetism and electricity	(2) Mechanical theory of heat (1) Kinetic theory of gases (1) Potential theory
1893–94	(4) Forces and motions, Force of gravity, Heat, Molecular forces, Sound	(1) Mechanical theory of heat, the elementary laws, the equation of state (3) Potential theory, main properties of electric potential (1) Theory of magnetism and electricity, electromagnetic theory of light
1894–95	(4) Light, Magnetism and electricity	(2) Principle of equilibrium with applications to

		mixtures of two substances (2) Maxwell's theory of magnetism and electricity
1895–96	(4) Mechanics, Force of gravity, The teaching of heat and of molecular forces, Sound	(2) Static electricity (2) Treatment of the first and second laws of the mechanical theory of heat, Applications to a single substance, The law of corresponding states, The principle of equilibrium
1896–97	(4) Light, Magnetism and electricity	(2) Thermodynamics (2) Electromagnetism, Equations of electromagnetic field, Electromagnetic theory of light
1897–98	(2) Elementary treatment of modern theories of electricity	(2) Electricity and magnetism, Thermodynamics, general properties of an equilibrium (2) Kinetic theory of gases, explanation of thermal phenomena

From this list one can deduce that there was a two-year cycle for the course on general physics. Such a cycle was needed to cover the first principles of the whole of contemporary physics. The contents seem to change little from year to year although the details were probably changed as science advanced.

The courses in mathematical physics were not so broad; they concentrated on two main themes — thermodynamics, which sometimes included the kinetic theory of gases, and electricity and magnetism. Here the contents would certainly change from year to year, particularly in thermodynamics where they would reflect the advances of his own research. Other topics such as mechanics and optics were left to the general course.

All the evidence [190, 194, 202, 221] shows that the lectures on general physics were clear, exact and lucid, and that they were illustrated by convincing demonstrations. The lectures on mathematical physics were not so clear. He allowed the audience to interrupt with questions or objections. Such permissiveness does not make the lecturer's task an easy one, but he never evaded discussion. Often the seminars were a continuous session of questions and answers, to the detriment of a systematic exposition but to the benefit of training students to think about physical problems, as Kohnstamm remarked [193, p. 161–62]. Another student recalled: 'He is master of his subject; he understands it perfectly and thoroughly. If there is an unexpected question or difficulty he goes at once to the heart of the matter, and his curt words show at once how he is totally absorbed by the problem and boiling over with ways of tackling it. His lecturing is enthusiastic, strong and clear, with informal digressions and questions, and illustrated with figures on the blackboard. He was tireless; often he kept on for two hours with only a fifteen minute break between the two periods, during which he did not even sit down' [190, p. 328].

His seminars were attended by only a few students. There were four in the audience for the lectures on thermodynamics in 1881–82; one of these, Jan Valkenburg, who died young, was van 't Hoff's favourite pupil to whose memory he dedicated his *Etudes de dynamique chimique*. The other three all left their mark in science; J. J. van Laar (1860–1938), G. Bakker (1856–1938) and Ch.M. van Deventer (1860–1931). Van Laar, who remembered these classes to the end of his days, has left an interesting study of van der Waals's style of lecturing. 'A lecture is conceived mainly in an historical spirit. Often van der Waals cannot resist the temptation to introduce material on which he is currently working . . . which absorbs him totally. His lectures, therefore, often threatened to become rather fragmentary because of the need to make later changes when his studies have advanced further. In compensation, however, his students often receive first drafts of his papers and in this way experience for themselves how research advances, and understand better the difficulties posed by problems still unresolved' [186d′, p. 37][5].

The recollections of Ernst Cohen, who attended the University from 1886 to 1892, are rather juvenile but in character; 'Van der Waals's lectures on thermodynamics were unforgettable. Those who wanted to understand them had no choice but to work intensively. The number of students never exceeded a dozen so van der Waals could question them continuously to make sure that they were still keeping up with him. If he saw that any were not working hard then he got very angry, and did not hide it' [*13*, p. 498]. Sometimes he became excited, raised his voice and stammered slightly: 'I, I, I . . . had to work from four in the morning to eleven at night. I was then as young as you are, and I toiled. But you twiddle your thumbs, and that's the only work you do' [252].

It is worth adding here Zeeman's recollection of van der Waals's way of guiding the practical work in the laboratory. He gave the students an introductory talk before the experiments, and afterwards if anyone had any difficulty, 'He put several questions which at once narrowed the field in which the difficulty might lie, and, at the same time, show also how far the man who asked the question understood the problem. Idlers did not like this method but it was very effective' [205, p. 110–11].

Practical work in the laboratory, which is not included in the lists above, made his teaching load excessive. As a rule there were laboratory classes each day [5, p. 50–51], thus for the year 1891–92 there were twelve hours a week

[*16*, p. 51]. There were, however, also years in which the numbers of hours were not fixed (*onbepaald*).

The head of a department was, automatically, also head of the laboratory, the offices, and any dependent institutions. (Thus Hugo de Vries was the director of the *Hortus botanicus*, the Botanic Garden.) So after the establishment of the new Physical Laboratory van der Waals had to appoint new staff to superintend the laboratory work. His first colleague was Ph.C. Kaz who had studied at the University from 1877 to 1881. While a student he had already assisted van der Waals particularly in his work on the principle of corresponding states for which he had made many of the calculations needed to check the principle, as was acknowledged by van der Waals in the German edition of *Continuity* in 1881 [$1d_1$, p. 157]. In the autumn of 1882 he became the assistant in the Physical Laboratory, a post he held until 1896. At the same time a junior assistant (*amanuensis*) F. W. de Zwaan was appointed who worked under van der Waals until the professor's retirement. There was a second junior assistant from 1887 and, at last, in 1891 an assistant (extraordinary) professor was appointed. This was W. H. Julius (1860–1925) from Utrecht, a talented pupil of Buys Ballot who later became well known for his work on the physics of the sun. A born experimenter, Julius opened his career at the University with an inaugural lecture on 'The methods of research in physics', and went on to develop laboratory experiments for postgraduate students. He became Co-Director of the Physical Laboratory in 1893, a position that explains the rather vague phrase in a biographical sketch of van der Waals [216]; 'Julius gave van der Waals some help.' This man of an 'artistically fine soul', as Einstein wrote [*17*], could not carry the heavy burden of these duties and five years later returned to Utrecht as a full (ordinary) professor. Probably the impulse for this move came from an incident described by Cohen forty-three years later [*18*].

In January 1896 Röntgen had sent accounts of his momentous discovery of X-rays to a number of well-known physicists, including van der Waals. On 10 January Cohen repeated Röntgen's experiments on his own initiative, and the next day these were reported in a newspaper under the trenchant headline 'Taking photos through a wall'. In a paragraph below the paper stated that the facilities for these experiments had been provided by Professor Julius. Van der Waals was himself deeply interested in Röntgen's discovery, and the experiments in the Physical Laboratory had been carried out without his knowledge or consent. He was very displeased, and did not hide his indignation. 'After my explanations, however,' Cohen wrote, 'the sky became clear again' [*18*, p. 354], and good relations between van der Waals and Julius were restored.

This resumption of good relations did not, however, diminish the heavy teaching load that fell to van der Waals. It is astonishing how he found time to resume his own research at a time when he had to create from scratch a new department and two courses in physics. The list of his publications shows, however, many works from the years 1877–81. The most important of these were on the equation of state and on the thermal properties of fluids; it was at this time that he discovered the important principle of corresponding states which will be discussed later (See Chapter 6, Section 1). There was also work on a variety of other problems, such as geometrical interpretation of the propagation of sound [9], Korteweg's theory of ponderomotive forces [14], and an interesting report on 'Electrical phenomena in spaces of a high degree of rarefaction of matter or the so-called radiant matter' [13], which deserves a fuller consideration. It was the subject of the second of van der Waals's two appearances in *Diligentia*, mentioned in Chapter 2.

Van der Waals gave his lecture to a packed hall and impressed his audience by his strong clear presentation and by the range of models, instruments, plates and experimental demonstrations that he used. The experiments were repetitions of those used by Crookes in his address 'On radiant matter' to the British Association on 22 August 1879, which was published in *Nature* on 28 August and 4 September. Van der Waals referred also to a special leaflet, containing the text, that had been distributed at the lecture[6]. The equipment for his experiments was made by van der Waals himself, and one can sense some of the pride he took in describing their details. His 'secret' consisted in pre-heating the evacuated glass tubes for several hours at 300–400 °C before sealing them. Tubes containing rarefied gases made in this way gave reproducible results for up to eight or nine months. There is no need to list all the beautiful and elegant experiments since these are to be found in the account of Crookes's well-known lecture but their repetition is evidence of van der Waals's experimental skill.

The main interest in his lecture is in its theoretical content for his views were quite different from those of Crookes. From the beginning he said that he could not accept Crookes's claim that there was a 'fourth state of matter'. He could not accept a state of extreme rarefaction as a new state of aggregation since even the liquid and gaseous phases should be treated as a single state of matter. He observed that at the high vacuum reached by Crookes (and also by him) each cubic millimetre of air contained about 5×10^{10} molecules, and there were, besides, many more adsorbed on the walls of the vessel. 'The phenomena are best compared with our picture of the solar system in which the air is not distributed uniformly but is concentrated at the surfaces of the planets'[7] [13, p. 47]. He made many experiments of heating and cooling evacuated vessels to show just how great a quantity of gas could be retained on the glass walls. He turned then to electrical phenomena in evacuated tubes and, having repeated Crookes's beautiful experiments, he went on to say: 'Is it possible to explain these phenomena? Crookes's answer to this question is unfortunate; he stood here, like Newton and Huygens before him in their search for the nature of light, and faced the two alternatives: is the cause material or immaterial?' Crookes decided in favour of the first and so was led to the name *radiant matter*, but there were arguments put forward by Goldstein and others for the opposite view, namely that a special state of the ether is the cause of the phenomena. The problem was unsolved, but there was no cause yet to speak of a fourth state of matter. If the rarefied state were to be regarded as a new state of aggregation then there would be a legion of such states. Such expressions as *radiant matter* 'are permissible for the advertising, à la Edison,' of spectacular electric lights, but they are too reminiscent of claims to have split all the elements, or to have shown that chlorine is an acid, all of which have vanished like mist. So, van der Waals concluded, 'they should honour Crookes's experimental talent, but the way in which he published his results certainly cannot help the advance of science' [13, p. 54].

So severe a criticism cannot be matched in any other of van der Waals's papers, and this lecture seems to be the only early evidence of his scientific ethics. His position is the exact opposite of Crookes's romanticism; the brilliant poetical lecture of the English scientist obviously irritated him – it might be poetical but it was not scientific.

As well as showing his experimental skill this lecture also tells us of his interest in electricity and magnetism, a field that was second in importance only to molecular physics. After *Continuity* his second and third publications were both on electric discharges in gases. The first doctoral work done under his supervision was also in this field, for Korteweg's was not really in this category, as is explained above. It was the thesis of H. W. C. E. Bückmann (1855–?), *On electrical discharge in rarefied gases* [*19*] which he defended on 28 October 1881. As Bückmann himself recalled in 1908, experimental work was carried out only with great difficulty because of the lack of space and the poor equipment available in the early years of van der Waals's professorship. We know from Kamerlingh Onnes that much of the experimental skill came from van der Waals. The author studied glow discharges and so was working at a higher pressure than in the repetition of Crookes's experiments. He followed the changes of colour and the transitions from band to line spectra, trying to show that the stratification corresponded with the distribution of gas density along the tube. He studied also the influence of magnetic fields on these phenomena. It is not surprising that he was quite unable to interpret his results and that the first of his conclusions was that 'no existing theory is capable of explaining an electrical discharge in rarefied gases' [*19*, p. 77]. His main contention that any theory would have to explain the fact that 'the gas itself behaves as a carrier of the electric charge' [*19*, p. 73] was, however, perfectly correct. Among Bückmann's other discoveries was a spectral band belonging not to free nitrogen but to a compound with carbon, possibly HCN. This observation recalls that of van der Waals [3] in which he showed that some lines attributed by Watts to carbon in 1869 were, in fact, due to CO_2. Although the thesis was undoubtedly written by Bückmann alone it is not likely that he would contradict his promotor's views and so we may safely ascribe to van der Waals also Bückmann's general proposition which he gives in the form of a quotation from a paper by Hittorf: 'It is not impossible that the interaction of gases with electricity, like that with heat, will reveal most easily the essentials of the phenomena, and so rid modern physics from the last of the imponderables, namely those of electricity.'

The second thesis written under van der Waals's supervision was also on an electromagnetic problem, *On light reflected by magnets*, by Ph. C. Kaz, who was mentioned earlier. For this work van der Waals invented a new way of measuring Kerr's magneto-optical effect, that is, the rotation of the plane of polarization of light reflected from a magnetized surface. He proposed that both the polarizer and the analyser be rotated simultaneously until there was no rotation of the plane of polarisation [20], a procedure that was described forty years later in a history of Dutch physics as a real experimental advance [*20*].

The best evidence for the constancy of van der Waals's interests in these problems can be seen in the contents of his courses on theoretical physics. When, after twenty years, he allowed himself to lecture only on selected topics, those he chose were molecular physics, thermodynamics and topics in electricity and magnetism[8]. Kohnstamm wrote that his exposition of Maxwell's *Treatise on electricity and magnetism* was 'the most excellent culmination' of all his lecture courses [187, p. 57]. It is, therefore, not surprising that the last lecture that he gave on experimental physics, after the course had been handed over to the second professor, was on 'New theories of electricity'.

The first four years of Amsterdam were the high point of van der Waals's long life. He was at the height of his powers and enjoying every moment of his constant activity[9] – organizing the new department, lecturing, experimenting, mastering the thermodynamics of Gibbs, working as a member of the committee studying lightning conductors at Delft, reading manuscripts submitted for publication, and working on the German edition of *Continuity*, copies of which reached the Netherlands in June 1881 and which was soon to make his name known round the world[10].

His children were growing up in the happy atmosphere of the comfortable house on the *P. C. Hooftstraat*, but then disaster struck. Shortly after leaving The Hague Anna Magdalena became ill with consumption (tuberculosis). The family correspondence of those years contains many hints and references to Mother's health. She remained, however, the life and soul of the loving home and, when the disease eased a little, took full care of her husband and children. It seems as if her husband refused to face the inevitable outcome, and, until the very last day, did not believe that she would die. On 28 December 1881 Anna Magdalena 'passed away quietly and calmly after long but patiently-borne suffering', as van der Waals wrote on the card that announced her death [232, p. 10]. She was only thirty-four years old.

'The heavy blow hit him hard and nearly broke him' [206, p. 14]. A month later he addressed to his wife verses full of love, pain, and hope of meeting hereafter[11]. Life had to continue, however; he had his children to care for and his duties as a professor could not be evaded.

Fifteen-year-old Anne Madeleine took over the household duties, helped in the first few months by Chantepie's wife and, as far as she could, she acted as mother to her brother and younger sisters. Her father wrote to her on 27 September 1882; 'Today is the anniversary of the day that Mummy and I were married. I have been rather depressed lately and do not know if you will ever see me fully recovered again. I find consolation in the children, however, and am glad that they are on such good terms with me. Janti [the son] has for a long time been so loving and cordial that I am continually surprised at it. And the others do everything I could wish for. I hope that God blesses them because now that Mummy has departed they must be satisfied with the little they can get from their father' [232, p. 11].

The grief for his loss never left him. Kohnstamm, who became a close friend of the family fifteen years later, remembered, 'Only one shadow lay over this family, a shadow that the attentive observer could see was never absent – the death of his wife whom he loved with an uncommon devotion and strength. The shadow was manifest even in the curtains of the front room which were kept drawn, as was noticed for many years by every visitor[12]. Its presence was more obvious in the portrait that stood on the armchair beside van der Waals's writing table, piled up with his books and manuscripts. To touch the portrait or to put something down on the chair would have been sacrilege, as was clear to everyone who came into the room[13]. However, although the sorrow that lived on in the home was never fully banished, there was no dejection or bitterness' [206, p. 5–6]. He had always been reserved in his manner, but after his wife's death he became, to everyone but his family and a few intimate friends, the unsmiling severe figure whose image is stamped on the memory of all his younger colleagues who had not known him before this tragic event.

In 1892 he received an offer of the chair at Leiden; Rijke was retiring and a new Professor of Experimental Physics was needed. Amsterdam held its breath while its most famous scientist made up his mind. In the end he did not yield to the 'singing of Leiden's sirens', as the Rector put it in his Annual Report [*5*, p. 82], but remained true to his adopted university. Why did he refuse the offer when he was known to value highly his alma mater? Was it because of the as yet unfinished Physical Laboratory, or to be close to his wife's grave? It is probable that the principal reason was an inherent quality of van der Waals that made him a man of few but constant loves – he remained true to the choice he had made. It is therefore not surprising that he later held firm to this decision and refused tempting offers from all other universities[14].

Notes

1. For more details about Korteweg, see [*14*].
2. In 1888 Chantepie de la Saussaye was elected as a member of the Academy and from 1908–17 was the President of the Section of Philology. In 1899 he left Amsterdam and became a professor at Leiden, where he stayed until his death in 1920. His departure had no influence on his friendship with van der Waals. See the biographical sketch [*15*].
3. The table that follows is compiled from the *Jaarboeken* of the University and presents the numbers of those entering the University and the Physics and Mathematical Faculty. For years from 1890–91 the *Jaarboeken* gave also the total number of students in the University and its faculties. These figures allow us to comprehend the size of student body in the University during its first twenty-five years.

Academic year	Number of students entering		Total number of students in the University	Total number of students in the Physics and Mathematical Faculty
	the University	the Physics and Mathematical Faculty		
1878–79	95	15		
1879–80	73	11		
1880–81	78	9		
1881–82	123	18		
1882–83	63	10	487	
1883–84	124	21		
1884–85	147	21	557	
1885–86	126	20	585	
1886–87	165	34		
1887–88	167	18		
1888–89	168	23	733	
1889–90	164	34	809	
1890–91	176	31	826	78
1891–92	182	37	966	103
1892–93	186	39	1019	151
1893–94	205	36	1039	120
1894–95	175	24	949	98
1895–96	162	24	917	101
1896–97	170	32	897	91
1897–98	157	29	850	107
1898–99	191	42	867	109
1899–1900	165	33	835	110
1900–1	169	40	883	120
1901–2	183	32	857	127

4. It is worth mentioning three of the first medical students who studied physics under van der Waals and became afterwards well-known scientists: C. Eykman, graduated *cum laude* in 1883, Nobel prize winner for his work on vitamins; H. Zwaardemaker, graduated *cum laude* in 1881, who did outstanding work on the physiology of human hearing and invented some medical devices, such as sound-proof rooms, etc.; Eugen Dubois, the discoverer of the fossil man of Java, later a colleague of van der Waals in the Physics and Mathematical Faculty.
5. Van Laar described Lorentz's lectures as follows. They were clear and complete, but 'a listener taken in by the clarity and perfection does not feel the exceptional difficulties of some problems and dreams that everything is simple. Only afterwards, during his own independent study does he realise that the field is full of problems' [186d′, p. 36–37].
6. This and some other details allow us to deduce that van der Waals probably visited England during the summer vacation of 1879.
7. Note once more van der Waals's astronomical interests.
8. To complete the information about van der Waals's courses we give here the list for his last decade. (The number of hours per week is indicated in brackets.)

1899–1900 *Molecular forces* (1): capillarity
Electricity and magnetism (2): the state of electric and magnetic fields
Thermodynamics (2): applications to the equilibrium of a single substance
Mathematical physics (2): Green's theorem and spherical harmonics

1900–1 *Mathematical physics* (2): magnetism, electromagnetic induction, propagation of an electrical pulse
Thermodynamics (2): mixtures of two substances
A detailed consideration of a *chapter of experimental physics* for intending *candidaaten* (2): thermal phenomena, caloric phenomena, two laws of thermostatics and their consequences, thermal conductivity

1901–2 *Chapters of physics* for intending *candidaaten* (2): static electricity, magnetism and induction
Mathematical physics (3): potential, static electricity, comparison of Laplace's and Poisson's equations, inversion of spherical harmonics
Thermodynamics (2): the first and the second laws of thermodynamics, the equilibrium principle, external forces acting on a substance

1902–3 *Special chapters of physics* for intending *candidaaten* in mathematics, physics and chemistry (2): heat
Thermodynamics (2): binary mixtures
Mathematical physics (3): magnetism, properties of electromagnetic fields, induction, electromotive and ponderomotive forces, Hertz's oscillator, electromagnetic theory of light, reflection and refraction, double refraction, focus, light source in a magnetic field

1903–4 *Mathematical physics* (3): static electricity, Green's theorem of the potential, spherical harmonics, electric charge images, potential and force lines in a two-dimensional space
Mechanical theory of heat (2): consequences of the second law of the mechanical theory of heat, the principle of equilibrium, influence of external forces
Special chapters for intending *candidaaten* in mathematics, physics and chemistry (2): static electricity, magnetism, induction currents

1904–5 *Mathematical physics* (3): electromagnetism after Maxwell's theory, Hertz's oscillations, propagation of electric vibrations, Lecher's problem, electromagnetic theory of light, principles of the theory of electrons
Thermodynamics (2 and 1): binary mixtures, thermodynamic theory of capillarity
Selected chapters for intending *candidaaten* (2): the chapter of heat concerning thermal and caloric relations for a one-component substance

1905–6 *Physics* for intending *candidaaten* in mathematics, physics and chemistry (2): static and flowing electricity, static and dynamic induction, alternating currents, wireless telegraph, magnetism
Thermodynamics for intending *doctorandi* in mathematics and physics (2): the laws of thermodynamics and the principle of equilibrium of a single substance, influence of external forces, capillarity
Theory of static electricity for intending *doctorandi* in mathematics and physics (2): Green's theorem and its applications, spherical harmonics, potential and force lines in a two-dimensional space

The equilibrium of a solid with a fluid phase (1): phenomena displayed when a solid phase separates from a binary system

1906–7 *Chapters of physics* for students in mathematics, physics and chemistry of the second and third years (2): heat, the essence of heat, equation of state, geometric locii of maximum and minimum pressure, minimum value of the product pV, boundary line etc., the equivalence of heat and work, Carnot's cycle, entropy, isentropic relations etc.

Theoretical physics and electricity for intending *doctorandi* (3): 'Maxwell's relations, electromagnetic theory of light, double refraction, light sources and focus, problems in the theory of electrons

Thermodynamics for intending *doctorandi* (2): theory of binary mixtures; course of p-lines, q-lines and potential lines, spinodal and binodal curves

1907–8 *Mathematical treatment of static electricity* (2): Green's theorem, induction, theory of images, spherical harmonics, Schwarz's theorem

Theory of a binary system (1): course of p- and q-lines and potential lines, mechanical theory of heat, electric displacement and force

Treatment of the chapter on electricity for intending *candidaaten* in mathematics, physics and chemistry (2): condensers, magnetic induction and force, problems of electric currents and induction

9. In 1879 van der Waals joined the Mathematical Society, just after its centenary. The Society had the motto *Tireless labour overcomes everything*, which was surely dear to van der Waals.
10. This German version of *Continuity* became the real origin of van der Waals's theory for most of his contemporaries and for the next generation of scientists, and so it was sometimes believed that van der Waals's equation appeared in 1881. See, for example, Max Born's classic *Atomic physics* (1935).
11. This is the best place to present an example of van der Waals's verses [232, p. 12].

Zou dan, toen gij werdt verheerlijkt
In een meer volmaakten staat
Uwe liefde zijn verdwenen,
Als een zon die onder gaat?

Neen voorwaar, dat is niet mooglijk.
Die uw harte heer gekend,
Die in vreugde en smarte beiden
Aan Use liefde was gewend

Weet, die min was onvergankelijk;
Die was met Use wezen één.
Die werd, toen gij werdt geheiligd
Zelfs nog reinder dan voorheen.

Waarom kan ik ze dan niet voelen
Als toen gij nog bij mij waart?
Wordt dan reine hemelliefde
Niet den mensch geopenbaard?

Kan de Heer dan niet gedoogen,
Dat gij mij Uw liefde toont?
Is het troosten dan verboden
Aan degeen, die bij Hem woont?

Maar op al dat angstig vragen,
Op dien twijfel, op die klacht

Klinkt mij nooit een stemme tegen,
Die mijn bitter leed verzacht.

't Graf blijft zwijggen – Van de groeve,
Waar Uw stoflijk kleed in rust,
Keer ik ongetroost steeds weder,
Dieper van mijn leed bewust.

Toch blijf ik, lieve! aan U hangen,
Blijf ik smeeken onverpoosd,
Dat ik door een enkel teeken
Van Uw liefde word getroost.

12. These mourning curtains were taken away only in 1902 when his son married [232, p. 11, 175].
13. According to his grandson's recollection, it was '*her* chair. Even his children would never lay a hand on the high back of the chair when they were speaking to him' [242].
14. We may note here a contrast with van 't Hoff, whose less stolid nature scarcely stood up to the enormous task of lectures and examinations. After he had refused several offers, he ultimately left the Netherlands for Germany in 1896, see [*21* and *13*].

References

1. Fromentin, E., *The Masters of Past Time*, London, 1913.
2. *'s Gravenhage*, Den Haag, 1923.
3. Bosscha, J., On secondary education in the Netherlands. In *The Health Exhibition Literature*, London, 1884, v. 16, Conference on Education, Section D, Training of teachers, organisation of intermediate and higher education, p. 197–201.
4. Willink, B., Origins of the second Golden Age of Dutch science after 1800: intended and unintended consequences of educational reform. *Soc. Stud. Sci.*, 1991, v. 21, p. 503–526.
5. *Jb. AU (1877-1884)*, Amsterdam, 1885.
6. Saltet, R. H., Barend Joseph Stokvis, *P. C.*, 1902, 4 Oct., 14 Jg., N2, b. 17–18.
7. *Levensber.*
8. Korteweg, D. J., *Over de voortplantingssnelheid van golven in elastische buizen*, Leiden, 1878.
9. Korteweg, D. J. and de Vries, G., On the change of form of long waves advancing in a rectangular canal and on a new type of long stationary waves, *Philos. Mag.*, 1895, v. 39, p. 422–443.
10. Korteweg, D. J., Ueber Faltenpunkte, *Wien. S.-Ber.*, 1889, Bd 98, Abt. 2a, S. 1154–1191; *A.N.*, 1891, t. 24, p. 51–98.
11. Korteweg, D. J., La théorie génerale de plis et la surface ψ de van der Waals dans le cas de symmétrie, *A.N.*, 1891, t. 24, p. 295–368.
12. Korteweg, D. J., Plaitpoints and corresponding plaits in the neighbourhood of the sides of the ψ-surface of van der Waals, *Proc.*, 1902–3, v. 5, p. 445–465.
13. Cohen, E., Chemie en chemische industrie te Amsterdam in den loop der eeuwen, *Ch. W.*, 1940, Jg. 37, b. 486–502.
14. Beth, H. J. E. and van der Woude, W., Levensbericht van D. J. Korteweg (31 Maart 1848–10 Mei 1941), *Jb. KAW*, 1945–6, b. 194–208.
15. Kuiper, K., Levensbericht van Pierre Daniel Chantepie de la Saussaye, *Jb. KAW*, 1920–21, b. 103–128.

16. *Jb. AU, (1891–1892)*, Amsterdam, 1892.

17. Einstein, A., W. H. Julius (1860–1925), *Astrophys. J.*, 1926, v. 63, p. 196–198.

18. Cohen, E., Vijftig Jaren Revolutie, *Ch. W.*, 1939, Jg. 36, b. 351–356.

19. Bückmann, H. W. C. E., *Over elektrische ontlading in verdunde gassen*, Academisch proefschrift, Amsterdam, 1881.

20. Kuenen, J. P., *Het aandeel van Nederland in de outwikkeling der natuurkunde gedurende de laatste 150 jaren* (Gedenkboek van het Bataafsch Genootschap der Proefondervindelijke Wijsbegeerte te Rotterdam ... 1769–1919), Rotterdam, 1919.

21. Cohen, E., *Jacobus Henricus van 't Hoff. Sein Leben und Wirken*, Leipzig, 1912.

CHAPTER 5

At the Royal Academy of Sciences

Van der Waals was a member of the Academy for almost forty years. Here he sent all his papers, and here he showed himself to be a leading organizer of science.

From the earliest days of his membership he was one of the most conscientious attenders in his Section, that is, the Physical and Mathematical Division (PMD). But after the meeting on 24 December 1881, when he had presented a report on the possible dependence on temperature of the internal pressure of liquids and gases [19], his name vanished for three semesters from the reports of the monthly meetings. For a year and a half a member of the Academy ignored its meetings, without even an apology for absence or explanation of why he would not attend. He ran the risk of expulsion, since a member who failed to attend for two years without a valid reason was expelled automatically. After the death of his wife, however, van der Waals had lost all interest in scientific gatherings. It was perhaps with the aim of bringing him back to active life that the Section elected him to be a Vice-President. This election *in absentia* took place in the spring of 1883 and seems to have achieved its purpose since van der Waals started to attend again in the autumn semester of that year. The next autumn he gave a report [20] on the magneto-optical investigations of his first assistant, Ph. Kaz.

This administrative post was, however, largely ceremonial, for during the thirteen years he held it he had to chair only twelve meetings. At first he deputized for Buys Ballot and later for van de Sande Bakhuyzen, so his duties were not too onerous. His organizational talents and his devotion to the Academy and to the cause of Dutch science were, however, first revealed during these years.

There is some evidence from his contemporaries (e.g. [186]) to suggest that van der Waals was an unworldly man, somewhat withdrawn from affairs. It is, therefore, easy to agree too hastily with this view and to depict him as a recluse who was helpless in practical matters. Such a representation bears no relation to the real van der Waals, who was a man of sound common sense, of unique self-discipline, of enormous capacity for work, and with a remarkable gift for organization. He probably first gained such experience in the HBS where he was in charge of twenty teachers and more than 200 pupils. Later he would not have been able to have run his department in the University without a good grasp of practical matters. It was, however, in the Academy where his talent for organization, combined with his aspirations for the renaissance of Dutch science, finally blossomed. It was here that he displayed his 'academic patriotism'.

A characteristic example is shown in the minutes of the discussion on the proper place for keeping the standard metre scale and kilogramme mass in 1887. This was the time of an international effort to promote the use of the metric system, and the Government proposed that the Academy should hand over these standards to the Polytechnic at Delft, a move initiated by Bosscha, the Director of the Polytechnic and the leader of Dutch metrology. He insisted that the standards should be kept where they were used, that is, in Delft. Van der Waals asked at once how often the standards were used by the metrologists at Delft, and Bosscha had to confess that this was seldom necessary, perhaps once every ten years. Then followed one of the few of van der Waals's contributions to such discussions that have been recorded in full – 'Mr van der Waals cannot support Mr Bosscha's view. He admits that the advantages claimed by Mr Bosscha for keeping the standards in Delft have some truth in them, but suggests, nevertheless, that they have been greatly exaggerated in several respects. Mr van der Waals himself prefers to consider the question under discussion from a purely scientific point of view. From this approach to the problem it cannot be seen to be desirable either that the Academy, as the supreme scientific body of the country, should be debarred from its function of holding the standards, or for the standards themselves to be withdrawn from the control of the Academy. He suggests that it is wrong for questions concerning the standards to be decided by the criteria of the techniques of measurement. Mr van der Waals therefore considers it inexpedient to keep the standards at a place where metrologists are taught. Until there is a special Faculty for Measurements in the country then, in his opinion, such a decision would not promote the maintenance of the quality of the standards' [150]. Thus did van der Waals oppose a colleague to whom he owed much, giving his views politely and tactfully but firmly. It was the cornerstone of his position that nothing should encroach on the primacy of the Academy in all matters of science; it was, for him, the key factor needed to promote the advance of Dutch science, and he took every opportunity to increase its rôle and authority.

This position explains why it was that van der Waals consented to become Secretary of the Section of Sciences (the PMD) in place of C. A. J. A. Oudemans who retired in 1896[1]. Almost at once he was elected as General Secretary of the Academy, following the arrangement already mentioned in Chapter 3 according to which the duty of General Secretary was laid on one of the two Secretaries of the Sections of the Academy. The Section of Philology had nominated its Secretary Bellaar Spruyt for the post of General Secretary, but in the ballot at the Annual Meeting on 25 April 1896 van der Waals polled 24 votes and Bellaar Spruyt 15 [*1*]. In 1901, 1906 and 1911 there were no other nominations and van der Waals was returned unopposed.

This post meant more work than van der Waals had had before since it was the General Secretary who ran all the business of the Academy; the scope of his duties was very wide. As Secretary of the Section he had to give notice in good time of forthcoming meetings in the official publication *Nederlandse*

Staats Courant, to arrange the publication, distribution and safe-keeping of all printed matter of the Section, and to read every letter and report sent from it. As General Secretary, he had to organize the Annual Session of the Academy, to prepare and read the annual Reports on all its activities, to supervise the Library (publishing lists of new accessions and reading the annual accounts), to distribute all the printed material of the Academy, taking care to provide copies for members of the Royal family, for the Chairmen of both Houses of Parliament and for all the members of the Academy, and finally to supervise the Academy archives and even its warehouse.

One cannot but wonder how van der Waals managed to get through all this work with the assistance of only a couple of clerks. Papers brought to him by the clerk were read and signed at odd moments during the practical classes in the laboratory. But when a matter was important he would spare neither time nor effort, as can be seen by his work on the committees of the Academy, which he often chaired. The rules required that a committee be set up each time the Government put a scientific or technical point to the Academy and wanted a recommendation. Particularly frequent were requests for advice on the design of lightning conductors for public buildings. Such requests seemed to be particularly common throughout Europe at this time. In Britain, Maxwell and later Ambrose Fleming and Oliver Lodge were called upon for advice, and in Germany Kirchhoff, Helmholtz and Siemens were all members of the Lightning Conductor Committee of the Berlin Academy of Science.

In Amsterdam the Lightning Conductor Committee had three members, of whom from 1878 to 1902[2] van der Waals was always one. At first his colleagues were Rijke and Bosscha; then C. H. C. Grinwis (1831–99), a physicist from Utrecht, replaced Rijke, then he was replaced by Lorentz, and finally Kamerlingh Onnes took Bosscha's place. Some of the buildings that were protected during van der Waals's membership of the Committee were the arsenal, powder magazine, cartridge factory and iron foundry at Delft in 1879, the mental hospital at Delft in 1882, the new Rijksmuseum, the chef d'oeuvre of Cuypers, in Amsterdam in 1887–88, and the Mauritshuis, the old palace at The Hague, which contained almost as rich a collection of paintings as the Rijksmuseum, in 1895–96.

The records of the Committee tell us something about how it went about its work. All aspects of the problem were given careful consideration, from the corrosion resistance of materials, to an analysis of general questions on methodology and the discussion of the scientific extent of our knowledge of the problems of lightning. Three features mark every investigation, first, a study of the building to be erected, usually with a visit to the site, second, a consideration of recent experience in other countries, and third, simple physical considerations that avoided too much theory and speculation. It is interesting to see that there was no tendency to play safe in spite of the burden of responsibility. On the contrary, the reports show an obvious intention of playing down fears excited by technical ignorance. An analysis of the problems of protection in Delft is illuminating in this respect. In 1879 the municipality had planned an extension to the arsenal but the principal Supervisory Engineer of The Hague and Delft protested in view of the danger to such an extension if adjacent houses were to be struck by lightning. After a full study, set out in its reports, the Committee found that they had to pay attention not only to the arsenal but also to factory buildings surrounding it, and that there was sufficient protection if the lightning conductor were installed on the roof of the watchman's house standing between the factory buildings.

Interesting examples of the work of the Committee can be found in the protection of famous Dutch museums, such as the Rijksmuseum which was finished in 1886. Since this was to house the national collection of paintings it was natural for the Academy to be called on for its expert advice. The instruction to them was signed by the Minister of Internal Affairs on 3 June 1887, and by the 25th Bosscha was able to read the report of the Committee at a meeting of the Section of Sciences. Assisted by Cuypers who had given them working drawings together with oral and written explanations, the members of the Committee studied in detail the ramified network of conductors proposed for the new building. They concluded that the system[3] would be effective but recommended some changes to give the installation a longer life [138].

For the new lightning conductors for the Mauritshuis in 1895 the Committee again approved the use of Melsam's system of protection but with some simplifications after the pattern of Oliver Lodge to make it less costly [140, 141]. This report was presented by van der Waals; the next year he put forward, on behalf of the Committee, proposals for a more rational organization of the service for protection against thunderstorms, which included a new post of inspector of lightning conductors, who should be an engineer or physicist with good qualifications in the field of electricity, who could inspect existing installations and control work in progress. The Academy, van der Waals said, in no way intended to abandon important practical problems but it should not be involved in every individual case; it should be able to investigate the really difficult matters of principle, and to be consulted in problematic cases where the inspector thought it necessary. After that report the Committee, now chaired by van der Waals, undertook only two further studies, in 1897 and 1902. In the preceding fifteen years they had made nine investigations. Such a reorganization of the method of working was typical of his contribution to the activities of the Academy.

He was involved in many other committees of the Academy but we need consider only two of them, which were both important and characteristic of his work.

In January 1896 the Academy received an unusual query from the Ministry of Justice. It had become apparent that prisoners in solitary confinement were able to exchange words, so the Minister wanted the Academy to find an inexpensive way of stopping this. The Chairman of the Section, van de Sande Bakhuyzen, proposed to set up a Committee comprising van der Waals, Kamerlingh Onnes, Lorentz and van Diesen. Van der Waals, unlike the others, asked for a few days reflection before taking on this ticklish job but, having consented, he then became Chairman. He ensured that there was a serious engineering investigation using what is now called a 'systems analysis' approach. The Committee met the staff of the Ministry, viewed all seven Dutch prisons, examined the state of the problem in France, Germany, Britain, Austria and Belgium, ordered special measuring instruments that were made in the Leiden physics laboratory, and consulted specialists in physiology and medicine. After a year's work the Committee produced a detailed report [146], which revealed that prisoners had been able to talk to each other through the ventilation openings, by tapping messages via the central heating tubes, and even simply by the transmission of sound through the walls. In both Dutch and foreign gaols the ventilation system was the most commonly used method. They confirmed this by direct experiment; it must have been a comic sight — the leading lights of Dutch science trying to speak from one cell to another via ventilation holes. The principal recommendation of the Committee was therefore a change from natural ventilation to a system of forced ventilation, a change for which, they said, there were also strong hygienic arguments. A section of the report contains an analysis of the problem of making a forced system impervious to sound. They discussed also the psychological effects on the prisoners of being forbidden to open the windows, a ban that they considered inhumane and admissible only for a short time as a punishment. They set out also other features of solitary confinement that had bad effects on the physical and mental states of the prisoners, and which should therefore be eliminated.

The whole report is an extremely interesting document covering a wide range of problems arising from the original assignment, beginning with thermophysics and ending with economics and ethics. Such a complex set of investigations had clearly required careful planning. It is significant that this is the only one of van der Waals's many reports for the Academy to be included in the *Catalogue of Scientific Papers*, the Dutch part of which was compiled by Korteweg.

It happened that doubts about the recommendations of this Committee were expressed in Parliament during a discussion on the budget in 1898, and published in an official document, a *Memorie van Antwoord*. This gives us an unusual opportunity to see how firm and tough van der Waals could be in a field that had little connection with physics. He replied to this document with a long letter from the Section of Sciences to the Minister of Justice in which it was argued convincingly that the Committee had carried out its task with great skill, and that the authors of the official document either did not understand their report, or did not wish to understand it. It is worth citing the preamble to the letter as an example of dignity and firmness. 'According to Rule 2 of the Regulations of the Royal Academy of Science, as approved by the Royal Edict of 23 February 1855, it is the first duty of the Academy to serve as an advisory body to the Government in the field of science, ... If its work is to be fruitful then the Government and its official employees must have no misconceptions about the interpretation of reports of the Academy. Whenever the Government has been in doubt and has applied to the Academy for elucidation of the meaning of a report, the Academy has considered it an honour to provide that elucidation. But when the Government, without making such an application, gives official evidence of its misapprehension of a report, then the Academy believes that it is its duty to correct that misapprehension, whatever the circumstances that gave rise to it. This unpleasant duty the Academy now discharges in appealing to Your Excellency' [147, p. 361–362]. The Minister of Justice asked the Academy to reconsider the question, but the Committee, supported by the Section of Sciences, merely repeated its recommendations and stated that it considered its work to be finished. Van der Waals was obviously much affected by this incident since he set it out in full in the Annual Report of the General Secretary [*3*], in contrast to the usual short mention of Committee's activities.

At almost the same time he was involved in the work of a Committee of quite a different kind. At the February meeting of the Section in 1896 there was read an important letter from the Minister of Internal Affairs on a matter of great concern to the Physical Laboratory of Leiden University. The Minister enquired about the danger from and permissibility of investigations that covered a wide range of uses of compressed gases. He attached a memorandum from a special Committee of the States General of South Holland (the province to which Leiden belongs) in which there was an appeal to the Government to forbid such investigations, since many people and some organizations had protested against the use of compressed gases; they had been frightened by rumours of the extreme danger of working with compressed hydrogen. The protesters were joined in their complaints by the municipality of Leiden, the Leiden section of the Society of Public Benefit, the University curators and Professor T. Zaayer, a member of the Academy and Director of the University Anatomy Theatre which was situated between the Physical and the Chemical Laboratories. Extensive explanations from Kamerlingh Onnes were attached to the memorandum, but these had not satisfied the plaintiffs. The Minister needed the opinion of the Academy before he could come to a decision, and a Committee was set up to formulate that opinion. Its chairman was van der Waals and the other members were Korteweg, S. Hoogewerff, Professor of Chemistry at Delft Polytechnic, and C. Lely, a leading engineer in hydrology who afterwards led the ambitious scheme for draining the Zuiderzee. The Committee immediately asked Kamerlingh Onnes what were the safety measures adopted in foreign cryogenic laboratories, so he wrote to the other leading experts on gas liquefaction – James Dewar in London, Karol Olszewski in Cracow and Raoul Pictet in Berlin.

Van der Waals was able to present the report of the Committee at the May meeting of the Section [145]. He said that the Committee had first studied carefully the materials used in Kamerlingh Onnes's laboratory, had then collated the information from London, Cracow and Berlin, had considered foreign practice and safety measures for the manufacture, testing and storage of vessels for use at high pressures, and had finally examined all foreign records of accidents that had resulted from the bursting of vessels containing compressed gas. From all this evidence they concluded with confidence that there were no grounds for banning the use of compressed gases in research at Leiden. Moreover, van der Waals reported, it was not strictly correct to use the word 'explosive' here since this word, with its connotation of extensive destruction and human casualties, should be applied only when there is a sudden release of chemical energy as, for example, when gunpowder or dynamite explodes. (Van der Waals is, no doubt, referring here to the notorious explosion of a ship on the canal, which was the original cause of the Leiden 'Ruins'.) When a vessel containing a compressed gas bursts it is only physical energy that is released and, for typical volume and pressures, the energy of expansion of the gas is equivalent to the chemical energy of no more than 3 kg of gunpowder. For such quantities there are no special rules or limitations either on its storage or carriage.

The Committee found that the safety measures required by Kamerlingh Onnes, and carried out in practice, were quite satisfactory. Van der Waals noted wryly that the length of this list of precautions was taken as evidence by the protesters of the extent of the danger, so that the scrupulousness of Kamerlingh Onnes had added to the number of protests.

Supporting letters from London, Cracow and Berlin were added to the report. Dewar's reply was particularly forthright; he wrote ('in haste'), 'It would be a terrible disaster for science in your country (and universal science) if the municipality of Leiden succeeded in carrying out any restrictions on your splendid cryogenic laboratory and the fine work you are doing. I cannot understand such a position. . . . I may say that I have made all my experiments with high pressure apparatus before the Prince of Wales and the Sister of your Queen Dowager the Duchess of Albany without the slightest hesitation and no suggestions of danger were even suggested' [145, p. 12].

It was noted in the report that five years previously one of the members of the Committee had had the opportunity to inspect closely Dewar's equipment in the Royal Institution in London. It was situated just a few steps away from Piccadilly, one of the finest streets in the British capital, and nobody was frightened by this situation. 'One of the members of the Committee' was in fact van der Waals himself [*4*].

The last, but not least, of the arguments in the report was that in densely populated and built-up cities there were factories and plants that were potentially much more dangerous than Kamerlingh Onnes's compressed gas cylinders. To meet the wishes of Professor Zaayer the Committee recommended that there was no increase in the number of cylinders near the Anatomy Theatre but noted that there was a much more serious and real danger to the Theatre from the Chemical Laboratory if there were to be a fire. 'We think,' van der Waals concluded, 'that the Academy may certainly recommend that the Minister gives unconditionally the permission asked for' (that is, for the continuation of the work of the Cryogenic Laboratory) [145, p. 10]. The report was put to the vote and approved, but then one of the members of the Section stood up and declared that he wished to dissociate himself from the decision just taken; it was Professor Zaayer, the Director of the Anatomy Theatre.

Van der Waals was of course not an unbiased onlooker at Kamerlingh Onnes's single-handed fight with the reactionaries of Leiden, for the work of the Cryogenic Laboratory was at the focus of his scientific interests, and he was on close terms, both professionally and socially, with its Director. There is, however, no doubt about the honesty and objectivity of his work as Chairman of the Committee. Its conclusion rests on the same principles as the other reports on protection against lightning, namely an analysis of experience elsewhere in Europe, a sensible evaluation of all the circumstances, and an avoidance of a policy of playing safe or 'double insurance'.

It was only in 1898, after many bureaucratic procrastinations, that Kamerlingh Onnes got official clearance for resuming and expanding his research. During the next decade the Cryogenic Laboratory was to become a world centre for the physics and technology of low temperatures. It is no reflection on Kamerlingh Onnes's abilities to say that none of this would have been possible had van der Waals's Committee come to any other conclusion. Many years later W. H. Keesom, the successor of Kamerlingh Onnes, wrote about the episode: '. . . during this period the laboratory had passed through a very serious crisis . . . which arrested the work for some years and which might have been fatal to the cryogenic laboratory' [5].

There was another and possibly more important side of van der Waals's work for the Academy, and that was in managing its publications. The *Verslagen* (Proceedings) of the Academy became his favourite vehicle for organizing science. Hitherto the Academy had published *Verhandelingen* (Memoirs), which appeared irregularly, and *Verslagen en Mededelingen* (Proceedings and Communications) which was a periodical, but was rather archaic in form and flabby in content. It contained accounts of meetings of the Section, scientific communications, reports on papers presented for publication and the papers themselves, reports of committees and, finally, in a section with a separate pagination, a list of books acquired by the Library. From the summer of 1892 this amorphous publication was transformed by van der Waals into an up-to-date and concise scientific monthly journal devoted to the publication of original scientific papers; only some short accounts of meetings remained to remind one of the old tradition. It is true that the old-fashioned *Verslagen en Mededelingen* are more appealing to the historian of science, with their many minute details of the life of the Academy, the lists of members present at each meeting, authors' abstracts of their papers etc., all of which have been excluded from the new version. But for those who wanted the latest scientific results or who wished to exchange such results with others, the new *Verslagen* sufficed, and these were the aims that van der Waals had in mind.

From 1900 the Academy began to publish also an English edition of the *Verslagen*. This edition was also organized by van der Waals and soon became known internationally as the *Proceedings of the Section of Sciences*[4]. It has played an important role in making known to the whole world the contributions of Dutch scientists.

He put much time and effort into the publishing of the *Verslagen*, and in 1902, when the volume of work became too much for him, he managed to establish a special post of assistant secretary to help with the publications. The first to be appointed was his son, J. D. van der Waals Jr., and two years later, when he had become a Professor in Groningen, he was succeeded by Kohnstamm. He also persuaded his youngest daughter, Johanna Diderika, a teacher of English, to take up publishing; she worked as a translator and proof-reader of the *Proceedings*, for which she received fl. 200 a year.

One final illustration can be given of van der Waals's concern for the *Verslagen* and his sound views on the organization of labour. In 1898 the question of whether telephones should be installed at the Academy was hotly debated at the Annual Meeting. This innovation would cost a lot of money and so aroused opposition. The General Secretary was a convinced advocate of technical progress; he argued that telephones, by reducing the time taken to call for aid, would diminish the risk of fire and that 'the publication of the *Verslagen* is held up by the large volume of correspondence that is now necessary but most of which would be avoided if telephones were installed' [*6*]. He insisted that the first telephones were installed in the Section of Sciences and in his house, so that he could do his editorial work at home.

It was hardly surprising that, when the Academy took its farewell of its retiring General Secretary, they chose as a gift for him complete sets of the *Proceedings* in both Dutch and English. Lorentz, the President of the Academy, addressed his old colleague and friend and reminded him how, sixteen years ago, he had fallen in with the unanimous wishes of his colleagues and taken up the post of Secretary. 'We asked ourselves then', Lorentz went on, 'if perhaps we asked too much from you. Dare we expect that a mind that could penetrate the mysteries of Nature, that the hand that had written *Continuity*, would bear the heavy burden of the Secretary's trials. Our apprehension that it might be too heavy for you turned out to be groundless. Your name lost none of its brilliancy. With youthful enthusiasm you undertook all the Secretary's duties and showed a talent for organization that is rarely found among scientists' [*7*].

Why did van der Waals voluntarily take on himself this 'heavy burden of the Secretary's trials'? He was totally devoted to science and quite free from ambition, so that he always declined the most senior posts; he never, for example, became Rector at any time during his thirty years at the University. The direct answer was given by Kohnstamm: 'The idea that the Academy must be the centre of the scientific life of the Netherlands, that it must be the focus of all advances made by Dutch scientists, but with full acknowledgement of the international character of science, and so lead to a revival and strengthening of national consciousness – this idea was what always guided van der Waals in the performance of his duties at the Academy' [197].

Notes

1. The Vice-President's post was occupied by Stokvis at this time.
2. Probably the Commission continued in being after this date but there is no indication of further activity [*2*].

3. Cuypers had used the system invented by the Belgian physicist, A. Melsam. Its principle is that of a Faraday cage; that is, that an electric discharge is impossible inside a metal cage.
4. These *Proceedings*, unlike the *Verslagen*, contain no minutes of the meetings of the Section of Sciences since these are on purely domestic subjects; only the bare texts of communications presented are published.

References

1. *Jb. KAW* voor 1896, b. ciii.
2. Archief Kon. Nederlandse Akademie, inv. no. 26 and 27. Tables of contents of the minute books for 1870–1909 and 1910–24.
3. *Jb. KAW* voor 1898, b. lxxvi–lxxvii.
4. *N.P.R.I.*, 1891, v. 13, p. 362, 478.
5. W. H. Keesom, *Leiden Comm.*, Suppl. No. 57, 1926, p. 11.
6. *Jb. KAW* voor 1898, b. cii–ciii.
7. *Jb. KAW* voor 1912, b. cxxxvi.

CHAPTER 6

Scientific work in the 1880s

6.1 Corresponding states

The German translation of *Continuity* was published in June 1881 [1d$_1'$] and became, for several decades, a scientific 'bestseller'. The work was rescued from the captivity of a 'restricted' language [*1*, p. 3] by Thomas Friedrich Roth who had taken his doctorate the previous year under the supervision of Gustav and Eilhard Wiedemann, with a thesis on an experimental study of the compressibility of gases [*2*]. In his introduction to the translation Roth wrote that he had been persuaded to undertake this work some time previously by Eilhard Wiedemann who had also given him a great deal of help [1d$_1'$, p.v]. The German translation might have appeared much sooner had it not been delayed by van der Waals himself [*2*, p. 22]. It was not so much that he wished to check the accuracy of the translation or to introduce corrections or additions (although these were made), but that he had made a fundamental advance in his ideas, and so the translation had to be supplemented.

At the meetings of the Section of Sciences in October and November 1880 he set out a fruitful development in molecular physics called the principle (or law) of corresponding states [12ac$_1$]. Two substantial papers in Dutch were published the same year [12,15]. Roth worked quickly; abstracts appeared in the *Beiblätter* in January and April 1881, and complete and even extended translations were incorporated into the German version of *Continuity*. Roth, one of the most active contributors to the *Beiblätter*, was for many years almost a 'personal abstractor' of van der Waals's work, and even drew attention to it in his abstracts of other papers in the field [*3*].

Van der Waals arrived at the principle of corresponding states by analysing the transition from gas to liquid represented by the loops in his isotherms, that is the curve BCDEFGH in Fig. 3.2. These loops are unobservable and the experimental points lie on the horizontal line CG. At what pressure must this line be drawn? In 1873 he had 'not had the good fortune to solve this important problem.... I have not succeeded in finding, in any of the properties of the saturated vapour, a characteristic one that I could use to determine where this line must be drawn through the isotherm' [1e$_2$, p. 232,233].

In 1875 in a lecture to the Chemical Society in London [*4*], Maxwell solved this problem. The line must be drawn so that the areas of the loops above and below the line are equal, that is

$$\int_{V_1}^{V_g} P\,\mathrm{d}V = P_{\mathrm{sat}}(V_g - V_1), \tag{6.1}$$

where P_{sat} is the saturated vapour pressure and V_g and V_l are the specific volumes of gas and liquid at the points G and C respectively. This is Maxwell's well-known 'equal-areas' rule. Clausius reached the same result by a different route in 1879, which he published in March 1880 [*5*]; it was from 'a kind communication' from van der Waals that he learnt that Maxwell had obtained this result five years before him [*6*, p. 280].

The locus of the end-points of the horizontal line, C and G, forms the border or binodal curve that marks out on the *P–V* plane the boundary of the region of liquid–gas coexistence. Maxwell had first sketched qualitatively the course of this curve in the first edition of his *Theory of Heat*, of 1871; its shape had long attracted the attention of scientists and of engineers working with steam engines. Knowing the course of this curve allows one to understand and to predict many of the scientifically and technically important properties of saturated vapours, such as the dependence of their density on pressure and temperature and (see below, section 6.2) their heat capacities. Some of these properties, such as the dependence of the saturated vapour pressure on temperature, are easily accessible to experiment but their interpretation was obscure. Thus in *Continuity*, van der Waals wrote: 'It is true that there have been many theoretical attempts at a law, but so far none has been successful. There is, however, no lack of empirical formulae; in the *Fortschritte der Physik* for 1847 there are already forty different ones, and sometimes two or three of these overlap in the same range of temperature for which they are claimed to hold' [1e$_2$, p. 232].

Van der Waals's equation, when supplemented by the Maxwell–Clausius equal-area rule, provides in principle a complete description of the gas and its transition to the liquid, including the shape of the boundary curve. There remained the problem of how to extract this information explicitly. This was first tackled for a variant of van der Waals's equation due to Clausius [*5*] which fitted better Andrews's measurements on carbon dioxide. In June 1881 a parametric solution to the problem of calculating the vapour pressure curve and the coexisting densities was published by the young Max Planck [*7*]. In his paper he makes it clear that he had read the original Dutch version of *Continuity* and not merely the German abstract of 1877. No doubt a copy of the thesis had been sent to his teacher at Munich, Professor Philipp Jolly. In August of the same year, 1881, Clausius sent to the *Annalen* a paper [*6*] in which he wrote that he had entered into similar calculations long before the publication of Planck's paper or Roth's translation of *Continuity*, and that he had delayed publication only because of the tedious computation required. Early in 1882 calculations by a method similar to that of Planck were made in Moscow by P. A. Zilov and his teacher A. G. Stoletov [*8*].

None of the authors was entirely happy with his results, which neither led to an explicit relation between vapour pressure and temperature nor fitted the experimental results closely, even if Clausius's equation, with three adjustable parameters, was used.

Van der Waals had also followed the same road. 'Over and over again I tried to find the equation of the [boundary] curve' he wrote in 1880. He refrained, however, from publishing his results, which were apparently similar to those of Planck and Clausius, 'not only because of the tediousness of the calculation and the complexity of the final equation' [1d$_1'$, p. 125–126], but also because he realized, probably better than his colleagues, that neither his equation nor Clausius's 'improved' version of it was to be trusted when the specific volume becomes as small as that of the saturated liquid at low temperatures. There was, therefore, little point in exact computations based on such equations and he, at least, had little inclination to embark on purely mathematical analyses.

But might there not be features of the boundary curve that do not depend on the details of its direct calculation from an equation of state? In seeking an answer to this question van der Waals returned to his old idea that one substance differs from another only by the values of two parameters, a and b. Could the equation not be rewritten in such a way that these did not enter explicitly? He knew that the critical constants could be expressed in terms of these parameters; $P^c = a/27b^2$, $V^c = 3b$, $RT^c = 8a/27b$, and so realized that the substitution

$$P = lP^c, \; V = nV^c, \; T = mT^c,$$

gives the universal equation of state,

$$(l + 3n^{-2})(3n - 1) = 8m. \tag{6.2}$$

The equation is universal since it contains no parameters characteristic of any individual substance, and so it is equally valid for all. 'All specifics have disappeared', as van der Waals succinctly expressed it [1d$_1'$, p. 127].

He called the variables l, n and m the 'reduced variables', and eqn (6.2) the 'equation of the reduced isotherms'. In these reduced coordinates his equation implied that all substances have the same isotherms. Since the equal-areas rule can also be expressed in terms of the reduced variables then it follows also that all substances have the same reduced boundary curve, and so it is no longer necessary to seek its explicit form since one can construct it for any substance from the experimental results for another substance. Thus in seeking to solve the problem of the boundary curve he found something more important — the law of corresponding states, which requires that the thermophysical properties of substances are the same in corresponding states, that is, states with a pair of equal reduced variables from the complete triplet of variables[1]. The law is a direct consequence of eqn (6.2) or, more precisely, it is a direct consequence of the existence of an equation of state with only two parameters characteristic of each substance, since the existence of such an equation implies that if two reduced variables are the same for a set of substances then the third is also the same throughout the set.

Van der Waals pointed out some of the consequences of the law and showed that they were supported by the experimental data. At a given reduced temperature all liquids have equal reduced vapour pressures, equal reduced specific volumes of the vapour and liquid phases, and equal products of the latent heats of evaporation and the molecular weight. At a given reduced pressure, all substances have equal values of the product of the molecular weight and the difference of the reduced volumes of vapour and liquid. At a given reduced pressure and temperature the compressibilities of all substances are inversely proportional to their critical temperatures.

Comparisons are of doubtful value, but it seems that at this stage of his career van der Waals had a deeper insight into molecular physics than any other scientist, except perhaps Gibbs. Clausius, Planck and van der Waals all tackled the problem of the boundary curve but the first two produced only formal solutions; it was van der Waals who extracted an important new generalization from his solution to the problem, a generalization whose value was recognized immediately. It is hard to over-estimate the utility of the law of corresponding states in bringing a whole mass of scrappy information on many different fluids into a coherent and intelligible picture. The way was now clear to estimate thermophysical properties that were inaccessible to experiment and a powerful impulse was given to the development of the technology of gas liquefaction. The discovery of the law led, moreover, to a new idea in molecular physics, that of thermophysical and hence of molecular similarity. For many physicists it served for a long time as a basis for their research programmes. James Dewar, one of the pioneers of gas liquefaction, said in 1902 that if one sought a physical principle of greater power than the law of corresponding states one would have to go back to Carnot's cycle [*9*, p. 29]. So with the appearance of *Continuity* in German in 1881, with not only the well-known equation of state but the new law also, van der Waals's fame quickly spread world-wide.

The first response to van der Waals's new result came from Kamerlingh Onnes, then Bosscha's assistant at the Delft Polytechnicum. 'I had reported on the law', van der Waals wrote in 1904, 'and already, before it had been printed, Onnes, who then knew of it only by word of mouth, produced a paper in which he deduced it from more general principles' [74, p. 86]. On 24 December 1880, Kamerlingh Onnes's derivation, based on a principle of mechanical similarity, was presented to the Section of Sciences by van der Waals [127]. This was apparently the first contact between these two scientists; it soon grew into a close personal and professional relationship. Kamerlingh Onnes's work confirmed van der Waals's intuitive feeling that the law is more accurate than his equation of state, from which he had originally derived it. Since it could be derived from more general principles it need not be tied to any particular equation with all its inevitable errors.

Nevertheless the law of corresponding states, for all its fertility, is not exact, as van der Waals realized after finding that his extensive calculations based on it never coincided perfectly with the experimental results. In each case the question arose as to whether the discrepancies were due to the inaccuracy of the law or to errors of measurement, since the quality of many of the experimental results left much to be desired. The fact that the law was found to be only approximately correct at first greatly depressed van der Waals; 'In the beginning I considered it proper to demand [for a physical law] an absolute exactitude if one is inclined to attach to it any value. I reasoned as follows; "What is inaccurate is useless". And from the moment I had to agree that the theorem is only approximate I began to doubt if it deserved to be published to the world. Later I realised that I had overlooked the fact that all our knowledge is but a sum of approximations, and that the essential merit of the theorem lies in its assertion that all substances belong to one genus, just as all people belong to the genus *Homo*' [102, p. 43]. This last observation is an original and deep observation and one can possibly discern in it a forerunner of modern ideas on approximate symmetries.

The law of corresponding states was clearly a favourite of van der Waals; on it he wrote one of two popular papers [73], a form of writing he otherwise eschewed. It was the only subject he allowed himself to 'philosophize' about. Many times he returned to the theme that it caught the essential idea of all substances being but different samples of a universal Matter. 'If we liken, say, the height to the temperature, the width to the pressure, and the length to the volume, then we could say that all substances are copies of one another' [155]. But if this is so then there arises an interesting problem that seems to have been overlooked by the philosophers. 'Realising that all people belong to the same human race, we wonder at the great variety of their intellectual abilities and their conditions of life. And when we admit that all substances belong to the same genus we are no less astonished at the huge differences that exist not only between their dimensions but their qualities also. One substance is useful and another is a lethal poison. Where does this diversity come from?' [102, p. 44]. His ruminations on the law of corresponding states give us a unique view of his innermost thoughts on the purposes of his reasearch: 'To discuss a unity in all the diversity of the material world. . . . the material world should be regarded not as a thing of chance, but as something created and living according to simple laws. . . . The physicist whom I would compare to an artist and who aims at the most scrupulous determination of the properties of a substance would find that this law was of no use to him. But there is also another whom I would compare to a philosopher amongst physicists and who is aware that the notion of "correspondence" allows him to discern a unity and to find harmony where it had seemed previously that chance and confusion had reigned' [73, p. 285]. Van der Waals clearly saw himself as a 'philosopher amongst physicists' but he was, perhaps, a 'latent' philosopher.

6.2 Thermodynamics and the theory of solutions

In the 1870s thermodynamics played little part in van der Waals's work. Only the eleventh chapter of *Continuity*, and two papers, *On the influence of pressure on the temperature of maximum density of water* [7], and *On the specific heat of a saturated vapour* [8], are applications of what was then still often called 'the mechanical theory of heat'. These are of interest mainly because of the way in which he clearly formulated the problems, the elegant geometrical constructions used, and the use of more than one analytical solution. The first paper included not only theoretical work but also some experimental work he had carried out in the HBS in 1876. He compared volumes of water in a glass piezometer in the temperature range 0.1 to 6.1 °C at ambient pressure and at 7 atm, which was presumably the upper limit that he could reach. His view was that these measurements were inadequate for quantitative results, and that was certainly true, but that his qualitative conclusions were correct; the temperature of maximum density decreases as the pressure rises. The modern value for this change is 0.02 K atm^{-1}. The second paper is the more significant and probably arose from reflection on the behaviour of vapour–liquid systems outside the critical region. It was known that for water and for a number of other liquids, the specific heat of the saturated vapour is negative, that is, heat is removed to take the saturated vapour to a state of higher temperature, and so of higher pressure. For ethyl ether, however, the specific heat is positive, and van der Waals had discovered the reason for this 'anomaly'; 'The specific heat of a saturated vapour is negative when, at a certain point on the line of the saturated vapour [in the pressure–volume diagram], the adiabatic line rises more steeply than the line of the saturated vapour, and it is positive when the adiabatic line rises more gently' [8, p. 172]. The paper contains general formulae and calculations of the thermodynamic properties of specific substances. From these calculations he predicted that chloroform vapour would behave like steam below 120 °C and like ether above this temperature, a result that was later confirmed by the experiments of Hirn and of Cazin, and discussed further by Chwolson [*10*][2].

We can see that van der Waals had an excellent command of thermodynamics as it was then practised in Europe but, at the same time, his attitude to it was reserved and unenthusiastic. Very typical of him are these words from *Continuity*; 'It is the boast of thermodynamics that its laws do not rest on any assumptions as to the structure of matter, and consequently embody truths which are in so far unassailable. If, however, we are prevented from making more searching investigation into the nature of bodies through fear of leaving the region of invulnerable truths, then it is clear that by so doing we wantonly cut ourselves off from one of the most promising paths to the hidden secrets of nature' [1, p. 113; $1e_2'$, p. 225–226].

A few years later, when considering two ways of doing research, he wrote: 'The first way, which conforms to the laws of thermodynamics, has a solid foundation and leads to trustworthy results. But this way has its darker side because it leaves uncertain the mechanism that gives rise to the phenomenon. Another way is to follow the principles of the kinetic theory of molecules since these lead directly to knowledge of the mechanism; thus those who follow this path set before themselves a much more exalted aim. But since the theory is still imperfectly understood they run the danger of finding the wrong mechanism' [10, p. 199–200]. Van der Waals obviously preferred the risks of the second way which promised to realize his 'much more exalted aim'.

In 1879 there was an event in his life which, although outwardly insignificant, was to change his attitude to thermodynamics. In that year van der Waals received from Gibbs both parts of the latter's *On the equilibrium of heterogeneous substances*, the first of which was published in 1876 and the second in 1878. (He received also a paper on the density of dissociating vapours, published in 1879, and later received Gibbs's works on optics, his memoir of Clausius and his book on statistical mechanics.) It is possible that Gibbs knew of van der Waals's name from the abstract of *Continuity* in the *Beiblätter* of 1877 or from Maxwell's earlier review in *Nature*. He may, however, have merely included his name in a distribution list that included most of the professors of physics and chemistry in Europe.

So van der Waals became one of the ninety or so possessors of Gibbs's papers on heterogeneous equilibrium, and was the only one who carried the subject further. At first his response was rather sceptical; he reported to the Section of Sciences that he considered the point of departure of Gibbs's deductions, that is, the principle of equilibrium[3] as 'open to doubt' [10ac]. He therefore compared calculations on dissociation in gases by the traditional methods of differentiation with calculations based on Gibbs's method. The agreement of the two methods is the conclusion of his paper *On the relation between pressure, volume and temperature at dissociation* [10]. This little known work, sent to the press in January 1880, is an important landmark in van der Waals's scientific development.

It is, first, the start of his 'collaboration *in absentia*' with Gibbs[4]. The previous criticism of thermodynamics is now omitted, and in the introduction it is pointed out that the thermodynamic method 'was applied very successfully, at any rate for mixtures of gases, by Willard Gibbs'. 'In an important work *On the equilibrium of heterogeneous substances* he has established an expression for the relation between pressure, volume and temperature for dissociating gaseous mixtures that has the same validity as the laws of Boyle and Gay-Lussac' [10, p. 200]. Secondly, the paper marks van der Waals's first moves into the realm of physical chemistry; the concept of 'dissociation' was one of the leading ideas in what was then called the field of 'chemical mechanics'. According to van der Waals, 'dissociation, in its general sense, includes phenomena belonging both to physics and to chemistry. To the domain of chemistry belongs the grouping of atoms into composite molecules, and to physics the grouping of those molecules into larger aggregates' [10, p. 199].

In conclusion, we note that the author realizes for the first time that, by following Gibbs's methods, a phenomenological approach can be carried much further than had previously been realized. It should be mentioned that he also presented his expressions in the form of the law of mass action of Guldberg and Waage, which had been published in its final form in German in 1879 and which he quoted with respect. For this transformation he introduced partial pressures calculated from Dalton's law.

All three aspects of this paper are evidence of van der Waals's scientific development, although it contained nothing that was really original. It did, however, introduce one new theme, and that is the idea that a purely thermodynamic approach is insufficient to solve the problems of equilibrium; it was also necessary to know the *P–V–T* relations of the mixture, and it was this proviso that led to the restriction, 'at any rate for mixtures of gases' in his evaluation of Gibbs's work. The idea that one must combine thermodynamics with an equation of state was one that remained with van der Waals for the rest of his life. In this paper he used an equation of state for the gas mixture that was the same as he had proposed for a pure substance, but with parameters a_x and b_x which depended on the composition of the mixture – in this case on the degree of dissociation. He added, however, that 'It is not yet known how the constants that enter the expression relating pressure and temperature, at fixed composition of the mixture, depend on the ratio of the constituent bodies' [10, p. 204].

This thermodynamic paper ends with a characteristic remark expressing his continual hankering after a kinetic treatment, reminiscent, perhaps, of Galileo's *E pur si muove*. 'Although the way in which dissociation changes with temperature is still beyond the scope of kinetic theory', he wrote, 'I do not believe that its derivation from that theory will be too difficult. I hope shortly to be able to derive the variation with temperature from the theory of molecular motion' [10, p. 217].

It was, however, a further ten years before he could return to this problem. According to the University statutes, each Faculty had the right, every other year, to put forward a subject for a prize essay. For ten years van der Waals did not include any topics of his own in those proposed by the Faculty of Physical and Mathematical Science, perhaps because of the absence of suitable students. In the academic year 1888–89, however, he proposed the topic, 'An investigation of the effect of dissociation on the laws relating the density and pressure of saturated vapour, the velocity of sound, the ratio of the heat capacities at constant pressure and constant volume, the latent heat etc., under normal conditions'. The gold medal, with the inscription *Nur der Starke wird das Schicksal zwingen*, from Schiller's *Das Ideal und das Leben*, was awarded to A. J. Swart, a candidate from Amsterdam University, who later developed his essay into a doctoral thesis. It is of interest that the chemical aspects of Swart's results were accounted for in detail by van 't Hoff in a paper in the *Zeitschrift für physikalische Chemie* [*11*]. The other aspect of the work, the molecular kinetic theory, remained only an attempt to build a kinetic theory of gas mixtures.

Little is known of van der Waals's studies during the 1880s. After the death of his wife there was a long silence, broken only by the issue of academic testimonials and similar documents. It is known, however, that his studies continued and were directed to mastering Gibbs's thermodynamics. Kohnstamm has drawn attention to the fact that this was the only case where van der Waals's studies were directly influenced by the theoretical work of another man [206], if we except the first 'revelation' of Clausius's work that he acknowledged in his thesis. Thirty years later van der Waals himself pointed out, 'It was of great importance to me to be acquainted with Gibbs's treatises on the equilibrium of heterogeneous substances, which he sent me immediately after their appearance' [103e, p. 263]. (The slight inaccuracy of the last phrase may be forgiven.)

In 1885–86 he read seven reports on the thermodynamics of phase equilibria at meetings of the Section of Sciences [21–26], but he prepared an abstract only of the first. Kohnstamm wrote of this period of his life: 'Depressed also by bodily sufferings, he had lapsed into a state in which the publication of the results of research had lost any attraction for him. Scientific work did not cease – to abstain from work was something he could not do. . . . But what need was there for him – so free from ambition and so deeply hurt that he was indifferent to all society – to take the trouble of making the results of his work available to others? Whatever it was, the danger that threatened him was eventually lifted, thanks to the initiative of Kamerlingh Onnes, and to his own strong sense of duty and discipline' [206, p. 24–25].

Van der Waals himself later told what had happened. 'Onnes had told me of his intention to start investigations into mixtures in his laboratory. One knew how to investigate a single substance, but for a mixture all was still mysterious. What were the problems here? Not only was there no theory to guide us, but such uncertainty prevailed that they did not know where to begin and how to go on' [74, p. 82].

Here we must interrupt the narrative to say that this was true not only for Kamerlingh Onnes but also for everyone else. The problems of gaseous mixtures, especially in the critical region, had exercized Maxwell, Andrews, Cailletet and Stoletov, but none of them could break out from the circle of ideas developed for single substances, and from the restriction of Dalton's law that the pressure was the sum of the partial pressures of the single components.

'I was amazed by this announcement,' continued van der Waals, 'and told him, thus amazing him also, that I had been long engaged in studying the theory of binary mixtures, and had, to a large degree, already completed it. I was confident about the principles, and although I had not yet worked out all the details I could tell him briefly some of the results. During this conversation, which lasted perhaps two hours or so, there was no time to discuss the theoretical basis; however the rule for coexisting phases, and even the occurrence of the loop that can explain critical phenomena, the plait point and the critical point(s) of contact, were mentioned. The precise theoretical foundations were omitted, and the set of assertions naturally seemed problematical to Onnes. . . . When I told him that if he were to see the basis of the calculations then his doubts would vanish, then he began to insist that I publish the theory. I was then passing through those years when the publication of new results was of no interest to me, and I was not certain that I would ever emerge from it. Onnes, however, put forward an argument in favour of publication to which I was susceptible; he did not want to undertake investigations that were guided by a theory which he could not refer to' [74, p. 82–83]. After this discussion there appeared in the programme of lectures at Amsterdam for 1888–89 the subject of *The molecular theory of mixtures* [*12*].

What was it that van der Waals had done? The essence of his theory was the combination or, more accurately, the synthesis of the laws of thermodynamics and the equation of state generalized for a mixture, so that the parameters *a* and *b* depend, in a defined way, on the composition. It was what Lorentz was to describe as a 'happy synthesis' [195, p. 308]. Van der Waals had managed to solve a problem that had beaten Maxwell.

The idea behind the generalized equation of state is the same as that in *Continuity*; in a mixture, as in a single substance, both liquid and vapour are composed of identical 'mobile permanent particles having the properties of size and attraction' [58, p. 583–584]. (Certain anomalous substances and chemically reactive mixtures were excluded from the theory.) 'Naturally all question's that were unanswered for a single substance remained unanswered for the mixture' [61, p.51].

The equation of state was created by van der Waals himself, and the thermodynamic framework was provided in an almost completed form by Willard Gibbs. According to his son, van der Waals thought that, except for Maxwell, no contemporary physicist was greater than Gibbs [*13*, p. 410]. Van der Waals acknowledged this respect in his own particular way; he used Gibbs's notation throughout his work[5]. Later he pointed this out in his Nobel Lecture; 'In his honour I named the equilibrium surface the ψ-surface. For the free energy whose significance for the equilibrium he was the first to recognize, he always represents by the sign ψ' [103, p. 11; 103e, p. 263].

Thermodynamic surfaces were invented by Gibbs in 1875–76. He discovered that the representation of the dependence of a thermodynamic function on its arguments (e.g. energy as a function of entropy and volume) allowed one to visualize the conditions of equilibrium between different states of aggregation, or between different phases in general, more easily than from an analytical treatment. If one constructed a plane that was a tangent to the thermodynamic surface at two points, then those points represented coexistent phases.

Gibbs's geometrical method found an ardent advocate in the person of Maxwell. In the fourth edition of *The Theory of Heat* in 1875, he discussed in detail Gibbs's work on the energy surface for a single substance. He proposed a simple way of detecting special lines and points on a thermodynamic surface; he greased a glass plate and rolled it over the surface, keeping it in contact with the surface at two points. At the points of contact the grease sticks to the surface and traces out the loci of the points of contact, which represent the coexistent states. The corresponding lines on the glass plate are a planar representation of the coexistent states. After receiving from Gibbs *On the equilibrium of heterogeneous substances*, Maxwell used this work as the basis of his address at the International Exhibition of Scientific Apparatus in May 1876, in the South Kensington Museum in London, and demonstrated there models of different thermodynamic surfaces [*14*]. Eight Dutch physicists attended this exhibition, including Rijke and Bosscha, the first of whom lectured immediately after Maxwell. It is therefore probable that van der Waals was told of Maxwell's lecture. Having read Gibbs's work it was, in any case, natural that van der Waals, with his geometrical habits of thought, became interested in the potentiality of geometrical methods for the study of the thermodynamics of mixtures. His synthesis of thermodynamics and the equation of state was expressed in the geometrical construction of the ψ-surface of van der Waals, that is, the representation of the free energy of a homogeneous binary mixture at a fixed temperature as a function of the two arguments volume V and mole fraction x.

The construction of this surface is possible since, from the equation of state, ψ is known explicitly as a function of V and x. Its shape describes the system in a simple way; those parts that are convex to the horizontal V-x-plane represent stable homogeneous states, and those parts that are concave (the so-called *plaits*) are unstable and break up into two coexisting phases. The slope of a tangent plane to the ψ-surface, that is $-(\partial\psi/\partial V)_x$, is the pressure of the system. If there is a plait then the tangent plane touches the surface at two points which represent the coexisting phases. The rolling of the plane describes the change of composition of these two phases with changing pressure, at the fixed value of the temperature. The ψ-surface can assume a great variety of shapes, depending on the temperature and on the parameters of the equation of state.

All these results were first publicly reported by van der Waals at the session of the Section of Sciences on 23 February 1889. His report was entitled, *Some results obtained during investigations into the theory of a mixture of two substances*, and was accompanied by a model of the ψ-surface of a system with two liquid phases and a vapour phase. The abstract of the report [29] is remarkable for the richness and density of its content — the generalization of the equation of state to binary mixtures, the derivation of an explicit expression for the free energy, the geometrical representation of the ψ-surface, the loci on this surface that separate stable homogeneous states from those in which the system breaks up into two or three phases — all these main topics of his theory were already present in this first report.

In a letter of 25 February 1889, van der Waals wrote (in English)[6]: 'I am occupied in studying a theory of a mixture of two different kinds of matter. Last Saturday I have given a brief exposure of it in the Royal Academy of Amsterdam.

'What I have written on this topic in my *Continuity* etc. is given when I had very restricted views on this subject and must be corrected. . . . a theory of two kinds of matter will correct some loose opinions I have given in the *Continuity*.'

Shortly afterwards he developed the report into a long paper for the *Archives Néerlandaises*. In his Nobel Lecture he said that 'I had written it in Dutch but my esteemed friend Bosscha undertook the difficult task of rendering it into French, a task all the more exacting as I had written it in an extremely concise form and the mathematical treatment led to particular points (plait points) and particular curves which had at the time rarely been examined in detail. Previously, however, my friend Korteweg, to whom I had communicated in broad outline the outcome of my examinations, had studied the mathematical properties of these curves, a study which I have often found of great use' [103e, p. 262]. Bosscha also made a few additions to the text and added some references to earlier works of van der Waals, notably *Continuity*.

Thus his important paper *Molecular theory of a substance composed of two different species* was presented to the world [31]. 'A remarkable work', wrote Kohnstamm, 'Almost on every page one feels the author overcoming the problems that prevented him from writing. If something can be said in ten words, he never uses twenty, but often only five or even two. That is why it is so hard to read, or, to be exact, so hard to understand fully. . . . For ten or more years I have read these fifty pages, perhaps twenty or thirty times, and each time, as my understanding grew, I have found a new hidden remark or a piece of subtlety that I had previously missed. Even now I am not sure that I have exhausted its contents' [206, p. 27].

Many years later, in his Nobel Lecture, van der Waals omitted entirely any mention of his personal problems. 'The reasons why I hesitated so long over publication were many and it would serve little purpose to enumerate them here. But one of the reasons of scientific importance was the question which I kept asking myself: is it any use, until the study of the equation of state has been completed, applying it to mixtures?' [103e, p. 262–263]. Kohnstamm comments, 'But this argument is not so convincing as it seems, since this doubt could be removed only by experimental corroboration or refutation, and this could be obtained, thanks to Kamerlingh Onnes's persistence, only after the theory had been published' [206, p. 26]. Nevertheless, van der Waals's doubts on this point were genuine, even if there was another, and probably more important reason, which held up many of his important papers. After having delivered the manuscript to Bosscha, he added in an Introduction, '. . . my work on several particular points is not yet complete, so some time will pass before I shall be in a position to treat the subject comprehensively' [31, p. 1]. It is clear how painful it was to the author to part with work that he thought was not yet wholly complete. The same happened again with the theory of capillarity, and with the publication of the lectures on thermodynamics and, we may guess, in other less important cases.

Let us now consider the other side of the question, and ask what gave the impetus to this line of research. In his Nobel Lecture, he wrote '. . . the consideration that even with b [held] constant my theory had not been unimportant in the case of a single substance gave me the hope that many phenomena could be explained qualitatively provided that a suitable value for a and b were introduced for mixtures' [103e, p. 263]. Besides his natural wish to expand his own original theory one can guess at another reason, not mentioned explicitly. Studies of mixtures allowed one to look from outside, as it were, at the equation of state of a single substance. In other words we can say that in the hierarchy of systems that we can study, binary mixtures are a supersystem with respect to those formed from a single substance. Indeed he later deduced values of a and b for the components from data on binary mixtures, and his papers on quasi-association use directly his theory of mixtures.

Finally, we must take into account some external influences that were important at the time of the publication of this work. The end of the 1880s and the beginning of the 1890s were a time of particular importance in the study of solutions, and one that was marked by the heat and intensity of the arguments – the time of the *Sturm und Drang* of the physical theories of solutions of van 't Hoff and Arrhenius. As always, van der Waals went his own way and his work deals with solutions of all concentrations, not just dilute solutions. His work went against the prevailing stream, or perhaps outstripped it. Nevertheless he was not entirely indifferent to the topic of the day for the last sections of his paper deal with van 't Hoff's theory. This discussion explains why van 't Hoff wrote to Arrhenius, from Amsterdam on 5 January 1890, 'In *Archives Néerlandaises* there is already in press or perhaps with the printer a paper of van der Waals in which he has worked out an equation of state for a mixture of two substances. He has told me that it contains also the osmotic laws for dilute solutions, and also the extension of Avogadro's law. I think that this will reassure many physicists' [*15*, p. 278]. It was due to van 't Hoff that this paper had already been published in March 1890 in a German translation by Ostwald in the *Zeitschrift für physikalische Chemie* [31d]. Van 't Hoff considered van der Waals's results to be an important demonstration of the correctness of his theory, and referred to them in several later papers, and in particular at a famous discussion on solutions at the meeting of the British Association at Leeds in 1890 [*15*, p. 282]. All this reinforces Kamerlingh Onnes's remark that 'van 't Hoff owes more than is generally known to the scientific influence of van der Waals' [233].

This is the appropriate place to consider briefly van der Waals's two papers on solutions of electrolytes. They are only a small off-shoot of his main line of work but are of interest both as documents of their time and as evidence of his contribution to the passionate debates on the theories of van 't Hoff and Arrhenius. As we have seen, van der Waals supported sympathetically van 't Hoff's theory of dilute solutions, but his view of Arrhenius's theory was less simple. The nub of the matter was that, in its original form, the hypothesis of electrolytic dissociation had been put forward without any consideration of what caused the dissociation, that is, without considering the interaction between the solute and the water. For van der Waals, who had entered the field as an investigator of non-ideal systems, it was self-evident that such consideration was absolutely necessary. He carried out a theoretical analysis of the vapour pressure of electrolytic solutions and compared his results with experiment, 'first of all with the aim of pointing out that it is wrong to deny an interaction between solvent and solute, as is often done at present' [32d, p. 314]. Van der Waals used his theory of mixtures to obtain equations for the dependence of the vapour pressure of a solution on its concentration and compared these with the experimental results of Regnault and Tammann, in particular the vapour pressure of dilute sulphuric acid. It is interesting that at first he deliberately used an expression in which no account was taken of the dissociation of the solute. He explained, 'Of course I should immediately take dissociation into account, but I wanted to test first to what extent this is forced on one by the behaviour of the pressure.' His main conclusion was that the 'observations are not in accord with the hypothesis of electrolytic dissociation if it is not simultaneously accepted that the solvent strongly influences the solute, and vice versa' [33d, p. 215]. Having reached this conclusion, van der Waals put forward three statements: '(1) recombination of ions dissolved in water must be accompanied by an absorption of heat; (2) the degree of dissociation of the same substance dissolved in various media may be very different; and (3) with increasing solute concentration the fraction of molecules that are dissociated must go through a maximum, a conclusion that explains the maximum in the conductivity that electrolytes exhibit at certain concentrations' [33ac].

It is noteworthy that these papers of van der Waals were one of the first constructive criticisms[7] of the primitive version of the theory of electrolytic dissociation, and the first that was founded on a serious quantitative analysis of the experimental results.

Van 't Hoff wrote to Ostwald about these papers on 18 May 1891; 'You have probably also received the most recent (Dutch) papers of van der Waals on electrolytic dissociation. I shall soon have translated them for the journal [i.e. *Zeitschrift für physikalische Chemie*] and expect to complete this in a few days. The results support recent views. Your expression for the degree of dissociation has been extended to concentrated solutions, although I have doubts as to the practical value of this.' Ostwald replied two days later, 'I had already asked van der Waals to translate his papers for me, and he has agreed; would you be so kind as to tell him of your intention so as to avoid both of you doing it.' A month later, on 18 June, van 't Hoff told Ostwald that van der Waals had completed his translation [*16*, p. 238–239]. He had made some additions also and these revised papers appeared in the July issue of the *Zeitschrift für physikalische Chemie*. They were later noted with approval by Ostwald in the chapter on electrolyte solutions in his well-known *Lehrbuch der allgemeinen Chemie*.

After this digression let us return to the paper on the molecular theory of mixtures and the events that followed its publication, most of which took place in Kamerlingh Onnes's laboratory. The issue of *Archives Néerlandaises* was no sooner published than Kamerlingh Onnes set as the subject for a prize for the students at Leiden, 'Observations are needed to test van der Waals's theory of mixtures of two substances.' The necessary work had already been started in 1889 by Johannes Petrus Kuenen, the favourite and most gifted of the students of Kamerlingh Onnes. According to van der Waals, 'Kuenen's masterly work was a real proof-by-fire for the theory' [74, p. 83–84]. Kamerlingh Onnes wrote in 1922 that 'The genius of van der Waals, then depressed by deep mourning, took a new flight when he was asked to work out in connection with Kuenen's measurements his theory of binary mixtures given before only in sketch' [*17*, p. 674].

Thus it was that in the laboratory at Leiden there was started a systematic experimental study of the ψ-surfaces of several binary mixtures. Van der Waals waited eagerly for each new set of results so that he could compare his theoretical conclusions with the experimental evidence. A telling detail is his mention of 'the plot sent by Mr Kuenen as soon as his measurements had been made' [48f, p. 91]. He took part at once in the discussions of the work, and the results of the first ten years of this cooperative effort were summed up in a report at the International Congress of Physics at Paris in 1900 [58] and, in a more complete form, in the second volume of the second edition of *Continuity* [61] which was dedicated to Kamerlingh Onnes with the words, 'The happy cooperation between theory and experiment, illustrated in the pages that follow, is principally due to you. The publication of the concise paper *Théorie moleculaire* in 1889 was due to your insistence. From then until the present day you have unremittingly exercised your rich talents for experimental investigations to produce results that are vital for the theory.'

6.3 Capillarity

In 1893 van der Waals published in the *Verhandelingen* of the Academy a characteristically long and dense paper of fifty-six pages entitled *The thermodynamic theory of capillarity under the hypothesis of a continuous variation of density* [37]. It was a paper whose value was fully appreciated only sixty years later.

As was discussed in Chapter 3, it was his reflections on Laplace's theory of capillarity that were the starting point of the work in his thesis. These reflections culminated in the triumph of his equation of state but did not lead to any deeper insight into capillary phenomena than was provided by Laplace's theory. This is not surprising since the development of a truly microscopic or molecular-kinetic theory of capillarity encounters great theoretical and computational difficulties that have only recently been overcome. A phenomenological theory could be formulated only in the language of thermodynamics, and with this van der Waals was insufficiently familiar – at least until he had studied Gibbs's fundamental papers. From these, however, he might have concluded that the subject was already dead since Gibbs's treatment of capillarity, published in 1878, not only covered all kinds of surface phenomena but also could be thought to be a model for any physical theory since it was irreproachable in its logic and was deduced rigorously from the minimum number of premises, namely the first and second laws of thermodynamics.

Van der Waals had fully appreciated the power and fruitfulness of Gibbs's approach but he believed that logic is not the best instrument with which to enquire into the laws of Nature. It seems that the cold and unrelentingly abstract methods of the New Haven recluse met in van der Waals the stubborn resistance of a passionate believer in an intuitive approach. This belief may lie behind the dubious compliment he pays to Gibbs's account of capillary phenomena in the introduction to his own paper: 'the author [Gibbs] takes more than a hundred pages to explain their nature and to discuss both well-known and less well-known applications' [37e, p. 201].

His real objection to the work of Gibbs (and of Laplace and others also) was that they supposed, or appeared to suppose, that there was a mathematically sharp dividing surface at which there is a discontinuous change of density from that of the bulk liquid to that of the vapour. Such a supposition is, however, contrary to the principles of thermodynamics which require that the density distribution within the two-phase system is determined solely by the pressure and temperature, and by any external fields such as gravity; one is not allowed to prescribe an arbitrary distribution. Moreover, simple physical reasoning shows that the density of the liquid cannot remain constant up to a sharp dividing surface since the molecular pressure or cohesion at the boundary is less than in the bulk. Gibbs himself was well aware that his dividing surface could not truly represent the structure of the interface and in the first few lines of his exposition wrote, 'That such is not rigorously the case is evident from the consideration that if it were so with respect to the densities of the components it could not be so in general with respect to the density of energy, as the sphere of molecular action is not infinitely small' [*18*, p. 219]. There was, however, no direct evidence then of the thickness of the surface or capillary layer (as van der Waals calls it); it was known only to be very thin. So Gibbs treated the problem in a way that was independent of any assumption about the thickness; he 'aimed at formulating a thermodynamic method independent of surface layer thickness' [*19*, p. 11].

It is clear from the title of his paper that the idea of continuity is again central to van der Waals's thought. He writes that 'circumstances which need not be mentioned here led me to conceive a different theory of these phenomena' [37e, p. 201]. It is tempting to suppose that among these was that of meeting the challenge to his commitment to continuity and a reluctance to bow down before the concept of a discontinuity in density which had recently been endorsed by Gibbs. These were not the only reasons, however, for later he writes that variations in density can, in general, be either continuous or discontinuous but 'the difficulty of obtaining an expression for the energy when there is a discontinuity of density has led me to attempt the calculation only for the case of a continuous variation' [37e, p. 207].

His first attempt, in his thesis, at solving the problem of capillarity led nowhere, but now having mastered Gibbs's methods he returns to the attack. This second assault was one of great boldness for it was the first time that thermodynamics had been applied to an inhomogeneous system, in this case a capillary layer in which the density changed continuously through the thickness. This layer is, however, extremely thin, so is it legitimate to use the familiar thermodynamic quantities, which are averages over space, when the physical properties are changing so rapidly? Is it even certain that the conventional thermodynamic equations remain valid for such a system? It seems that the answers to such questions were at first not clear to van der Waals and so, having conceived his theory as early as 1888, he held back from full publication and contented himself with a short note communicated to the Section of Sciences on 28 May 1888 [27]. This apparently passed almost unnoticed although Lorentz prepared an abstract of it for the *Fortschritte der Physik* [27c].

His hesitations had vanished by the time of the meeting of the Section of Sciences on 25 February 1892, when Kamerlingh Onnes reported [*20*] on the work carried out under his supervision at Leiden for the thesis of E. C. de Vries. This comprised precise measurements of the variation with temperature of the surface tension of ether. It had been known for many years that surface tension vanishes at the critical point but it was generally accepted that it did so linearly with temperature. De Vries was the first to measure it with sufficient accuracy near the critical temperature to be able to show that it vanishes more quickly; that is, at the critical temperature the tangent to the curve lies along the temperature axis. Van der Waals took the floor immediately after his Leiden colleague [35]. He pointed out that the supposition that the density changed discontinuously at the surface between a liquid and its vapour leads inevitably to a linear decrease of the surface tension near the critical point. It follows that the results obtained by de Vries are evidence that this supposition is false, at least for liquids near their critical points. On the other hand the new results are in agreement with the predictions of his theory which is based on the supposition that there is a continuous variation of density through the capillary layer.

Van der Waals bases his theory on the second law of thermodynamics, as Gibbs had done, but he does not forget to note that this 'is already found in essence in the work of Clausius' [37e, p. 201]. 'A given substance distributes itself in a given space, at a given temperature, in such a way that $\int\rho(\varepsilon - \tau_1\eta)\mathrm{d}k$ becomes a minimum' [37e, p. 202], where the integrand is the free-energy density and the integral is over all the available space. This is the most convenient form of the law for his purposes. He emphasises that if we are not to lose all capillary phenomena then we must take into account the fact that in a highly inhomogeneous system the free-energy density at a given point depends not only on the local density but also on the density at neighbouring points; indeed it is this dependence that gives rise to capillarity. For a planar liquid–gas surface normal to the z-axis he limits himself to the two leading terms of a Taylor expansion:

$$F[\rho(z)] = f[\rho(z)] - \tfrac{1}{2}c_2(\mathrm{d}^2\rho(z)/\mathrm{d}z^2), \tag{6.3}$$

where $f[\rho(z)]$ is the free-energy density of a hypothetical homogeneous fluid of uniform density $\rho(z)$, that is, a substance identical to the fluid in question, under the same external conditions and obeying the same equation of state, but without the capillary layer. Such a fluid is what James Thomson (see Chapter 3, Section 3.7) had in mind for the unstable part of what we now call van der Waals loops and, indeed, Thomson was the first to suggest that it might play a rôle in capillary phenomena. The coefficient c_2 is given by

$$c_2 = 2\pi\int x^2\psi(x)\mathrm{d}x, \tag{6.4}$$

that is, it is the second moment of $\psi(x)$ which is the function whose derivative with respect to the intermolecular separation x is the intermolecular potential.

The zeroth and first moments of this function play a fundamental rôle in Laplace's theory, see Chapter 3, Section 3.1, eqns (3.10) and (3.11).

The conditions that the total mass be fixed, $\int \rho(z)\mathrm{d}z$ = constant, and that the free energy be at a minimum with respect to all possible local changes of density, allow him to determine the shape of the density profile $\rho(z)$, and the surface tension σ:

$$\sigma = c_2 \int (\mathrm{d}\rho(z)/\mathrm{d}z)^2 \mathrm{d}z. \tag{6.5}$$

By expansion in a Taylor series near T^c, he showed that the dependence of surface tension on temperature followed the relation

$$\sigma \sim (1 - T/T^c)^{3/2}, \tag{6.6}$$

and hence, as de Vries had found, there is a horizontal tangent at $T = T^c$, and not a non-zero slope as Laplace's theory would require from a discontinuous change of density at the liquid–vapour boundary.

It is clear that van der Waals had had all the essentials of his theory in his mind during the five years that elapsed between his short communication of 26 May 1888 [27][8], and the publication of the full details in his classic paper of 1893 [37]. There were many points in his treatment with which he was not entirely comfortable. These included not only the general uncertainty, mentioned above, about the applicability of standard thermodynamic equations to an inhomogeneous system, but also more technical points. Is it correct to assume, for example, that the entropy density is strictly a local property, that is, one independent of the density gradient? He made this assumption and noted that 'Gibbs himself evidently inclines to the same view' [37e, p. 211; *18*, p. 221]. Furthermore, is it correct to assume that the Taylor expansion used in calculating the 'inhomogeneous' part of the free energy can be restricted to the first two terms?

One curious point emerges. Van der Waals's treatment relates all capillary phenomena to the integral c_2, the second moment of $\psi(x)$, and not c_1, the first moment, eqn (3.11) of Chapter 3, which arises from Laplace's treatment. He writes, 'It is certainly striking that the term thought by Laplace to be characteristic of capillarity is just that missing from the equation for the energy, and which therefore should be missing also in the subsequent development of this theory of capillary action. The term which should account for these phenomena is c_2, and this quantity is as many times smaller than c_1 as c_1 is smaller than a. This circumstance could easily lead to the belief that we should find from the hypothesis of a continuous variation of density that the capillary energy would be so small that it disagreed with the value found experimentally, at least unless we ascribe to the sphere of attraction dimensions which, in their turn, will disagree with other observations. We show below, however, that these difficulties are imaginary' [37e, p. 210]. A full, rigorous and quantitative resolution of this paradox requires the resources of modern statistical mechanics [*23*].

Notes

1. For example P^c and T^c for ether are respectively 36 atm and 468 K, and for benzene 48 atm and 563 K. So when ether is under a pressure of 3 atm and at a temperature of 234 K, the state corresponds with that of benzene at 4 atm and 282 K ($l = 1/12$, $m = 1/2$).
2. The modern way of discussing the problem is to show that the heat capacity of the saturated vapour is positive if the heat capacity of the vapour at constant pressure exceeds the entropy of evaporation.
3. Thirty years later he wrote of the principle of equilibrium: 'Gibbs restricts himself to formulating the law but gives no proof, which is scarcely possible in the present state of our knowledge. The law is to be understood as a wide generalization, the only proof of which that we can seek, at any rate for the present, is in the thoroughness of the experimental confirmations of deductions from it' [113, p. 116]. The stipulation 'at any rate for the present' was related to the advances in statistical mechanics made by Boltzmann and Gibbs.
4. Next year his derivation of Maxwell's rule by Gibbs's method was the first real result of this 'collaboration' [$1d_1$, p. 92].
5. ε energy, η entropy, ψ free energy, ζ free enthalpy (or Gibbs free energy), μ chemical potential.
6. Letter to R. Threlfall, the translator of *Continuity* into English, in Sydney University Archives [157(b)].
7. Even earlier were criticisms by the Russian chemists I. A. Kablukov and V. A. Kistyakovskii.
8. It is almost certain that his work was independent of that of Karl Fuchs, Professor of Physics at Pressburg (now Bratislava, in Slovakia) who made a similar expansion of the energy density in terms of the derivatives of the density with respect to height, and went on to use these to discuss the pressure and the surface tension. Fuchs's first paper [*21*] was published in an issue of Exner's *Repertorium der Physik* that was distributed in April 1888, and so its appearance may have prompted van der Waals to get put his own ideas into print as soon as possible. Fuchs's work, and similar work by Rayleigh [*22*], was however mechanical in essence, not thermodynamic, and so missed one of van der Waals's main points.

References

1. Volevich, I. V. (И. В. Волевич), *Sovremennaya literatura Niderlandov*, Moscow, 1962.
2. Roth, F., Ueber die Zusammendrückbarkeit der Gase, *A. Ph.*, 1880, Bd 11, S.1-36.
3. Roth, F., Abstracts of papers by W. Ramsay and J. B. Hannay, *Beibl.*, 1881, Bd 5, S. 422, 784.
4. Maxwell, J. C., On the dynamical evidence of the molecular constitution of bodies, *Nature*, 1875, v. 11, p. 357–359, 374–377.
5. Clausius, R., Ueber das Verhalten der Kohlensäure in Bezug auf Druck, Volumen und Temperatur, *A. Ph.*, 1880, Bd 9, S. 337–357.
6. Clausius, R., Ueber die theoretische Bestimmung des Dampfdruckes und der Volumina des Dampfes und der Flüssigkeit, *A. Ph.*, 1881, Bd 14, S. 279–290, 692–704.
7. Planck, M., Die Theorie des Sättigungsgesetzes, *A. Ph.*, 1881, Bd 13, S. 535–543.
8. Stoletov, A. G. (А. Г. Столетов), Notes on the critical state of bodies (in Russian), *ЖРФО*, 1882, t. 14, c. 167–175.

9. Dewar, J., President's address, *BAAS Rep.*, 1902, v. 72, p. 3–50.
10. Chwolson, O. D., *Traité de physique*, t. 3, 3.Fasc., p. 795–802, Paris, 1911.
11. Swart, A. J., Úber die Dissociationsgesetze in Gasen, umgearbeiteter Auszug van 't Hoffs, *Z. ph. Ch.*, 1891, Bd 7, S. 120–139.
12. *JB. AU (1888–1889)*, Amsterdam, 1890, b. 48.
13. van der Waals Jr., J. D., Gibbs' invloed op de mengseltheorie, *Ch. W.*, 1926, Jg. 23, b. 410–411.
14. Maxwell, J. C., *On the equilibrium of heterogeneous substances*, in South Kensington Museum Conferences held in connection with the special loan collection of scientific apparatus: Physics and mechanics, London, 1876, p. 145–150. Reprinted in *Philos. Mag.*, 1908, v. 16, p. 818–824.
15. Cohen, E., *Jacobus Henricus van 't Hoff: Sein Leben und Wirken*, Leipzig, 1912.
16. *Aus dem wissenschaftlichen Briefwechsel Wilhelm Ostwalds*, Bd 2, Berlin, 1969.
17. Kamerlingh Onnes, H., Prof J. P. Kuenen, *Nature*, 1922, v. 110, p. 673–674.
18. Gibbs, J. W., *Collected Works*, v. 1, New York, 1928.
19. Rusanov, A. I., The centennial of Gibbs' theory of capillarity. In *The modern theory of capillarity: To the centennial of Gibbs' theory of capillarity*, ed. F. C. Goodrich and A. I. Rusanov, Berlin, 1981, p. 1–18.
20. de Vries, E. C., Metingen over den invloed van de temperatuur op de capillaire stijghoogte bij aether, tusschen den kritischen toestand en het kookpunt van aethylen, *V.M.*, 1892, d. 1, b. 156–160; *A.N.*, 1895, t. 28, p. 210–219.
21. Fuchs, K., Ueber Verdampfung, *Repert. Phys.* 1888, Bd 24, S. 141–160, and later papers, S. 298–317, 614–647; *Sitzungsber. Kais. Akad. Wiss. Math.-Naturwiss. Kl.*, 1889, Sekt. IIa, Bd 98, S. 740–751, 1362–1391.
22. Rayleigh, Lord, On the theory of surface forces. II. Compressible fluids, *Philos. Mag.*, 1892, v. 33, p. 209–220, reprinted in *Scientific Papers*, Cambridge, 1902, v. 3, p. 513–523.
23. Rowlinson, J. S. and Widom, B., *Molecular Theory of Capillarity*, Oxford, 1982, chap. 3 and 4.

CHAPTER 7
An assessment

The middle of the 1890s marks a new period in van der Waals's life. He was approaching his sixtieth birthday, his main scientific work was behind him, and so one might have expected him to rest on his laurels. This was not the case, however; it seemed rather as if he had got his second wind.

The publication of the *Molecular Theory* and then of *Capillarity* marked his return to the international scientific community. His election as General Secretary of the Academy brought about a further change in his life and made it hard for him to carry out properly his University duties at a time when the number of students was growing steadily. He sought and obtained the establishment of a second (extraordinary) professor to take over the lectures in experimental physics. In 1897 a student of Kamerlingh Onnes's, Remmelt Sissingh (1858–1927) was appointed to this post. He specialized in optics and later revised part of Bosscha's textbook into a two-volume treatise called simply *Light* (1902–7). He was quickly followed by another student of Kamerlingh Onnes's who joined van der Waals as a lector. This was Pieter Zeeman (1865–1943), who was appointed the year after he discovered the remarkable splitting of spectral lines in a magnetic field. Zeeman soon took over the supervision of the practical classes although at first, as he recalled later [205, p. 110], van der Waals had merely wanted him to assist in the classes that van der Waals himself was conducting. Zeeman also lectured on certain topics, for example the optics of crystals, and continued his research with better equipment than had been available to him in Leiden. In 1902 he and Lorentz, who had developed the theory of the Zeeman effect, were the first Dutch physicists to receive Nobel Prizes.

With his assistants in place, van der Waals could restrict himself to the teaching of theoretical physics and thermodynamics, but naturally all the administrative problems fell on him as the Professor and Director of the Laboratory[1]. He had passed through some hard times but now it seemed that he was entering a period of stability. The restoration of his equilibrium may have been helped by the appearance on the scene of Philipp Kohnstamm who entered the University in 1893 and soon became van der Waals's closest pupil. The rapport between the two men started even when Kohnstamm was an undergraduate; he was to recall later that van der Waals 'could be gentle, like a father, kind and consoling. I experienced this myself when, as a young student, depressed by illness and other troubles, I asked him to come and speak to me "not as a professor with a student, but as a teacher with a pupil"' [206, p. 5]. Conversely, we can see how much van der Waals confided in Kohnstamm when he spent ten of the twenty minutes devoted to a test in thermodynamics telling him instead of his own recent work on the subject of the test [187, p. 58].

Kohnstamm's elder son wrote of his father that, 'his bonds with van der Waals were very deep-rooted. Van der Waals's position as a convinced Christian, who combined exact physical thinking with a deep Christian conviction, was an important factor in my father's conversion to Christianity' [241]. Kohnstamm himself wrote in his autobiography that van der Waals's influence 'was not only scientific, it went much deeper. . . . I came into close contact with a natural philosopher of the highest rank and with a character as pure as refined gold who would not accept any hint of compromise with the truth' [*2*, p. 18]. His writings about van der Waals [187, 193, 197, 199, 206, 218] are the most important source of information on van der Waals's life, and his work in physical chemistry was the natural continuation of van der Waals's. At the same time he is an important and independent figure in Dutch science and culture, and for that van der Waals also deserves some of the credit.

Philipp Abraham Kohnstamm[2] was born in Bonn on 17 June 1875, the second child of a white-collar worker for a Dutch company [*2*]. His first years were spent partly in Germany and partly in the Netherlands where his mother's family lived; Professor B. J. Stokvis (see Chapter 4) was his mother's brother-in-law. He attended first a private school in Amsterdam and then a three-year HBS and, from 1890, a five-year HBS. (In the first of these one of the teachers was a brother of van der Waals but Kohnstamm does not tell us his name.) At fifteen he had to choose between German or Dutch nationality and, although his parents had had to return to Germany, he chose to be Dutch and stayed in Amsterdam. He preferred the freedom and democracy of the Netherlands and feared the increasing anti-semitism in Germany.

At the University he did not at first make up his mind about the field in which he wished to specialize. He later told van der Waals, 'The decision came when I attended for the first time your lectures for *doctorandi*. Beyond the scientific context of your lectures I perceived two statements: that there is only the one truth, and that we have to seek it out. It was then that I realised that my further development had to take place under your guidance' [218, p. 38].

In 1900 he became an assistant at the Physical Laboratory and the next year he defended his doctoral thesis on the pressure–temperature–composition relations for binary liquid–vapour systems. This work is remarkable both for the great effort put into the critical evaluation of different methods of studying phase equilibria, and for the range of useful experimental results. At the same time it was also evidence of the breadth of his interests; many of the *Stellingen* are on philosophical and cultural problems. The acknowledgement in his preface to the late Bellaar Spruyt was no mere formality. To van der Waals he wrote, 'For us who have the privilege of seeing you almost daily this contact means infinitely more to us than the gift of a highly refined theory or more ingenious experiment. This is neither the place, nor have I the right, to explain my meaning adequately, so let me content myself with a word of gratitude for all that I have received from your abundant presence' [*3*, p. ix].

After his 'promotion' *cum laude* Kohnstamm stayed on as an assistant in van der Waals's department and, during 1907, tried his hand as a lecturer in logic – he was chairman of the Philosophical Association at the University. The next year, after van der Waals's retirement, he became a Professor in the Physics Department and lectured on thermodynamics[3]. His inaugural lecture was on 'Determinism and science'. He wrote many papers on thermodynamics and phase equilibria and on similar problems in the field of van der Waals's work, including a series of studies of the effects of high pressure on the mutual solubility of liquids. These were started with Jean Timmermans (1882–1971) from Brussels, and continued with A. M. J. F. Michels, who was first his assistant and later his successor[4]. From 1919 he became also Professor of Paedagogics but continued to teach thermodynamics until 1928, after which he devoted himself purely to social problems. He summed up his work in thermodynamics in his editing of van der Waals's lectures in a two-volume *Textbook* [122,123], which is discussed in the next chapter, and in a review on 'The thermodynamics of mixtures' for the *Handbuch der Physik* of 1926 [163]. His textbook on *Heat* (*Warmteleer*) went through several editions in the Netherlands. He wrote also about a hundred books and papers on education, psychology, philosophy and religion. His closest friend was Paul Ehrenfest (1880–1933)[5] and he wrote a moving essay, 'Paul Ehrenfest as the spokesman of his times' after the latter's suicide in 1933. After the Nazis came to power in Germany he wrote a series of books against fascism – *The psychology of anti-semitism* (1934), *National socialism as a spiritual danger* (1936), *Democracy, dictatorship and education* (1937), and *Democracy under fire* (1938). He survived the Second World War and, when he died on 31 December 1951, the Dutch radio and television programmes interrupted their New Year revels to announce his death.

Van der Waals's sixtieth birthday was in the autumn of 1897. No notice was taken of the occasion. In the spring of the next year, however, came the jubilee of his thesis and this occasion was not allowed to pass without ceremony although the celebration was entirely without encouragement from him. Bosscha prepared an address that was signed at first by a few scientists and then distributed for signature to a wide circle of persons and institutions. It began with the words: 'The day of 14 June 1873 is a significant one for Dutch science. On that day in Leiden there was the public defence of the famous thesis of Johannes Diderik van der Waals, *On the continuity of the gaseous and liquid state*.' The bulk of the address was an appeal for donations to establish a Van der Waals Fund for the benefit of science and as a mark of gratitude to van der Waals for his services to the Netherlands. The fund was to 'supply monetary support for about the next ten years for experimental research related to the theories put forward by the eminent scientist and to be conducted at Amsterdam under the supervision of the founder of the theory' [5, p. 47]. The Council of the Congress of Dutch Naturalists and Physicians, of which Bosscha was the chairman, contributed fl. 1000, a payment ratified by the General Assembly of the Congress who contributed a further fl. 200 in 1899 and again in 1900 [*5*, p. 47–48]. Individual contributions of fl. 5 a year were solicited later [*4*, p. 3]. The fund was established by the anniversary date of 14 June 1898.

On that day the post flooded in to his house in the *P. C. Hooftstraat*. The address from Leiden came early in the morning and van der Waals was touched to see that it was signed even by the aged Rijke. The first telegram from abroad came from the Naturalists' Society of Moscow[6]. American colleagues sent congratulations and good wishes; a letter of 20 June from van der Waals, expressing his thanks, was found among Gibbs's papers [*6*, p. 239]. It is the only record of any correspondence between the two physicists [248]. An apposite telegram from the Physical Chemistry Institute at Leipzig read, 'Dem Forscher der Continuität wünschen wir Continuität der Forschung'. (To the researcher of Continuity we wish the continuity of research.) It was signed by Ostwald and thirty-six other scientists from several countries who were then working in Leipzig [244].

Van der Waals returned polite letters of thanks[7] to each new round of congratulations (all written, it seems, on Friday evening, 20 June), but succeeded in seeing that there were no overt celebrations, although he was compelled to listen to official speeches of congratulation at the Section of Sciences of the Academy and at the University. He flatly refused to sit for his portrait.

Four years later there was another jubilee, the twenty-fifth anniversary of his professorship[8]. The same procedure was followed: formal congratulations, the avoidance of fuss, and the refusal of a portrait[9]. He answered the Rector's address 'with a few words thanking the Council for their tolerance of his shortcomings in formal matters' [*7*].

At the first silver jubilee in 1898 a few friends, colleagues and students had gathered at his house and, as Kamerlingh Onnes has related, the total absence of ceremony emphasized the significance of the occasion. Julius handed over the *Van der Waals Fonds*, explained its intended use to support experiments and said that its disposal was at van der Waals's discretion. 'We wished', Kamerlingh Onnes wrote, 'to do all in our power to relieve him of monetary difficulties in the development of his ideas' [233][10]. In his reply van der Waals told his guests about the scientific path he had followed and the main influences on his work. It was then that he confessed that it had been his original aim to be a mathematician and it was only after he had come under the influence of Rijke and, later, of Bosscha, that he had turned instead to physics [233].

It is appropriate at this point to attempt now a preliminary assessment of van der Waals as a man and as a scientist. A useful starting point is the fulsome address prepared by Kamerlingh Onnes on behalf of the University of Leiden [244], since it reveals much of the character of both men.

Dear colleague van der Waals,

On the occasion of the silver jubilee of your 'promotion', we feel compelled to offer you our hearty congratulations and express our feelings of pride, gratitude and sympathy.

Having defended your thesis in our University you chose to support Leiden, so that it became famous for excellent work of the kind that establishes the reputation of a university.

It is seldom in the history of science that there are jubilees like today's, when it is hard to know which of your achievements we should note with most respect, your masterpiece written 25 years ago, or the new field of science it opened up, or the subsequent work that made you yourself the leader of this field.

We know how little you care for outside recognition but it is appropriate today to assure you that the University of Leiden is proud of its connection with your internationally known name through the doctor's degree it conferred on you. We remember with gratitude that your thesis, in the introduction to which you acknowledged our Faculty and, in particular, your Promotor, has enriched the human spirit by ideas of extraordinary fruitfulness. It is the greatest duty of a university to encourage and support such endeavours.

A work such as yours can arise only when a mind of the greatest power is willing to use its noble gift to reach the highest goal and to allow itself to be guided by a single-minded aim of penetrating the essence of things as far as is humanly possible. Who can foresee or measure the benefits that your work will bring in the remote future now that you are trying your luck in extending your investigations and directing further work with renewed strength towards the same goals. Both the outcome and the value of the fruits already achieved are incalculable, and [your] satisfaction in work that has shown such internal strength should be great indeed. But you believe strongly that you will be able to go further in realising your ideals.

When observations of Nature were few, you have always dared to arrive at the truth intuitively and to present it with such a depth of vision that it could only later be understood by others also. You believed that 'we should not be hindered from more searching enquiries into the nature of bodies through a fear of leaving the realm of incontrovertible truth since by so doing we should deliberately cut ourselves off from one of the ways that lead to new truth'.[11] You succeeded, however, in finding laws that revealed experimental gaps, which led to better measurements which confirmed those laws and thereby showed them to contain new incontrovertible truths.

If we recall that the law of corresponding states is sketched in a few words near the end of your thesis, then we can be in no doubt that there will be many other places in your works where ideas already present in your mind await elucidation, and that in still other places you have written in the firm conviction that you will eventually make your way to further truths although, for the moment, these are still hidden from you by a seemingly impenetrable wall.

We wish you the satisfaction of arriving at these truths also.

We congratulate you on the acceptance of your ideas by all scientists; they are now a basic part of science and are quoted in almost every issue of the journals since they are needed in current research and are the inspiration of the pioneers in many fields of science.

We congratulate you on being able to celebrate this jubilee in the plenitude of your creative powers.

Finally, we congratulate you on the facts that a great number of scientists all over the world try to develop your ideas, that your followers, who seek your help and example so often, are gathered sympathetically and respectfully ever more closely around you, and that there are now so many who have followed the wish at the end of your thesis that "the motion of planets and music of the spheres will be neglected for a while in admiration of the maze in which the tiny atoms run"[12].

The theme of this address seems to be that van der Waals had, in his thesis and later, concentrated all his effort on one great problem. He was, in fact, a single-minded man and this problem was that of the nature of matter. To put this in concrete terms we can define it first, as a main theme, the deduction of the properties of matter – gases, liquids and their mixtures – from the properties of molecules and, as a supplementary theme, to obtain the properties of molecules from first principles. This second theme could not then be tackled but it was probably this problem that lay behind his continuing interest in electromagnetism. This point can be illustrated by some passages from a little-known paper of 1903 [69] that shows his interest in atomic physics.

'First, there exists the material atom, that is, ponderable matter. This fact must be underlined. Secondly, the electricity in the atom is driven [on irradiation] into an oscillatory motion. . . . Thirdly, we believe that the electricity is atomically distributed in ponderable matter and hence that there are indivisible particles of electricity as well as of ponderable matter. We have been driven to such a conclusion by the phenomena of electrolysis. Fourthly and finally . . . the electricity does not move on its own in an atom but is connected with a small part of the atom of ponderable matter. This means that a ponderable atom is not indivisible, as has been thought previously, but that there are particles smaller than atoms' [69, p. 508–509]. 'The fact that so many kinds of light can be emitted simultaneously is, however, still quite inexplicable. Our concept of the nature of a radiating atom is fundamentally inadequate, it lacks some idea which, when it is found, will allow us to interpret all the present complexity as the outcome of a single and simple cause' [69, p. 511–512].

In spite of minor digressions such as this he never strayed from his real interest at the centre of which lay his molecules, 'small bodies with a real volume'. His belief in these could be almost childish, as when he said angrily to students who harboured any doubts about the existence of atoms: 'You will never achieve anything in the molecular field' [247, p. 103]. He would tolerate no discussion of this point; molecules were his truth and he believed in them passionately. At the age of seventy-three he said, of the problem of accounting for the size of the molecules in the equation of state, 'I still wonder whether there is a better way. In fact this question continually obsesses me, I can never free myself from it, it is with me even in my dreams' [103e, p. 256]. If he felt like that in his old age one can only guess at his passion for molecular science in his youth. 'There was something very one-sided in his greatness, but how could he not be single-minded when he was so great', a younger friend wrote [204, p. 153].

Perhaps it was this obsession with his own ideas that made him impervious to outside disturbances; he was able to detach himself from his surroundings. Zeeman was astonished to find him unperturbed by the sound of hammers and the noise from the machine shop behind the wall of his study. 'I am busy with my work', van der Waals explained [205, p. 111].

This absorption in his own ideas seemed to hinder his acceptance of the ideas and theories of others. Zeeman recalled: 'The originality of his spirit made him unwilling or, at least, reluctant to follow arguments put forward by others. I had a striking example of this trait when I told him of Balmer's famous formula which he had not then encountered. He said, "Don't think that you can tell me that a law of nature can be discovered by making wild guesses." I failed to convince him. After a few months I returned to the subject and set out the evidence from the beginning but in a different form. He said then, "You seem to be totally committed to this formula. Why, I'm believing it a little myself now."' [205, p. 111–112].

The same characteristic was displayed in an episode recalled by Kohnstamm: 'Once, after a lecture that visibly tired him, I asked him what was wrong. It appeared that he was struggling with the derivation of an expression and had been at it since the previous evening. I could not hide my surprise for the expression was quite outside his field and both it and its derivation were to be found in textbooks. "You think", he answered, "that I should stand up and get a book from the shelf and search for it. Why, I am sure that it would be better for me if I were to derive it for myself".' Kohnstamm added: 'One can understand how such an attitude affected the lectures of such a man. Nothing was borrowed, everything was his own. His first demand on his students was for them to do their own work, and their own thinking' [206, p. 30–31]. It was because 'he could work out only what he had discovered for himself' [206, p. 29] that outside scientific influences on him were negligible, except for the first 'revelation' of Clausius's paper and for the works of Maxwell and Gibbs.

His early publications reflect his good command of the contemporary literature in physics and physical chemistry, but in later life he read little; in the tactful words of van Laar, 'he only takes information from the literature of his field to a limited extent' [186d′, p. 42]. Nevertheless he always glanced over the latest issue of each journal to see if it contained new experimental results; the discussions and calculations of others he passed over. 'He acquired an unusual skill in such scanning', wrote Kohnstamm, 'he generally knew in a few minutes if there was something in a paper "that could be used". This skill was obtained in part by hard work; he knew the field better than anyone else from his long calculations' [206, p. 30]. It is, therefore, not surprising that his papers contain few references to and little discussion of other men's theories. This lack was probably connected with his dislike of polemics, noted by van Laar [180d, p. 43] whose inclination was always to argue a point. Van der Waals preferred to work in his own way and not to argue, but there were times when he protested strongly if he thought that his ideas were under attack. We have seen (Chapter 4) his sharp attack on Crookes's belief in 'radiant matter', and later discuss his views on the size of the parameter a_{12} in his equation of state for mixtures (Chapter 12, Section 3.1). A further example is his discussion of the problem of connecting thermodynamics with an equation of state, a problem to which he returned many times (Chapter 12).

The central tenet of his scientific thought, the existence of atoms and molecules, was, however, too important a matter to be discussed. Kohnstamm tells that he never saw van der Waals so angry as he was after a technically excellent lecture in Amsterdam in which Ostwald had denied the existence of atoms.

When Ostwald was later forced to abandon his views, after Perrin's work on Brownian motion, van der Waals was no less indignant since he could not forget that Ostwald had refused to accept the equally convincing evidence from the kinetic theory of gases [206, p. 4; 218, p. 37]. These emotions, however, never appeared in his writing when he was always reserved, even in argument.

There is an interesting example of his way of discussing problems, and of his style of writing, that we owe to the happy circumstance that Boltzmann asked him to publish in the *Proceedings* his manuscript 'On the characteristic equation of v.d. Waals'. Boltzmann wrote that he did this, 'because I think this is the best way to attain my purpose, viz. to incite those who are interested in this question to a discussion which might be useful to science' [*8*, p. 398]. Van der Waals communicated the paper and an extract from this letter at the next meeting of the Section, on 25 March 1899 [*8*]. At the following meeting, on 22 April, he gave his answer, the preamble of which is the example we seek. 'In the proceedings of the former meeting Prof. Boltzmann has inserted a communication, accompanied by a letter. In this letter Prof. Boltzmann has expressed the wish that this communication would give rise to a discussion. As the result for Prof. Boltzmann's values of the coefficients of b/v, b^2/v^2, etc. which appear in the characteristic equation, differs from that which I have obtained, I consider the invitation to discussion as addressed also to me. And though as a rule I prefer to leave the discussion between different results, obtained in two different ways to the gradual development and extension of our views, which is brought about in course of time, I will not abstain from complying with his request, in the hope that this discussion may be "useful to science".

'I am perfectly aware of the difficulties attending this discussion. Prof. Boltzmann's *Vorlesungen* form a logical coherent whole, and the different results agree so well, that we may be sure that they contain a perfectly correct solution of the problem, as Prof. Boltzmann conceives it. On the other hand, I am convinced that also my solution, leaving some unsolved questions and some minor points out of account, gives a true explanation of the problem as I see it. . . . As our results differ so much we cannot but have considered two different problems', and the aim of his paper is 'to get some certainty on this point' [56e, p. 768].

This passage gives some idea of van der Waals's style, one aspect of which deserves some attention. Each paper opens with a detailed and often pedantic explanation of why he has decided to tackle the problem indicated in the title and why he now publishes it in its present form. When, after all this explanation, he passes on to his subject, the same style is maintained, so that it seems at first that the thoroughness of the exposition would make the actual reading an easy task. This is not so, however, for, as with Gibbs, the leisurely narrative is combined with a dense concentration of ideas. A. Rakovskii, the editor of the Russian translation of the van der Waals–Kohnstamm *Textbook* wrote that 'the book is written in a very compressed style and in difficult language.

... The main difficulty of translation was that a sentence could contain several, sometimes hidden, ideas and so became clear only after reading later chapters' [122r, p. 8]. Clearly van der Waals was anxious to consider a problem in all its aspects, but did not wish to expose it clearly to his readers; they would appreciate the problem more fully as they learnt more of the subject. Perhaps van der Waals understood this short-coming when he wrote, 'when someone has made a theoretical discovery he is quite convinced himself of its correctness but, until it is generally accepted as true, he may prefer, when addressing others, to describe it from their point of view' [69, p. 511]. Such a position was not one he found easy to adopt and this difficulty is one of the reasons that he wrote only to two popular articles[13]. One of them, on the principle of corresponding states [73], we have discussed in Chapter 6, the other [69], which he wrote after Lorentz and Zeeman had received the Nobel Prize, is as important for our understanding of his thought and philosophy. The reasons that prompted him to write were set out in an article whose lighter touch can be contrasted with that of his scientific papers: 'Our newspapers devote long columns to many topics and tell people of developments in politics and of social events. We learn much about sport and something about art. But of science we hear nothing, or are told that we should heed the proverb, "a word is silver, but silence is golden." ... When the editor of *De Gids* asked me: "Would it be possible to explain to the educated public the significance of the Zeeman effect?", I took up the challenge with enthusiasm, although I was doubtful if it was possible to answer this question. Since we are unaware of what the educated public knows of modern views of nature – and how can we be aware when the subject is never raised in the press? – this question cannot be answered. I wanted to try, however, to see to what extent it would be possible and I leave it to the editor of *De Gids* to decide if this article is suitable for the educated public' [69, p. 493–494].

If we move from the external style of his writing to the deeper matter of the internal style of his thinking, then the first point to note is how pictorial is his expression of his ideas. His molecules, his curves and his surfaces, all these he saw rather than described by formulae. This pictorial cast of thought explains why he was not worried by minor lapses in mathematics and even occasional lack of logical rigour in his theoretical derivations, as long as these did not obscure the main picture[14]. It follows naturally that he often used geometric methods (see e.g. [8,9,42]), particularly in his theory of mixtures where he enthusiastically described complicated constructions on thermodynamic surfaces. He displayed here an extraordinary power of visualizing space; geometric forms lived in his mind, they formed his principal mode of thinking.

Perhaps there is a link between thinking pictorially and the power of intuition; certainly the two came together in van der Waals. Lorentz once remarked that it would have been impossible to discover the equation of state without 'a vivid power of imagination' [195, p. 311]. This intuition was displayed first and most clearly in *Continuity*. Boltzmann was certainly correct when he noted that the equation of state was found 'by inspiration, as it were' [*9*][15], and the force behind Maxwell's criticism of the way in which the correction *b* was introduced did not really disturb van der Waals. Kohnstamm wrote that, 'I once asked him how it had happened that he had knowingly made a false assumption. "You see", he answered, "I certainly knew when I wrote it that what I was doing was not strictly correct, but I knew that the result was right and had no wish to search for a long roundabout way to reach it"' [206, p. 16–17]. It would be hard to find a better example of intuiting.

Pictorial thinking and particular attention to geometric methods reflected also another side of his scientific outlook; the structure or the qualitative aspect of a subject was more important to him than numerical relations. He was angry if physicists engaged in pure calculation and forgot the essence of the phenomenon; he called them 'ciphermen' (*cijferaars*) and, in spite of his own extensive calculations, he would become cross with students who fell into this error [218, p. 36].

Another way of expressing these features of his scientific thought is to say that it was physical rather than mathematical. He did not underestimate the importance of mathematics, but his way of reasoning was physical. Of those who treated all gases and vapours as ideal he wrote (in *Continuity*) that 'the questions proposed and answered by those who adopt it have no reference to the facts of nature, but only to the arbitrary assumptions which gave rise to the questions' [$1e_2$, p. 225]. In discussing Korteweg's work on the geometry of ψ-surfaces he posed the question: 'Are the relations between the parameters that are needed to produce a second fold on the surface for mixtures of normal substances really present in nature, or are they simply mathematical relations with no physical meaning?' [58, p. 611]. Such questions either do not arise or at least are of only marginal interest to a pure mathematician. In conversation with Kohnstamm he expressed the same point more vividly: 'A mathematician is always inclined to remake nature according to his own ideas. We need to have a hundred experimentalists to hold one mathematician on the leash' [206, p. 9][16]. He had good reason to make such remarks. As Ya. Frenkel once said, a theorist's view of nature is like a platonic love of a woman, but this was far from van der Waals's view.

It is appropriate here to rebut the conventional view that van der Waals was purely a theorist. Van Laar was in error when he wrote that van der Waals did no experiments except for a few determinations of critical temperatures [180d, p. 32], but this statement has been repeated to this day, even in the Netherlands. Thus in an interesting paper mentioned above van der Waals is described as 'a theoretician who occasionally supervised an experimental PhD thesis' [*4*, p. 2]. In fact he did much experimental work early in his career: in *Continuity* he mentions his experiments with aerosols [$1d_1'$, p. 35; $1e_2$, p. 159; see Chapter 2]; later he tackled two purely experimental pieces of work, a precise study of the line spectrum of carbon [3], and a measurement of the dependence on pressure of the temperature of maximum density of water — he found that the temperature decreased with increasing pressure [7][17]. He carried out significant experiments on mixtures: determinations of the gas-liquid critical temperature of mixtures of carbon dioxide with air and with hydrogen chloride [1d$_1'$, p. 143], and a study of the effects of pressures up to 300 atm on the critical solution temperatures of liquid mixtures[18] [1d$_1'$, p. 145–146]. Finally, we recall his experiments on electric discharges in rarefied gases and his invention in magneto-optics (see Chapter 4).

It is probable that he stopped his experiments after the death of his wife but, as Kohnstamm has written, he kept a keen eye on the experimental work of others and often knew better than the man doing the work which method would be successful and what obstacles he would be likely to meet [187, p. 58].

His close interest in the experimental work of others was in part enforced, since he had no assistance and so, as well as giving lectures, he had to supervise all the laboratory work. But, as with his reading of the experimental papers in journals, this duty was not the main reason for his concern. Zeeman wrote, 'He was always full of interest whenever I showed one experiment or another. He listened readily to the work of others, and just as willingly he described his own. Many times I heard from him a short account of new work that he had prepared for his lectures or for communication to the Academy. All was made clear as he spoke by his characteristic scribbling and sketches' [205, p. 111].

He did not study experiments and experimental results for their own sake but because they provided rich food for thought. Kohnstamm emphasized that experiments gave him a sense of a firm grip on reality and that this sense was 'a powerful driving force' for his thinking. 'He knew that each of his constructions was only an approximation, that each of his steps should be checked by experiment which had the power again and again to show his short-sightedness, and to which he had to submit with resignation' [206, p. 9,10]. Van Laar confirmed that, 'Never, or at least very rarely, was he fully satisfied with his results and he works tirelessly to do better himself and to bring them to perfection' [186d′, p. 41]. We can understand why his papers are full of scattered doubts, confusions, hesitations and wishes to think a problem through again. Thus, in discussion a special case of three-phase equilibria and the possible existence of a cusp in the corresponding T,x- and P,x-curves, he writes: 'I am, however, not perfectly convinced of the existence of this cusp – and the surest proof of this is certainly that I so often return to this subject. Repeatedly the thought has occurred to me, that after all the P,x-curve might be rounded in this point. I have only reluctantly assumed the existence of this point' [107e, p. 668]. Such confessions are of particular interest since he was so reserved in other respects. Here he was frank, and could write also, for example, that 'Mr Kohnstamm corrected [my] errors in particular in the forms of the curves that I named q-lines' [112, p. iv].

The understanding of his dependence on this sense of reality is tacitly present in all his works, but once he expressed it openly. 'Although an application of an already-known phenomenon is important, the discovery of a new phenomenon is more important to the physicist, and not only to him. Every new phenomenon is, for a physicist, a means to test the correctness of our ideas on the essence of nature, and to see to what extent they need to be expanded and improved' [69, p. 500–501.].

Let us return to the discussion of his mode of thinking and note that his strongly visual imagination was necessarily a mechanical one since mechanics underlay the whole scientific philosophy of the last century[19]. He said once that he belonged to those 'who consider the whole of inorganic nature as a grand mechanical problem – and what else is it other than an unbounded series of equilibria and motions' [61, p. 51].

His intuition prevented him from taking this philosophy to extremes[20]. We notice a Cartesian limitation of mechanics to inorganic nature, thus excluding living beings and the world of the spirit from the realm of mechanics. It is more interesting, however, that even within physics itself he saw problems that apparently could not be brought within the scope of mechanics, the most important of which for him were the gravitational and molecular attractions themselves. His difficulties with these had appeared first in his Thesis where, in the sixteenth 'Proposition', he quoted Newton's confession of an internal contradiction in his system (see Chapter 3). Van der Waals wrote later: 'Whereas the French revolutionaries could rejoice that thrones were swept away and altars overthrown, Newton went much further; he threw away almost the whole of the contents of the Universe. For him, the Universe was almost empty; it contained only ponderable matter. But then he had to attribute supernatural powers to this ponderable matter, to the Sun and the planets, and so he himself deprived the ponderable matter that he had admitted of the character of physical bodies' [69, p. 502]. The refilling of empty space began, van der Waals supposed, with the work of Young and Fresnel, proceeded with that of Faraday and especially with that of Maxwell, and had now been carried further by Lorentz and Zeeman. 'Physicists are nevertheless, only reluctantly and with obvious resistance, restoring what Newton had thrown out' [69, p. 503]. Notwithstanding these advances we still cannot explain Newtonian attraction and have to speak of 'the riddle of gravitation' [69, p. 511].

A similar riddle for him was the origin of molecular attraction (see also Chapter 11, Section 2). He was convinced of its essential difference from Newtonian attraction – the parameter a was certainly not proportional to the molecular weight [92]. He could even oppose the existence of another kind of interaction, the repulsive molecular force postulated by Maxwell, a postulate which was, he said 'a hypothesis whose improbability is not to be denied' [92e, p. 150]; he preferred to ascribe a fixed volume to a molecule[21]. Neither this kind of force nor the electrostatic models proposed by his son could be responsible for molecular attraction and to the end of his days he could only hope, 'Perhaps we shall sooner or later form a picture of a molecule which requires that they necessarily attract each other, and learn how to compute the value of that attraction' [92e, p. 151]. Such thoughts allow us to maintain that he was 'a philosopher among physicists' [73, p. 285], albeit a 'closet' one, as was described in the last chapter.

In the Netherlands he was often contrasted with Lorentz who 'inherited the great clarity and classic calmness of Helmholtz, whereas van der Waals displays rather the waywardness and probing of Maxwell. Lorentz is easier to understand . . . Van der Waals is not an Olympian like Helmholtz or Lorentz, rather he is a philosophising dreamer, a searcher, but he is not inferior to Lorentz in the profundity of his insight or in the sharpness of his vision' [186d′, p. 34,36].

Zeeman once remarked [205, p. 102] that 'a Plutarch in the history of science' would be interested in a comparison of van der Waals with Gibbs; both were born at the end of the 1830s, van der Waals published his main works when he was 36, 43 and 52 years old, Gibbs when he was 34, 39 and 63. There the similarities end, for one worked mainly by intuition and the other had a powerfully logically mind. If we are to avoid the superficial analogies of a Plutarch then we find more striking resemblances between van der Waals and Maxwell, as has been noted already by Kohnstamm and van Laar [186d′, p. 33–35; 218, p. 36]. Both are characterized by powerful intuition, a pictorial way of thinking and the conviction that experiment takes priority over theory. Both were deeply religious men, and both were fond of writing verse. It is of interest that, initially, they both took rather a cool attitude to thermodynamics, but were led, partly by the influence of Gibbs, to undertake work of their own in this field[22].

As has been mentioned, van der Waals placed Maxwell first among contemporary physicists, and Gibbs second, but at almost every mention of Maxwell's name he is critical of weak points in Maxwell's argument [248]. We cannot know the reason for this; perhaps it was a lingering subconscious resentment of Maxwell's criticism of *Continuity* in his review of 1874, or perhaps a degree of jealousy for a man of so similar physical ideas and methods[23]. The fact remains that even in a popular article [69, p. 510] and in his Nobel Lecture there are ripostes against Maxwell, although the last of these has an affectionate air about it: 'Maxwell stated, although he erroneously considered to have proved it, [that] there is a law for the distribution of the velocities known as the Maxwell law . . .' [103e, p. 254].

He was profoundly confident in his belief of the independence of nature and its laws from human concerns; this belief underlies all his general statements, such as those quoted above. It is the belief that Einstein called 'religious'[24] although van der Waals would not have restricted this word to so abstract a meaning. Characteristically, he spoke little about his intimate thoughts but once wrote, in concluding a discussion on the nature of atomic radiation, 'This complicated phenomenon will certainly be explained in a simple way. This is awaited by most physicists, at least by those who see a work of God, which is all-embracing but indivisible in the whole of nature in all existing reality' [69, p. 512]. It was this feature of his *Weltanschauung* that had been noted by Boltzmann, according to Kohnstamm's account: 'When one speaks with van der Waals about a physical hypothesis one receives the impression that it is almost a religious truth for him' [206, p. 3]. This was in fact so, for 'science was for him not an ideal, but a means of service to God and for His glorification' [206, p. 6]. He looked on his scientific activity as his duty to God.

So we arrive finally at perhaps the most important feature of the portrait of a scientist, the motive of his work. Einstein wrote in a famous article on the motives for scientific research [*14*] that the temple of science would be almost empty if those who engaged in research for reasons of ambition or personal profit were to be expelled from it. 'Nevertheless there would be a few, from both the past and the present, who would remain within.' Max Planck was the man Einstein had particularly in mind but surely van der Waals would have been there also, among those who strive in science to see 'predetermined harmony', and for whom 'the spiritual effort resulting in such hard labour is also religious and passionate' [*14*]. These words apply equally to the passion van der Waals brought to his scientific work. Van Laar wrote: 'When he is engaged on something new, it absorbs him completely; his whole being is devoted to the problem and with such thoroughness that he knows no peace. At such times of high stress even a night's rest counts for nothing' [186d′, p. 42]. There is no reason to doubt Kohnstamm's assertion that van der Waals's search for truth was his way of serving God, but it is hard to disentangle his religious beliefs from what was simply an almost fanatic determination to discover the truth[25].

Van der Waals himself wrote once about his motives with an unusual candour; his article that marked the award of the Nobel Prize to Lorentz and Zeeman concluded with the paragraph: 'Those who appreciate scientific research for the sake of science itself were delighted by the theoretical work of Lorentz and the experimental work of Zeeman although, at that time, Nobel Prizes had not been heard of. These prizes can be given as a subsequent reward but, it is my firm conviction, they cannot stimulate scientific work of the highest standard. He who is not driven by an irresistible internal drive could never accomplish such work for the sake of a future material gain' [69, p. 512].

Here we can see again how impossible it is to separate the scientist from the man; as Kamerlingh Onnes wrote, 'It would be completely against the character of van der Waals if we were to speak of his private life' [233]. It is scarcely possible to penetrate further into the inner life of a man who wrote (albeit of the work of another man): 'Every discovery, either theoretical or experimental has its intimate history. How the first idea was born, what were the external influences that affected it, what disappointments arose in the early days, and what helped it to survive — all these remain unknown to the outside world. For the author himself they are all unforgettable, but they remain unspoken since we keep in our heart what has stirred us most. I am far from saying that I know the intimate history of this discovery [the Zeeman effect], but if I knew I should think it improper to speak of it' [69, p. 509–510].

Nevertheless a few facts and recollections from the later part of his life give us some picture of van der Waals as a man. The first impression is of a dull and dry-as-dust pedant. A punctual and reticent man of small stature, monotonous in his way of life, he was precise in all matters, his lectures and classes fell in the same hours, 9 to 10 and 10 to 11, for many years, he kept steadfastly to his established routine; each day he awoke, ate and went to bed at the appropriate time. The only uncommon feature in his appearance was his bright blue child-like eyes, and those who knew him well associated this with his straightforwardness: 'Dissembling and falseness were totally alien to him. . . . To say something that he did not think was physically impossible for him' [204, p. 154]. This trait made it hard for most people to associate easily with him. He was too artless, too independent of common conventions, and this aspect of his character led often to difficulties. For example, on the death of a member of the Academy, against whom he felt some antipathy, someone said to him that the Secretary had to attend the funeral. He replied at once: 'If I have to do that then I shall no longer remain as Secretary' [204, p. 154]. He could be silent, stubborn in his opinions and in his likes and dislikes, and often seemed to be gloomy and to bear ill will. 'Van der Waals was the last man that you could call soft or kind-hearted. There was something in him of the strength of the prophets of Israel, and, like them he could be severe and sharp' [206, p. 4]. His stubborn dislike was directed, amongst others, at van Laar, one of his most talented pupils, who complained to Lorentz, in a letter of 18 August 1899: '. . . another person whom I esteem highly, I mean Prof. van der Waals, is totally impossible in normal (scientific) contact. Never have I had any help or support from him when I was having difficulties with scientific work. Among one of the less pleasant characteristics of Prof. van der Waals is certainly the fact that he does not very much appreciate the work of other people (I do not mean my work, but work of much greater men, men with excellent reputations). *A* is a third-rate scientist, *B* is fourth-rate etc. Putting himself only for one moment in their place is not his strongest point. This accounts for the fact that he never gives any encouragement to others. He can sometimes be extremely cold' [*16*, p. 28]. Thus van Laar was calling on his own experience when he wrote in his essay on van der Waals: 'All his being emits a stubborn and self-sufficient power which may produce, together with his sharpness and reticence, an impression of ill will in those who do not know him well' [186d′, p. 15]. Cohen added, however, that 'His apparent sullenness was, as with many others, a result of his diffidence' [*17*, p. 498] but the matter was obviously not so simple. Fame changed neither his behaviour nor his habits. 'He lived as if on an island, in solitude and silence, with his daughters' [204, p. 151]. 'This man moved away from the world into his quiet study [186d′, p. 45].

The house on *P. C. Hooftstraat* was, however, by no means the dark cell of an anchorite. 'Old Jan', as they called him at home, 'was not the armchair-scientist whom you never see because he sits in his study buried behind his books and who, if you do catch a glimpse of him, is essentially somewhere else, absorbed in his own thoughts. There were few men who were so active at home, who participated so fully in all family troubles, or who took so full an interest in the petty details of everyday life' [204, p. 157].

The hard life he had led when young made him very modest in his needs, and thrifty, even perhaps tight-fisted, although he helped his younger brothers when he had become independent. There is a well-known legend in the family, quoted by one of his great-grandsons: 'Jan senior had heard the news that he had won the Nobel Prize and his grand-children went to his house to congratulate him. Coming home one of them said "*Grootvader smijt met 't geld*" ("Granddad is throwing his money about"). He had given each of them a *dubbeltje* (a 10-cent piece)' [242].

Although meticulous in his duties he was careless with the letters he received. 'He kept no letters except for those in the van der Waals Laboratory. When a letter had a blank page he used it as scribbling paper for his calculations. He left these sheets on the table in the middle of the room where, in the course of time, they accumulated to a depth of 5 to 6 cm. When the time came for spring cleaning, his eldest daughter, Annie, who was responsible for his household, urged him to visit a friend or a relative, or she took her chance if he went to a meeting. The top layer of 1 to 2 cm was then spread on the floor, the rest was put into the waste-paper basket, and then, as well as she could, she re-arranged the papers from the floor on the table as she had found them' [242].

The rules at home were firm; when his wife was alive he was strict with his children and only became more easy-going after her death [232, p. 11]. His principle 'if you do anything, do it well' was imposed on the children also, and it was irksome for them at first. Nevertheless the atmosphere of the home was not only morally high and clear, but could also be merry. 'Ah, what lovely days we had then with their seriousness, but also with joy, laughter, humour and childish tricks, when he, "a world celebrity", became the butt of all the jokes and he tried in vain to defend himself' [206, p. 5].

Psychologists note a correlation between intellectual ability and a sense of humour. Van der Waals's case confirms this, for although his scientific papers are without any trace of humour, he could unbend when he was merely writing about science, as is shown above. At home those who knew him well said that it was impossible to talk completely seriously with him. 'When van der Waals was spoken to, he never stayed serious for long; the humour soon broke through. . . . earnestness and humour always go hand-in-hand with van der Waals' [204, p. 154,152]. This was particularly so on holiday.

He usually rented a cottage for the summer although some years the family spent their holidays travelling abroad, to the Tirol in 1896 and to Switzerland in 1897 and 1899. More often they stayed in the Netherlands; from 1889 to 1891 they lived in the summer in a house near Putten that belonged to the Oudemans family. From 1892 they often rented a house at Groesbeek, in the south-eastern part of the country, which was near the villa of Chantepie de la Saussaye. It was here that he received Boltzmann in the autumn of 1898.

'I remember vividly', Zeeman recalls, 'two days spent at St. Jansberg near Groesbeek, where van der Waals had rented a house. The summer weather was marvellous, the lime trees were in flower, the bees hummed continuously, and van der Waals, sitting under the trees, looked quite unlike his usual appearance in his laboratory or study. I saw how susceptible he was to the beauty of nature and with what readiness he responded to jokes and laughter. Although he was already past sixty, he remained a good walker. He pointed out the beauties of the landscape, showed in which direction Nijmegen lay, in which Kleef, and in which Cuyk, where the end of the Mookerheide was, where the railway ran, and where the Maas flowed. On such walks he usually led the way, and then suddenly gave the signal for five minutes' rest for the party, and everyone sat down. He alone remained standing and, five minutes later to the second, he gave the signal for the walk to be continued' [205, p. 112].

When in the country he liked to play patience in the evenings, but sometimes disappeared into the study to write something down 'because his work carried on within him even when he was at rest' [204, p. 153]. Billiards was another of his recreations and he noted the differences between the collisions of the ivory balls and those between polyatomic molecules which were complicated by an inelastic redistribution of energy among the internal vibrations. During the academic year walking was his principal recreation. How much he was pressed for time we may deduce from his letters to Richard Threlfall, the translator of *Continuity*, who visited him in February 1889. Before his visitor's arrival van der Waals asked: 'As I am much occupied, it will be convenient if I know exactly beforehand the day of your arrival. I hope you will grant me the favour of having you for my guest. I cannot offer you very much but that little will be offered with great pleasure.' After his departure van der Waals sent a very short additional comment on binary mixtures and concluded: 'I have but little leisure, therefore be so kind as to excuse this writing . . .' [157(b)].

His more general cultural interests are not known to us. The paper quoted above, on the Zeeman effect, shows that he had a full understanding of the historical development of our scientific view of the universe, and that he was acquainted with the irony of Voltaire and of Heine, 'in whose peculiar style one has to know how much that is serious is to be found in his cynical jokes' [69, p. 502]. In the same paper he used musical analogies to discuss unsolved problems in the spectra of radiating solids. A piano stood in his house, which his daughters certainly played, but we do not know if he used it himself. Perhaps under the influence of his brother, who was a Sanskrit scholar, he was fond of Indian literature: 'The *Wisdom of the Brahmins* by Rückert[26] and similar treatises on Indian philosophy and poetry are his favourite reading' [186d′, p. 45].

The head of the family's attention to his children and relations was as great as to his science, and when grandchildren were born he liked playing with them even more than working in the laboratory, as he once confessed to Kamerlingh Onnes [204, p. 152].

One can be sure that his pride and hope was his son, not only his youngest child and only male heir, but also his scientific heir. At first he was taught at home by his father, and then he entered the Physical and Mathematical

Faculty in 1892. It was a great day for both father and son when he defended his doctoral thesis on 21 December 1900. It was entitled *A Statistical Treatment of Radiation Phenomena* [*18*] and is interesting to historians of science as evidence of the unease at that time about the theory of radiation. The crisis had not been resolved immediately by Planck's hypothesis of the quantization of radiant energy. One of the 'Propositions' (*Stellingen*) read: 'The true theoretical formula for the distribution over wave length of the black-body radiation has not yet been established. It is impossible to decide with certainty which of the formulae of Lummer and Pringsheim or of Planck[27] can be chosen as an empirical one.' The author obtained new results in the thermodynamics of radiation and put forward some neat suppositions about the nature of molecular interactions and their connection with radiation. The Academic Council accepted the thesis *cum laude*, a decision that was not taken only out of respect for his father.

There are some personal details in the thesis. As was customary, the text was preceded by a dedication: 'To my father, to the memory of my mother, and to my fiancée', and by a preface with thanks to his teachers and others who had helped him. These concluded in an unconventional way: 'But my gratitude belongs most of all to you, my father. All who know you must realise how much I am indebted to you. The strongest feeling which encourages me in defending my thesis is the earnest desire to be worthy of you, not only in science but also in my whole life hereafter.'

From 1898 he worked in the Physical Laboratory as an assistant to Sissingh and, after his 'promotion' lectured on theoretical physics as a *privaat docent*, mainly on the kinetic theory of gases. In the autumn of 1903 he became Professor of Theoretical Physics at Groningen and five years later returned to Amsterdam to succeed his father. He undertook numerous pieces of research in theoretical physics, including some on the equation of state, and was elected a member of the Academy of Sciences. His work is not to be compared with that of his father in its scientific value, but he became a significant figure in the Netherlands. His professional and scientific activity included the supervision of the Van der Waals Fund for many years and, beyond these duties, he sat on the editorial boards of social, political and literary journals, such as *Onze eeuw* (Our age) and *De Gids* (The Guide) and took an active part in the public life of the country. It is a, pity, however, that he told us so little about his father.

From 1901 he lived with his own wife and family while his sisters remained at home with their father, where they were known familiarly as Annie, Line and Jo. The eldest resembled her mother with her soft and even temper, and it was she who took over the running of the household, shunning the limelight up to the day of her death in 1937. Each of the two younger sisters had a strong intellect and a good sense of humour; they were close friends. Later, when 'young Jan' came with his children they arranged marvellously fantastic games. Both qualified as teachers in primary schools but only Jo taught English language and literature for most of her life. Line became a writer although she also taught history for a while in schools at Doorn and Bloemendaal. She was probably his favourite daughter. She had rather an unhappy youth with a difficult relationship with her father who, after the death of his wife, did not know how to treat his daughters. She described this time in the first chapter of her novel *Noortje Velt*, but later a very intimate relationship was established between her and her father [232, p. 14]. On the day before her thirty-fourth birthday, 25 June 1902, he wrote a letter to her.

All could be and much is different from what we should like to see. We often say, like those who went to Emmaus[28], "But we hoped that etc." Only then are we in the frame of mind that God has the right to demand of us, when we believe that everything must go on as it always has done. But I realise that it is difficult for us always to see things this way.

I write this to you not because, at least as far as I know, you are often in the mood that forces us to ask if we can trust in God, but at least for me, and probably for many others, the outcome of this war[29] is so disheartening that we are full of sorrow for the fate of our kith and kin, and are forced again to ask the question, "Does a loving and just God govern earthly affairs also?"

But God's voice must come down to enlighten us, as it did those who went to Emmaus, and make us see that all must go on as it is, and not be otherwise! And in the absence of this voice we shall nevertheless perceive in our hearts, owing to God's help, that He knows all things better than we can and that He is full of love, and all this shall suffice for our personal blessing.

My beloved Line! I wish that you may trust fully God's loving guidance and I wish that the new year of your life may be passed under His protection.

Good-bye until tomorrow

Your Father.

This letter is kept by his family [242] and was published in a biography of Jacqueline van der Waals [232, p. 25]. It is wholly in character and reflects faithfully the relationship between father and daughter.

He had a good appreciation of poetry, and wrote verse easily, often for his daughters, and his influence may have contributed to Jacqueline's becoming a novelist and a poet. Her works remain in print and, to most Dutch people, her name is probably better known than that of her father.

Her life was not rich in externals but it was inwardly full with her literary work, with the study of philosophy, history, aesthetics, art and foreign languages. English, French and German were obligatory for her, but she had also a good command of Norwegian and Danish. Like her father, Jacqueline was fond of botany; when she was training to be a teacher she collected a herbarium of more than 400 plants each properly identified and described. This album is now in the Hugo de Vries Laboratory in Amsterdam. Most of her life was spent in her father's house and among a small circle of his friends united by their common intellectual interests. These included the families of Zeeman, Kamerlingh Onnes, Kohnstamm and Chantepie de la Saussaye. The last was most sympathetic to her; he was the first to believe in her literary gift and it was his authority and persistence that drew publishers' attention to her work. His much-discussed lectures 'On confidence and doubt', which she attended in the winter of 1893–94 were a stimulus to her.

Her work has been described by critics as 'true and original poetry' [*19*, p. 370]. It includes a novel, four books of poetry and some essays of which the best known is one on Kierkegaard. Her *Last Verses* rank among the best in Dutch poetry and 'are a better witness than any spoken word to the nobility of her inner life' [206, p. 6].

The same could be said of van der Waals himself; we have seen how firm was his belief both from his letter to Jacqueline and in his confidence that all nature was a design of the Creator. He belonged, however, to no church and rarely attended services. A younger friend of his, Pastoor Leendertz, wrote that once an Englishman described at great length to van der Waals his involvement in church affairs and asked whether van der Waals did anything similar. The answer was 'I've read in the Gospel a similar story about a pharisee and a publican, and I should like to be as the publican and share his wish[30]' [204, p. 154].

He showed a similar reserve in his social activities, at least after the death of his wife. He never participated in the Dutch Congress of Naturalists and Physicians although his friends Bosscha and Onnes, as well as Lorentz and Roozeboom took active parts[31]. He avoided most other social activity and he only voted in a parliamentary election once in his life, a lapse that later gave rise to many jokes.

In spite of this lack of interest in public affairs he was 'a conscious and fiery Dutchman' [206, p. 31]. The feeling of national pride, common to all Dutch people, became particularly intensive at the time of great social advances at the end of the nineteenth century. The opening address by Stokvis before the first Congress of Naturalists and Physicians is an example of this. His theme was that the time had now come to resurrect the world fame of the science of the Netherlands, which had fallen after the 'Golden Age'. His reasons were of two kinds; the first, rather metaphysical, was that nations, like organisms, develop non-uniformly so that periods of stagnation are followed by those of 'Sturm und Drang', and the second, more practical, that the conditions in the Netherlands were now favourable for a renaissance. He cited the freedom of thought, of teaching and of research, and, above all, the development of the system of education. 'Let us do our duty', Stokvis appealed, 'We are gathered ... to announce before the whole nation, that we will do all in our power to promote the flourishing of science in the Netherlands' [*20*, p. 31].

The ardent patriotism of van der Waals stood out even against such a background although he seldom spoke of it, as was his custom with other intimate matters. Perhaps the most obvious manifestation was that almost all his work was originally published in Dutch, and only later in foreign translations. There were, however, two or three occasions when he expressed his sentiments in public. The article on the Zeeman effect, quoted above, opens with the words: 'The award of the 1902 Nobel Prize for Physics to two Dutchmen caused universal rejoicing in our country. This is to be welcomed in so far as it serves as a sign that our national spirit is still alive and has not degenerated into a flabby cosmopolitanism. Nevertheless, the fact that the excitement is so intense shows the extent of the ignorance of the high level that physics has reached in our country during the last decade . . . and shows also that our people are insufficiently informed about remarkable achievements in religious, intellectual and scientific domains.'

A year later he wrote in a paper given on the occasion of the twenty-fifth anniversary of Kamerlingh Onnes's doctorate: 'The motives that led Onnes to choose the subjects in which he wanted the Leiden Laboratory to work were mainly a desire that our fatherland should take its own proper part in the development of physical sciences. . . . Of course one can assert on good authority that science is cosmopolitan. We can, however, also adduce weighty reasons to support the assertion that a certain part of science is linked, more than others, to the special gifts inherited by a nation and to certain features of the national character, [and] that one part of science, more than another, is ripe for development since it fits in well with the training provided by the principal system of education[32]. . . . A look at the works on physics produced in the Netherlands in the last twenty-five years gives one the impression that one can already speak of the physics of our country' [74, p. 73–74].

On 14 June 1898 he had first appealed to his colleagues and friends for them 'to work together so that they could again speak of Dutch science, and in particular, of Dutch physics' [187, p. 56]. He told them that this was the goal he had set before himself in his youth, and indeed, it was the publication of *Continuity* that marked the beginning of the second Golden Age of Dutch science.

The publication of the German translation of *Continuity* gave van der Waals a European reputation which grew throughout the 1880s[33]. There was however, one country in which there was some opposition, and that was Britain. One of the reasons was Maxwell's criticism (see Chapter 3) and another a jealous protection of Andrews's contribution to the field, even after his death in 1885.

It was a young Cambridge physicist who had emigrated to Australia, Richard Threlfall, who prepared the first English translation [$1e_1'$] in 1890[34]. Its appearance led to controversy [*23*]. J. T. Bottomley, Andrews's first assistant and a nephew of Lord Kelvin, wrote a critical letter to *Nature* on 5 March 1891. He began with the doubt 'as to how much there is of real importance in these much-discussed memoirs' and formulated his attitude to them in these words: 'I do not propose to criticize the papers, though I do not think that either the thermodynamics or the conclusions will bear examination; but I cannot avoid the task, however ungracious, of pointing out that they do not show a proper appreciation of the work of Andrews.' On the S-shaped isotherms, taken by van der Waals from Maxwell's *Theory of Heat* (see Chapter 3), he wrote: 'This is to me unintelligible. The curves seem certainly not taken from Maxwell, but are somehow obtained from original curves of Andrews.' His concluding claims were that 'neither the idea, nor the proof of continuity is in any sense whatever due to him' and, 'I can find no vestige of a novel idea on the subject of continuity in the essay of van der Waals' [*23*, p. 415–416].

A week later, A. W. Rücker, Professor of Physics at the Royal College of Science, showed by quotation from the English edition that van der Waals did not claim priority for any idea that was not his own, that is, for the identity of the gaseous and liquid molecules, and for the theoretical derivation of the form of the isotherm, the shape of which had, as he acknowledged, already been suggested by James Thomson. His letter was followed at once by an admission by Bottomley of his mistake: 'I beg leave to withdraw absolutely the words used in my letter, "The curves seem certainly not taken from Maxwell",' but he insisted that, 'My criticism of Van der Waals's essay is in no way altered, however, unless perhaps it is a little strengthened' [*23*, p. 437].

This exchange concluded with a letter from R. E. Baynes at Oxford, an authority on thermodynamics and molecular physics, who had referred to *Continuity* in his own work as early as 1880. He added some significant points to those made by Rücker: 'It is not a point with him [van der Waals] to discuss the question of continuity except as bearing on his characteristic [equation]; but this continuity is doubtless taken for the title of the thesis as being the most important deduction from his theory. There is no question of priority: the author gives full information as to where the experiments bearing on the subject are recorded, and only claims to have shown this continuity as a consequence of known law.' He showed, using quotations from the Dutch original, that the only reproach which could be laid upon van der Waals was insufficient praise of Andrews's work in the preface; 'but it may possibly point to an earlier date to Prof. Van der Waals's theoretical conclusion than that of his thesis' [*23*, p. 487–488].

On 4 May 1891 van der Waals was among twenty-one foreign scientists elected as honorary members of the Royal Institution of Great Britain to mark the centenary of Faraday's birth. They included such men as Berthelot, Helmholtz, Pasteur, Gibbs, Cannizzaro, Mendeleev and Thomsen. The election of van der Waals may have displeased Ramsay because four days later he gave a popular lecture on liquids and gases before the Institution [*24*] in which there was no mention whatever of van der Waals's work; it looked like a deliberate snub.

Van der Waals made no comment but perhaps his feelings may be guessed from the extremely polite formality of his reply to the Institution, on 16 May.

> Sir
>
> I have the honour of accusing [acknowledging?] the receipt of your missive containing the communication that the Royal Institution of Great Britain has conferred upon me the honorary membership.
>
> Let me express the high satisfaction I feel, that the Institution, which at all times has taken the lead in the domain of science, has done me the honour of associating my name with herself. That this takes place at a moment that you remember your august compatriot, makes it particularly agreeable to me.

It is with sentiments of profound gratitude that I accept the honour, and I take the liberty of asking you to be so kind as to bring my thanks into the Institution.

If my health, which lately has somewhat suffered, permits it, I will be present in London to be admitted to the Institution.

I have the honour to be, Sir, your obedient servant,

J. D. van der Waals [154].

He went to London for the celebration of the Faraday centenary, and wrote to Jacqueline, not without irony: 'I cannot understand how it is that I have fallen among a few very famous men, mostly towards the end of their lives. It is probable that they wanted someone from Holland to place alongside those from France, Germany, Italy and Denmark' [232, p. 25].

It was the celebration at the twenty-fifth anniversary of *Continuity* that set the seal on his reputation. The next year Ramsay gave a lecture on the noble gases at the seventh Congress of Dutch Naturalists and Physicians [*25*], and this time he did mention van der Waals, although we do not know with what reluctance, if any. Further recognition came when van der Waals was invited to write the article on 'Condensation of gases' for the *Encyclopaedia Britannica* [64].

In 1900 he was elected a foreign member of the Academy of Science in Berlin, on the nomination of Planck and with the support of van 't Hoff. The same year he became a corresponding member of *L'Institut de France* after nomination by Poincaré; ten years later he became one of the twelve foreign members[35].

A more weighty manifestation of his world-wide fame came on the twenty-fifth anniversary of his professoriate when he was presented with an album, 'as heavy as lead', that contained the photographs of 170 physicists from all over the world [231, p. 152], and a unique collection of books and reprints on his field which was placed with the papers of the Van der Waals Fund [*7*; *4*].

In August 1900 he came to Paris for the First International Congress of Physics. The city welcomed 800 physicists from all over the world and its reports, which filled four large volumes, reflected the whole of the physics of those times. Van der Waals's report was on *The statics of fluids* (*mixtures*) [58] and was a condensed version of the second part of the second edition of *Continuity* [61], in which he described the problems he was then concerned with. It was too long to be read in full and a shortened version was given to the Congress by Berthelot on 11 August. The whole week, 6–12 August, was full of activity – personal discussions, visits to laboratories (including the famous Radium Laboratory of Madame Curie), a visit to the World Exhibition and even an ascent of the Eiffel Tower. For van der Waals it was an opportunity to make his work on the theory of mixtures known outside the Netherlands.

Thus during the second half of the 1890s there were some important changes in his life. An increased stability led to the restoration of his scientific activity that we noted at the start of this chapter. His growing international reputation enabled him to extend the scope of his work; he had more time and better resources. Numbers of publications are inadequate evidence for a biographer but the figures are striking; he published 19 papers (excluding translations and official documents) in the fifteen years after his wife's death, and 60 in the next fifteen years, of which some of the most interesting were on ternary systems [47;65].

We conclude with a list of the theses completed under his supervision during the last decade of his professoriate.

1898 N. J. van der Lee, The influence of pressure on the critical temperature of complete mixing.

1900 H. Hulshof, A direct derivation of the value of the molecular constant *a*, considered as a tension in the surface.

N. Quint, A determination of the isotherms for mixtures of hydrogen chloride and ethane.

E. H. J. Cunaeus, Measurement of refractive indices as a method of investigation of the compositions of coexisting vapour and liquid phases.

J. D. van der Waals jr, A statistical treatment of radiation phenomena.

1901 Ph.A. Kohnstamm, Experimental studies concerning the theory of van der Waals's *P–T–X*-surface: Preliminary studies and methods.

1902 J. J. Hallo, Magnetic rotation of the plane of polarization near an absorption band.

1904 C. H. Brinkman, Determination of isotherms for mixtures of methyl chloride and carbonic acid.

G. C. Gerrits, Determination of *P–X*-lines for binary systems.

1906 B. M. van Dalfsen, Saturation surfaces for ternary systems.

1908 C. Dorsman, Isotherms for mixtures of hydrogen chloride and carbonic acid.

G. van Rij, Apparent association: a contribution to the explanation of the divergencies of experimental results from the van der Waals equation of state.

Only the thesis of Hallo is on an unusual subject and serves to remind us of van der Waals's interest in this field and of the results recently obtained by Zeeman. It was Hallo who also made a further contribution to van der Waals's work by taking down his lectures in shorthand. These records were later used by van Rij in preparing his thesis (as van der Waals said in his Nobel Lecture) and by Kohnstamm in compiling the textbook on thermodynamics.

The list above is not a complete account of the work done under his supervision but even so incomplete a list leads us to one of the main achievements of his life's work — the creation of a Dutch school of physics and physical chemistry.

Notes

1. When electric trams first passed near the *Plantage Muidergracht* they disturbed the laboratory instruments. With the support of the Burgomaster, van der Waals obtained in 1907 a sum of fl. 20 000 from the city to develop equipment to shield the instruments [*1*].
2. Kohnstamm still, we believe, awaits a biographer. Some material can, be found in Poggendorff's biographisch-literarisch Handwörterbuch, Bd 5, S.663–664; Bd 6, S.1364–1365; Bd. 7b, S.2516–2521, in *Levensber.*, an autobiographical essay [*2*], a letter from his son [241], and in the Amsterdam weekly journals *Ch. W.* and *P.C.*
3. The relevant passage from his son's letter reads: 'My father got an appointment as *extra-ordinarius* especially for thermodynamics, to continue experimental studies on phase-equilibria as a continuation of van der Waals's work in that field, but van der Waals's son, also J. D. van der Waals, was appointed *ordinarius* in theoretical physics. My father's field was limited because; (a) there was already another professor of experimental physics – the famous P. Zeeman, (b) my father did not wish to devote himself exclusively to physics, his ambitions went partly into politics (1910–1917), far stronger however into philosophy, and later (after 1918) into education, psychology and theology' [241].
4. A. M. J. F. Michels (1889–1969), a skilled experimentalist, became the European authority on high-pressure physics and technology. See [*4*] and Poggendorff (note 2 above), Bd 7b, S.3264–3267.
5. Again, from his son's letter: 'One of his best friends from 1915 was Professor Paul Ehrenfest, Professor of Theoretical Physics at Leiden. . . . Ehrenfest was famous among European physicists as the originator of a monthly international colloquium of theoretical physics in Leiden after 1920, for which inter alia A. Einstein and N. Bohr regularly came to Leiden. Ehrenfest and Einstein often used to come the following day together to my father's home in Amsterdam. I still remember some of those visits vividly' [241].
6. It is printed at the end of Chapter 9.
7. As an illustration, we give here the text of such an answer, previously unpublished.

   Amsterdam 20 Juni 98
   Hochgeehrter Herr,
   Für den schönen Glückwünschen, den ich von 'der physico-matematischen Facultät des Kaiserlichen Universität Jurjew' empfangen habe, fühle ich mich zu tiefstem Danke verpflichtet. Glauben sie mich, wenn ich versichere dass diese Dankbarkeit stets rege in mir sein wird und dass ich die heilen Wünsche für Ihre Facultät hege.

   Ihr ergebenen
   J. D. v d Waals

   Herrn Loewinson-Lessing, Decan der phys.-mat. Fac. Universität Jurjew [157(d)].

   Jurjew, or Yuriev, is now Tartu, in Estonia.
8. The anniversary coincided with that of the University which was celebrated on 15 October with due pomp. Van der Waals was one of only five of the original professors still in post. All five jubilees were recorded in *Propria Cures*, and of van der Waals's, a reporter noted that it was celebrated in a 'very calm and homely way' [*7*].
9. His strong reluctance to be photographed or painted is shown by a note before an article [186] on him in 1900 in *Mannen en vrouwen van beteekenis in onze dagen* ('Men and women of note in our times'). Each article in this series was traditionally preceded by a portrait of its subject. In the case of van der Waals, however, there was a note that read 'Unfortunately the undersigned cannot put a portrait of van

der Waals before this biographical essay because Prof. van der Waals has told us that he has, for good reasons of his own, refused to pose for a portrait. H. M. Service has no picture more recent than one taken about 30 years ago. – The Editors'.

10. The Van der Waals Fund has prospered. From 1904 it was used to support studies of the effect of high pressures on fluid phase equilibria, under the direction of Kohnstamm. The first two papers that acknowledged its support were by Timmermans and Kohnstamm and appeared in 1910. The total is now about 330. See the article by Levelt Sengers and Sengers [*4*].
11. This is a quotation from *Continuity*, Section 72 [1d$_1'$, p. 119]. The English version is slightly different [1e$_1'$, p. 446; 1e$_2$, p. 225].
12. These are the final words of *Continuity* [1, p. 128; 1e$_2$, p. 239]. They are, however, not those of van der Waals who was quoting William Thomson who, in turn, was quoting Fleeming Jenkin [183, p. 14].
13. Most Dutch professors wrote dozens of popular articles and gave many public lectures; van der Waals did not. A contemporary compared him with H. G. van de Sande Bakhuyzen, President of the Academy and Chairman of the Section of Sciences until 1908, who retired at the same time as van der Waals. Van de Sande Bakhuyzen, as a true student of Kaiser's, knew how to address the public; van der Waals did not and the value of his scientific work was thus underestimated by the public [189].
14. This may be illustrated by a letter from the German physicist Waldemar Voigt to Lorentz, of 11 May 1899 (kindly lent to us by Professor A. J. Kox).
'I got into a scrape with Mr van der Waals because I made some objections about Boltzmann's understanding of molecular theory. I expressed myself thus: to reach a *perfect picture* of the mechanism of a process, which is explicitly *expressed* in equations, we and others add elements from our imagination that are not expressed by formulae and which can therefore be contrary to reality, although the formulae themselves are confirmed by experiment. Mr van der Waals was so obviously irritated by this opinion that I decided to leave in a hurry.
On the one hand it is a pity that I made so unfavourable an impression on a scientist I so much admire, but on the other, it was interesting for me to see his passionate adherence to his "picture". We lose our tempers most easily just when there is no possibility of coming to a strict conclusion' (italics by Voigt).
15. It was this 'vision' that Kamerlingh Onnes mentioned in his address. On Boltzmann's remark he had commented rather tartly: 'Such inspiration comes only to great scientists after ceaseless effort and unimaginably laborious explorations in all directions which have each time come up against an impenetrable wall of darkness' [196].
16. Compare a similar statement by Maxwell: 'There may be some mathematicians who pursue their studies entirely for their own sake. Most men, however, think that the chief use of mathematics is found in the interpretation of nature' [*10*, p. 248].
17. See Chapter 6, Section 2.
18. He found that in a mixture of ether and water the meniscus between the two phases became flatter at high pressures, thus suggesting that the mixing temperature fell with increasing pressure, a result later confirmed by one of his students.
19. This belief was quite general; an example of its expression is Bosscha's evaluation in 1899 of van der Waals's work: 'The courageous and successful endeavour of van der Waals to penetrate more deeply into the world of molecular motions, and the persistence with which he pursues it, are further evidence of how fruitful

it is to apply the principles of mechanics to explain physical properties of state and such phenomena' [*11*, p. 11].

20. The farthest he went in his published work was probably the expression of his opinion that: 'Most probably the molecule must be considered as belonging to the solid condition of matter . . .' [1, §27, p. 39; 1d$_1'$, p. 35; 1e$_1'$, p. 365; 1e$_2$, p. 160], and 'For helium the limiting density (m/b) is probably equal to that of the heavy metals. From the supposition that it is formed by splitting off heavy metals this follows already with a certain degree of probability' [83e, p. 528].
21. The reason for his choice was as follows: 'How and why the attraction at somewhat larger distances is converted into such a repulsive force is a question that was probably never put by him [Maxwell], and at all events was not answered by him' [92e, p. 150–51].
22. Maxwell, however, did not live to finish and publish his work on thermodynamics, see [*12*].
23. It is of interest that his description in *Continuity* of Maxwell's *Theory of Heat* as 'a little book which is certainly in the hands of every physicist' [1, p. 81] was omitted in the German translation of 1881 [1d$_1'$]. (It should have been but was not, reincorporated into the most recent English translation [1e$_2$].)
24. 'I have no better expression than "religious" for this belief in the rational nature of reality which is, at least to some degree, open to human understanding', letter of 1 January 1951 [*13*].
25. It is appropriate here to quote from an article on philosophy in the Amsterdam students' weekly journal concerning the motivation of scientific work. 'Prof. Korteweg, a contributor to the *Revue de Metaphysique et de Morale* speaks, I think, both for himself and his teacher, Prof. van der Waals, when he writes: "Such motivation, as the name indicates, the knowledge which we need to know things, and not that which we need to do something. It is an aesthetic thirst, the thirst for a peculiar beauty implicit in everything that is true, that is intricate and that drives science forward"' [*15*].
26. Friedrich Rückert (1788–1866) was a poet and Professor of Oriental Philology at the Universities of Erlangen and Berlin, known for his translations from oriental languages. The six volumes of *Die Weisheit der Brahmanen* ('The Wisdom of the Brahmins') was published in 1836.
27. He meant the first of Planck's formulae. The paper that introduced the quantum was submitted on 14 December, a week before the 'promotion' of van der Waals Jr.
28. St. Luke, Chapter 24.
29. The Anglo-Boer war had ended on 31 May 1902.
30. Literally, 'God be merciful to me a sinner' (St Luke, Chapter 18).
31. Nevertheless there was one society that he never ignored – the Holland Society of Science at Haarlem (the HMW, see Chapter 1), to which he had been elected as early as 1878. He was for many years a judge of the prize memoirs, and it was he who presented the gold medal to Zeeman in 1892 for his work on magneto-optics. From 1889 the HMW supported much experimental research to test van der Waals's theory of mixtures.
32. In a recent Russian book there was written: 'A further goal of the underlying investigation was to search for national traits. It is suspected that every nation has its own mentality, its own system of categories or a particular correlation of concepts that are themselves common to all nations. There is something which makes a nation reason in its own way in the course of solving its own problems' [*21*, p. 80].
33. The original Dutch edition of *Continuity* was exhausted shortly before this trans-

lation appeared. After the appearance of the abstract published in 1877 [1c₂] it was impossible to meet readers' demands. Thus in 1879 the abstractor for the Physical Society in Berlin wrote that his abstract [1c₁] had been compiled from Maxwell's account of the thesis, since the original was unobtainable. In 1881 an Amsterdam bookseller wrote to van der Waals to say that 'We continuously receive demands for your book from England, Germany and America, and cannot fulfil them' [231, p. 149]. Among the early owners of the Dutch version who emphasized its value were Wurtz in France [*22*] and Stoletov and Mendeleev in Russia (see Chapter 9).

34. For further details about the preparation of this translation, see [183, p. 90].
35. This was also on Poincaré's nomination, who wrote thus: 'He is, indeed, one of those men who make things happen because they think' [209, p. 793].

References

1. *P.C.*, 1906, Jg. 18, b. 4,18.

2. Kohnstamm, Ph., *Hoe mijn 'bijbelsch personalisme' ontstond*, Haarlem, 1934, 43 b.

3. Kohnstamm, Ph., *Experimentele onderzoekingen naar aanleiding van de theorie van van der Waals: Het P–T–X-vlak: voorstudies en methoden*, Amsterdam, 1901, 219 b.

4. Levelt Sengers, J. M. H. and Sengers, J. V., Van der Waals Fund, Van der Waals Laboratory and Dutch high-pressure science, *Physica A*, 1989, v. 156, p. 1–14.

5. *NNGC 7*, Haarlem, 1899, b. 46,47–48,51,81.

6. Wheeler, L. P., *J. W. Gibbs: The History of a Great Mind*, New Haven, 1951, 264 p.

7. *P.C.*, 1902, Jg. 14, b. 80.

8. Boltzmann, L., Ueber die Zustandsgleichung van der Waals, *Versl.*, 1899, d. 7, b.477–484; On the characteristic equation of v.d. Waals, *Proc.*, 1899, v. 1, p. 398–404.

9. Boltzmann, L., *Vorlesungen über Gastheorie*, Bd 2, Leipzig, 1898, S. 155; *Lectures on Gas Theory* trans. S.G. Brush, Berkeley, California, 1964, p. 356.

10. Maxwell, J. C., Introductory Lecture on Experimental Physics. *Scientific Papers*, v. 2, Cambridge, 1890, p. 241–255.

11. Bosscha, J. Openingsrede, *NNGC 7*, Haarlem, 1899, b. 1–20.

12. Garber, E., J. C. Maxwell and Thermodynamics, *Am. J. Phys.*, 1969, v. 37, p. 146–155.

13. Einstein, A., *Briefe an Maurice Solovine*, Berlin, 1960, S. 102; *Letters to Solovine*, New York, 1993, p. 118–119.

14. Einstein, A., Motiv des Forschers. In *Zu Max Planck's 60. Geburtstag: Ausspra-chen in der Deutsche physikalische Gesellschaft*, Karlsruhe, 1918, S. 29–32.

15. Klaverweiden, F. B., De Filosophische Studentenvereeniging, *P. C.*, 1906, Jg. 18, b. 126–128.

16. van Emmerik, E. P., *J. J. van Laar (1860–1938). A Mathematical Chemist*, Doctoral Thesis, Delft, 1991, 228 p.

17. Cohen, E., Chemie en chemische industrie te Amsterdam in den loop der eeuwen, *Ch. W.*, 1940, Jg. 37, b. 486–502.

18. van der Waals, J. D. Jr., *Statistische behandeling der stralingsverschijnselen*, Amsterdam, 1900, 93 + iv b.

19. Rispens, J. A., *Richtingen en figuren in de Nederlandsche letterkunde na 1880*, Kampen, 1938, 436 b.

20. Stokvis, B. J., *Nationaliteit en Natuurwetenschap. Rede tot opening van het Eerste Nederlandsche Natuur – en Geneeskundig Congres*, Haarlem, 1887, 32 b.
21. Gachev, Georgii (Г. Д. Гачев), *On national images of the world*, Moscow, 1988, p. 180 (in Russian).
22. Wurtz, Ch.A., *La théorie atomique*, Paris, 1879, 248 p.
23. Bottomley, J. T., Rücker, A. W., Baynes, R. E. [Letters to the Editor, concerning van der Waals's 'Continuity of the liquid and gaseous states'], *Nature*, 1891, v. 43, p. 415–416,437,487–488.
24. Ramsay, W., Liquids and gases, *N.P.R.I.*, 1891, v. 13, p. 365–374.
25. Ramsay, W., Recent Researches on the Gases of the Atmosphere, *NNGC 7*, Haarlem, 1899, b. 189–192.

Plate 1 The house in Leiden, now a butcher's shop, in which van der Waals was born. (Photograph 1984.)

Plate 2 Anna Magdalena van der Waals. (Photograph from her grandson.)

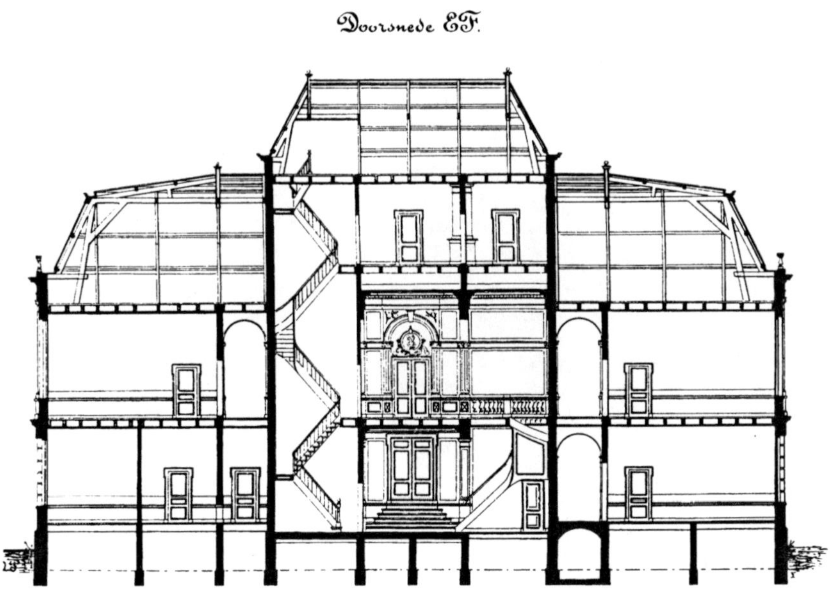

Plate 3 Section of the Laboratory at Amsterdam, as planned in 1880. The building erected did not follow this plan exactly. (Reproduced by courtesy of University Museum de Agnietenkapel, Amsterdam.)

Plate 4 Rear view of the laboratory showing the two wings added later. This building was a store in 1991 when this photograph was taken.

Plate 5 His home in P. C. Hooftstraat, Amsterdam. (Photograph 1991.)

Plate 6 Ph. Kohnstamm. (Photograph from his son.)

J. D. van der Waals en H. Kamerlingh Onnes bij den heliumliquefactor in het Kamerlingh-Onnes-Laboratorium te Leiden. 1911.

Photo Mr. A. H. Kamerlingh Onnes.

Plate 7 J. D. van der Waals and H. Kamerlingh Onnes by the helium liquefier in the Kamerlingh Onnes Laboratory at Leiden, 1911. This photograph was taken by Kamerlingh Onnes's son.

De Amsterdamsche oud-hoogleeraar in de Natuurkunde J. D. van der Waals, wien deze week den Nobelprijs van ruim 190.000 francs toegekend werd, door een onzer verslaggevers jongstleden Zaterdag genomen.

Plate 8 Van der Waals in a street in Amsterdam. This photograph was taken without his knowledge and was published in the weekly newspaper *Het Leven* [194].
The caption reads: 'The former Professor of Physics at Amsterdam, J. D. van der Waals who has this week been awarded the Nobel Prize of 190 000 francs, taken by one of our reporters last Saturday.'

Plate 9 Stamp issued as one of a set of three Dutch Nobel Prize winners on 7 September 1993. (Copyright is retained by PTT Post BV.)

Plate 10 Van der Waals at the time of the award of the Nobel Prize, December 1910.

Plate 11 Model of the Ψ-surface for two liquids and a gas, made by Van der Waals.

CHAPTER 8
The teacher and his school

'However great he is as a scientist, he is, perhaps, no less great as a teacher,' wrote Kohnstamm [206, p. 3]. 'In him, wrote Zeeman in the first lines of his obituary, 'the Netherlands has lost perhaps the best of its teachers' [205]. Van der Waals was first and foremost a teacher in the eyes of his countrymen. Formally this was so throughout his whole career, from being a teacher in a primary school through to becoming a university professor. He maintained a vivid interest in secondary education in the natural sciences all his life and used his official positions to ensure a high standard of teaching, especially in the HBS. When in The Hague he was for many years a member of the examination committee for secondary education [186, p. 133], and in Amsterdam he was chairman of a supervisory committee. A former student recalled [224] that van der Waals thus stayed true to his ideals, making heavy demands on both the teachers and their pupils. He opposed the simplification of lessons, or making them easier so that they were more palatable, believing that it was the rôle of the middle schools to select their best pupils for entry to the higher schools. He tried to help gifted young men to enter a university, but did not want to open the way for mediocrities who would subvert the high standards he sought to maintain.

In the Academy of Sciences he functioned as a teacher in the most general and widest meaning of the word. As a professor and as a member of the Academy who presented the papers of his colleagues for publication, van der Waals associated continually with his juniors, an association that was probably invaluable to him. Often, in his papers, he has observations such as, 'induced by questions of Dr Smits on subjects in which suchlike particularities occur, and at his request, I will discuss briefly at least the principal cases' [78e, p. 194]. (The question was on the equilibrium between a solid and fluid phase in the critical region.) It is impossible to say how many discussions of this kind took place, but it clearly cannot be fewer than the number of papers he presented to the *Proceedings*, since almost every one of them contains acknowledgements of his help. Among the dozens of papers presented are not only works of physicists, but also of chemists and even of biologists, when they had used physicochemical methods. An interesting example of this little noticed but stimulating influence of van der Waals is a biochemical work on adsorption which contains a discussion of the possible form of the attractive intermolecular potential [*1*]. An earlier example is Zeeman's paper on *An experiment on the so-called anomalous spreading of waves* [*2*], in which the author says that van der Waals had drawn his attention to the effect and had given his support during the experiments.

The rôle of van der Waals as a teacher of Dutch physicists and chemists is vividly exemplified by his collaboration with two eminent scientists who were themselves to head schools of physics and chemistry respectively – H. Kamerlingh Onnes and H. W. Bakhuis Roozeboom.

Chronologically, and because of his important scientific rôle, we take Kamerlingh Onnes first. He and van der Waals had a strong mutual influence. In 1882 Rijke retired and was succeeded by Kamerlingh Onnes who, until 1923, was head of the Department of Experimental Physics at Leiden[1]. The general direction of his work is shown by the title of his inaugural lecture, 'On the significance of quantitative investigation in physics', with its principal motto, *Door meten tot weten*[2]. Lorentz has characterized Kamerlingh Onnes's work by the words of Helmholtz, 'When a science is in an advanced state the only one who can make fruitful experiments is one who is deeply versed in the theory and knows therefore how to frame the questions, and, conversely, the only one who can theorise successfully is the one who has great practical experience of experimental research' [7, p. 292][3].

Kamerlingh Onnes met these criteria and, moreover, displayed uncommon business-like qualities – persistence, purposefulness, and a practical grasp of the management of a new physical laboratory. He intended to use his laboratory especially for investigations at low temperatures so as 'to simplify as far as possible the conditions under which the phenomena occur via extensive suppression of thermal motion' [9, p. 308]. In his Nobel lecture he said, 'What has given a character of its own to the Leiden work from the very beginning is that I allowed myself to be led by van der Waals's theories, particularly by the law of corresponding states which at that time had just been deduced by van der Waals. This law had a particular attraction for me because I thought to find the basis for it in the stationary mechanical similarity of substances[4] and from this point of view the study of deviations in substances of simple chemical structure with low critical temperatures seemed particularly important. I was greatly drawn to making the necessary precision measurements at low temperatures' [9, p. 306]. Kamerlingh Onnes's work on corresponding states was already under way by 1881 (see above, Chapter 6, Section 1).

From 1890 onwards, when the experimental technique of the Leiden laboratory had become sufficiently established, and when van der Waals's theory of binary mixtures had just been published, Kamerlingh Onnes embarked on a long series of investigations entitled *Contributions to the knowledge of van der Waals ψ-surface.* He remembered afterwards how, just after the monthly meetings of the Section of Sciences he went regularly to see van der Waals for 'a monthly privatissimum' as he called it. 'I found van der Waals always sitting at a table spread with papers covered with formulae. . . . His conversation gave me the opportunity of realising the enormous extent of the calculations that he made but never published. The greater part of them served only for the preparation of his scientific papers, but some contained valuable information. It was such calculations that led him to happy simplifications, and so were the means by which he built his theories. From his long experience of these calculations he could often interpret features from one or another plot of the results of the research carried out at Leiden. Time and time again he pointed out that some peculiarity on a plot was an error, and almost every time further investigation showed that his criticism was well-founded' [201].

The series of investigations that Kamerlingh Onnes had planned made a good start with Kuenen's thesis of 1892, *Measurements concerning the van der Waals surface for mixtures of carbonic acid and methyl chloride*. It was in this study and in later work by Kuenen that van der Waals's theory of mixtures received its first experimental corroboration. 'My work', Kuenen wrote in the paper [*10*] in which the contents of his thesis were presented in more detail, 'is entirely connected with the above named theory, but it is nonetheless insufficient for its thorough confirmation. At one important point I managed to show experimentally the truth of consequences deduced from the theory (§§40–47)' [*10*, p. 354]. These sections concerned the condensation of binary mixtures and contained the discovery of what he had called 'retrograde condensation'. This is the most brilliant of Kuenen's achievements, 'the way to which', he wrote, 'was directed by Mr van der Waals himself' [*10*, p. 377]. Everyone who has read the last pages of van der Waals's paper *Molecular theory of a body composed of two different substances* [31], in which there is a diagram of volume versus composition, can see that the idea of retrograde condensation was deeply hidden in van der Waals's work, and can see how great was the merit of Kuenen's work in which he was able to comprehend and develop the idea.

'I still hold in vivid remembrance', Kamerlingh Onnes wrote in 1922, 'how Kuenen, putting in action his magnetic stirrer, the simple but fundamental contrivance by which he succeeded in eliminating retardation [i.e. hysteresis], had the satisfaction of demonstrating to van der Waals the retrograde condensation, and of seeing van der Waals looking in deep reflection at the beautiful phenomenon which at once put his theory beyond any doubt. An admirable interaction of Kuenen's experiments and van der Waals deduction followed' [*11*, p. 679].

Kuenen's first paper was presented by Kamerlingh Onnes at a meeting of the Section of Sciences, its closing words were, 'The results of other experimenters, especially those of Andrews who described his experiments most exactly, are in thorough accord with the explanation proposed here. That such an experimentalist as Andrews could not see exactly how things were was a consequence of his ignorance of the theory of van der Waals without which retrograde condensation could hardly have been discovered' [*12*, p. 19]. The last sentence was diplomatically omitted in the English version published in the *Leiden Communications*, No. 4, and indeed was not entirely fair to Andrews who had published a brief and less precise description of the phenomenon.

Van der Waals followed Kuenen's work with the keenest interest. Kuenen told him of his results as soon as they had been obtained[5] and van der Waals commented on them. For example, in Kuenen's paper of 1893 we can read,

'Mr van der Waals has made, in a private communication, the interesting observation that Konowalow's laws are not obeyed on the portion of the binodal curve between points C and P' [*10*, p. 379][6]. Several years later in the second part of the second edition of *Continuity* van der Waals discussed Kuenen's investigations in detail and wrote, amongst other things, that it was he 'who first explored this phenomenon, confirmed it experimentally . . . and named it retrograde condensation' [61, p. 128][7]. Two years later Kuenen wrote in another paper: 'The idea of considering the sign of $(\partial^2 v/\partial x^2)_p$ on the plaitpoint I took from a letter from Prof. van der Waals written on the occasion of the communication of my paper to the Physical Society of London, May 24th' [*13*, p. 12].

It is for this work that Kuenen is now mainly remembered and it has thrown into the shade another investigation of no less importance. It happened that during the 1880s investigations of critical phenomena were plagued by subtle experimental mistakes that arose mainly from the slowness of diffusion, and so the difficulty of reaching equilibrium in the critical region. These errors led to misconceptions about the identity of the two coexisting phases at the critical point[8]. These problems were resolved by some fine experiments by Kuenen in which the electromagnetically operated stirrer mentioned by Kamerlingh Onnes was used to hasten the approach to equilibrium. These results were of great importance in defending the classical picture of the critical point which, he said, 'we can designate the Andrews–van der Waals theory' [*15*, p. 278][9].

The monumental range of investigation instigated by Kamerlingh Onnes included the development of experimental techniques for working under pressure over a wide range of temperature, precise measurements of P–V–T relations and of phase equilibria, especially in the critical region, a sound mathematical treatment of the results obtained, and a theoretical discussion of them along the lines laid down by van der Waals. Only a few examples can be given here to illustrate the contacts between the Leiden group and van der Waals.

The first model of the ψ-surface for a binary mixture, made by van der Waals himself, was only schematic. Kamerlingh Onnes proposed creating plaster models of the ψ-surfaces of the real systems studied in Leiden. This laborious job was carried out in collaboration with M. Reinganum, a German physicist who studied at Leiden and Amsterdam from 1899–1900, and who was later killed in the First World War. The final fruit of their efforts was an impressive object measuring 40 × 60 × 80 cm. Even when hollowed out this weighed 80 kg but 'it proved to be highly satisfactory for several constructions' [*17*, p. 280].

Such models proved later to be useful tools in the study of the system H_2 + He for which unusual and paradoxical properties were discovered. In November 1906 Kamerlingh Onnes presented to the Section of Sciences a short paper entitled *A gas that sinks in a liquid* [*18*]. The starting point of his reasoning was simple: 'If we have an ideal gas and an incompressible liquid without vapour tension, in which the gas does not dissolve, the gas will gather above the liquid under the action of gravity, if the pressure is sufficiently low, whereas the compressed gas will sink in the liquid if the pressure is made high enough.' In reality the picture just given is over-simplified because in practice the gas will dissolve in the liquid and the liquid in the gas, so that with change of pressure both the composition and the density of the two coexisting phases will change. Hence, 'every experiment in which a gas is compressed above a liquid is practically an application of the theory of binary mixtures of van der Waals.' At first Kamerlingh Onnes could not predict the theoretical conditions needed to bring about the supposed phenomenon, but later he realized these conditions in an experiment with a mixture of one part of helium and six of hydrogen at -253 °C.

On compression in a glass tube 'up to 49 atmospheres the liquid hydrogen was seen to deposit from the gas mixture, bounded by a distinct hollow meniscus against the helium. At 49 atmospheres the helium, or properly speaking the gas phase consisting chiefly of helium, went down just as water through oil, and remained on the bottom as a large drop. With further compression to 60 atmospheres and decrease of pressure to 32 atmospheres the volume of the bubble appeared to follow the change of pressure as that of a gas. At 32 atmospheres the bubble rose again. By changing the pressure the bubble was made to rise and descend at pleasure' [*18*].

Van der Waals responded immediately; at the next meeting of the Section of Sciences he put forward a general criterion for the occurrence of the effect discovered by Kamerlingh Onnes [83]. At the same time Keesom and Kamerlingh Onnes also discussed the theory of this phenomenon, which they called the barotropic effect. A consequence of this theoretical study of the effect was that they were led to a revival of van der Waals's prediction of the demixing of a gaseous system [*19*]. It is interesting to note that the original Dutch version of the paper did not contain van der Waals's prediction, which was added only in the English version of the *Proceedings*; the addition is clear evidence of the discussion of the problem with van der Waals after Kamerlingh Onnes had presented the work to the Section of Sciences. A preliminary calculation of the ψ-surface for the system helium + hydrogen had already convinced the authors that '*two different phases may be in equilibrium which must be both considered as gas phases*. Then the two substances which are the components of these mixtures are *not miscible in all proportions even in the gas state* [*19*, p. 786]. (The italics are those of Kamerlingh Onnes.) How these predictions have been fulfilled will be told in Chapter 12.

This discussion should conclude with an account of the liquefaction of helium, which was one of the most famous and brilliant achievements of the Leiden Laboratory. Kamerlingh Onnes has described how van der Waals's theory was his guide, and this description reads 'like an adventure novel' (Zeeman [198, p. 702]). There is no room for this novel here; the reader must consult Kamerlingh Onnes's original paper [*20*][10]. It is important to remember, however, that, according to Lorentz, it was a long-term programme 'accomplished after a well-considered plan, absolutely in the spirit of van der Waals' [195, p. 314].

The first and principal question that had to be settled was the value of the van der Waals parameter a for helium; only if this were positive was there possibility of liquefying helium. This point was settled by precise measurements of the P–V isotherms at temperatures down to that of liquid hydrogen, and the theory was then used to calculate the Boyle point and the Joule inversion temperature for helium. Only after all this work had been done was it possible to affirm that helium could be liquefied. 'Without the guidance of van der Waals's theory the great undertaking of Kamerlingh Onnes would have failed from a lack of satisfactory knowledge' [198, p. 703].

On 26 September, Korteweg, then Vice-Chairman of the Section of Sciences, said at the opening of the meeting following the summer vacation, 'Gentlemen, some words are needed before we begin our work. After our last meeting there has appeared what we might call a Vacation Issue of our Proceedings, as a supplement to the Proceedings of the June meeting. This unusual event had an even more unusual cause. There is no one here who will not recognize with great joy and pride, the scientific victory of our fellow member Kamerlingh Onnes in liquefying helium in the Leiden laboratory. We know what a combination of scientific understanding, experimental skill and persistence was necessary to gain this victory. After previous disappointments it shines before us in glowing light. A new Dutch name will for ever be associated with one of the most significant chapters of physics, and with the physics of the condensed states. Alongside the name of van der Waals, who founded the theory of this phenomenon, stands that of Kamerlingh Onnes as the conqueror of the most resistant gas, and as the man who has made a great advance in approaching closer to the point of absolute cold than anybody before; he will occupy forever one of the leading places in this branch of physics. Words of congratulation and honour befit such an achievement' [*21*].

Van der Waals particularly drew attention to the liquefaction of helium in his Annual Report as Secretary at the General Meeting of the Academy. As for Kamerlingh Onnes, he said repeatedly that 'The study of the equation of state related to the monumental theories of van der Waals has always been a main item of research in the Leyden laboratory'. And again, 'how happy I was to be able to show condensed helium to my distinguished friend van der Waals, whose theory had guided me to the end of my work on the liquefaction of gases' [*9*, p. 316,325]. After the Nobel ceremony in 1913 Kamerlingh Onnes added in his banquet speech: 'This day brings particular joy to my respected friend van der Waals. I know that today he is full of enjoyment just as I was happy to tell him that according to his predictions helium had been liquefied' [*22*][11].

It is not surprising that the reserved van der Waals intruded only very privately on the thermodynamic work at Leiden[12]. 'Although I was seldom present in the laboratory during these researches, and was only fortunate enough to be able occasionally to observe them directly, they were not unknown to me. For many of the investigations I knew not only that they had begun, but was also kept fully informed about their aims. I knew what was the question that Kamerlingh Onnes was trying to answer by experiment. I knew what were all the difficulties that were foreseen, the ways in which the first successes were achieved, the new points of view that opened up during the course of the work, and lastly I knew what results were achieved. All this I knew from conversations or from numerous letters or else from the detailed communications presented, which gave all the numbers and were accompanied by drawings and lantern slides. Moreover – and I refer now to the thermodynamic investigations, using the word thermodynamic in its widest sense – these investigations were closely related to questions to which I was seeking or had sought answers. Frequently it was the most recent result that Kamerlingh Onnes tested again and I awaited the outcome of these experiments with no less anxiety than he himself. I was never in any doubt that he would arrive at a completely impartial assessment. I often felt that I myself was also a participant in the investigations in the Leiden laboratory. This close collaboration with Kamerlingh Onnes on topics that were central to my interests, and this skilful and intense testing of ideas conceived on Dutch soil, were, I considered, one of the greatest pieces of good fortune that have fallen to my lot. I could only partially express my gratitude to Kamerlingh Onnes by dedicating to him the second part of *Continuity*' [74, p. 72–73][13].

Whereas the scientific cooperation of van der Waals with Kamerlingh Onnes was entirely in the field of physics, that with Bakhuis Roozeboom (1854–1907) was in the borderland between physics and chemistry. Hendrik Willem Bakhuis Roozeboom[14] left the HBS at Alkmaar in 1872 ranked first in North-Holland, that is, he was the best of the school-leavers in Amsterdam, Haarlem and Alkmaar. His first teacher in chemistry, and the Director of the HBS, was Dr J. D. Boeke who had an enthusiasm for chemistry that he passed on to his pupil. He recommended this gifted youth to J. M. van Bemmelen, his brother-in-law, who was then the director of the agricultural middle-school in Arnhem. He drew Bakhuis Roozeboom into his investigations of soils and later, when Professor of Chemistry at Leiden, helped him get a good job in a chemical factory. Bakhuis Roozeboom had not the means to afford a university education after leaving the HBS, but in 1878 van Bemmelen managed to appoint him as a lecture assistant at Leiden. Here he worked 'at high speed', spending his days in the laboratory and his nights at his books. In February 1881 he had taken the external degree[15] at Leiden as a *candidaat* in mathematics, astronomy, physics and chemistry, and in 1884 he was awarded his doctor's degree for a thesis, *On the hydrates of sulphurous acid and of hydrogen chloride, bromide and iodide*[16]. Bakhuis Roozeboom had come to these problems through his studies of dissociation; his first paper was on the dissociation of tertiary butyl bromide. He tried to resolve the confusing evidence on the dissociation of hydrates by seeking relations between three states of aggregation – crystals, solution, and vapour – at different temperatures and pressures.

One of his main results was the discovery that if the pressure on a solution were higher than the dissociation pressure of the hydrate, then crystals of the hydrate were precipitated from the solution. He presented the relations between temperature and dissociation pressure by means of graphs. After the defence of his thesis he found for the hydrate of hydrogen bromide a strange form of P–T plot, with a minimum and a maximum dissociation temperature. This meant that there was a range of temperature in which, on isothermal increase of the pressure on the solution, crystals of the hydrate were first precipitated, then dissolved, and then precipitated again.

Van Bemmelen, a member of the Academy, regularly communicated Bakhuis Roozeboom's papers to the Section of Sciences. The report of the strange P–T plot was read on 27 December 1884, and van Bemmelen recalled later that van der Waals had listened to him most attentively, but had asked no questions. Two months later, in February 1885, van der Waals claimed to have found an explanation of 'this startling phenomenon discovered by the recent research of Mr Bakhuis Roozeboom' [21, p. 379]. Van der Waals's communication contained an approximate differential equation, similar to that of Clapeyron and Clausius, but with the term on the right depending on the composition of the solution x, and that of the hydrate c. The slope of the P–T plot is determined by the sign of the difference $(c - x)$, and it was changes in this sign that explained the complicated course of Bakhuis Roozeboom's curve.

In further research Bakhuis Roozeboom obtained results that he could not bring into accord with van der Waals's equation. He therefore wrote directly to the author of the equation, so starting a correspondence that was not published for a further fifty years [220]. Although the letters from Bakhuis Roozeboom have been lost (the reader may recall how van der Waals used to treat the letters he received), seven of those to him have survived, and these are sufficient to enable us to understand the matter. In the first letter of 13 December 1885, van der Waals said merely that he hoped to take up the problem during the Christmas holidays. The second, of 22 March 1886 (probably stimulated by others from Bakhuis Roozeboom), contains a description of the normal course of a three-phase P–T curve, and an indication that this course may be disturbed by the appearance of a second solid phase. Three days later, on 25 March, came a very detailed letter (published in full [220, p. 57–60]), in which the equation for the curve was analysed in detail and its various consequences set out. Van der Waals suggested that, as well as the dihydrate of hydrogen bromide which had been described by Bakhuis Roozeboom, there was also a solid monohydrate. In the next three letters, 5, 12 and 31 October, van der Waals analysed the experimental results as they were being obtained. Finally in the seventh letter of 30 December, he responds to the manuscript of the paper in which Bakhuis Roozeboom had summarized his results. Van der Waals writes that he has 'read the manuscript with great satisfaction, and although something here and there gives rise to questions and reflections, he readily recognizes that Bakhuis Roozeboom has produced a work of high quality which he is confident is correct' [220, p. 64].

The next day, 31 December 1886, Bakhuis Roozeboom's paper *On the conditions of equilibrium according to van der Waals of two substances in the three phases, solid, liquid and gaseous* was sent to the press. In the introduction he wrote, 'I much regret that Mr van der Waals has not yet made up his mind to publish his views in full since these have several interesting applications. Since, however, the uses of his equation are not clear without any preceding account of the theory, he has kindly permitted me to present the necessary information here. For this permission I give my very sincere thanks' [*31*, p. 335]. (Van der Waals had published his formula without the detailed derivation given in his letters to Bakhuis Roozeboom. This was a difficult time for him after his wife's death when he shunned public appearances.)

Between March and October 1886 Bakhuis Roozeboom met van der Waals in Leiden, a meeting of great significance in the history of science. Van der Waals had gone to Leiden to give a lecture on Gibbs's thermodynamics of phase equilibria. His audience was small – van Bemmelen, Lorentz, Kamerlingh Onnes and Bakhuis Roozeboom. He paid particular attention to the phase rule and pointed out that Bakhuis Roozeboom's results suggested that the emergence of a second solid phase would explain the apparent conflict between theory and experiment. Van Bemmelen remembered later that van der Waals's arguments had been 'a revelation' to Bakhuis Roozeboom [*24*, p. 260]. It led him to the study of Gibbs's papers, 'On the equilibrium of heterogeneous substances' and of notes of van der Waals's lectures on thermodynamics. It is of interest also to record van 't Hoff's testimony – 'Roozeboom's life-work was settled when van der Waals, in some ways a man of similar nature, turned his attention to Willard Gibbs's famous work and to the phase rule formulated in it' [*32*]. From that day Bakhuis Roozeboom's research, which was systematically planned to set out the relations between different states of aggregation, or phases as they may now be called, had gained a theoretical foundation, as he himself pointed out a quarter of a century later [*33*, v. 1, p. 7]. As to van der Waals, 'he has discovered in Roozeboom's researches a connecting link between the phase theory of Gibbs and a field in chemistry that had been entirely untilled before Roozeboom' [*24*, p. 260][17]. This class of problems was, in fact, explored by Bakhuis Roozeboom and his followers and pupils in Leiden, particularly by W. Stortenbeker and F. A. H. Schreinemakers. Bakhuis Roozeboom soon gained the reputation in his own country of a first-rate chemist; in 1890 he was elected a member of the Academy on the nomination of van Bemmelen.

Van 't Hoff left Amsterdam for Berlin in 1896, after some dramatic events[18], and the chair of chemistry became vacant. It was van der Waals who insisted that it be offered to Bakhuis Roozeboom. The new professor gave his inaugural address on 27 April 1896 and, as tradition required, he concluded by paying his respects to the Curators, the Rector Magnificus, and the Professors; 'It is with particular feelings that I address you, my respected van der Waals,' he said. 'I never expected to become your colleague ten years ago, when you first took notice of my work and so gave a powerful impetus to my scientific development. . . . Without that I would not be standing here. I appeal now for further collaboration with you, since the future development of physical chemistry must rest on theoretical physics, and to whom, if not you, could I turn for its elucidation. The memory of the past leads me to hope that in the future also you will be unwavering in your benevolence' [*38*, p. 25–26]. The ceremonial circumstances influenced the form but not the substance of this appeal. Bakhuis Roozeboom said what he meant, and his hopes were fulfilled.

His years in Amsterdam proved to be particularly fruitful; here he gathered around him many pupils and co-workers to form his school of chemistry, and here he conceived and, in part, carried out his plan for his fundamental treatise on *Heterogeneous Equilibrium from the Standpoint of the Phase Rule* [*33*]. Here also he carried out his most important theoretical investigations including his classic paper on *Solidification points of mixed crystals of two substances* in which the phase diagrams were derived from plots of thermodynamic potentials [*39*].

There is no doubt that the blossoming of his work in Amsterdam had its origin in his contacts with van der Waals. There was a curious demarcation at that time in the Physical and Mathematical Faculty; those who worked in Bakhuis Roozeboom's laboratory on the heterogeneous equilibrium between solid and liquid phases were called chemists, while those who worked in van der Waals's laboratory on phase equilibria between liquids and gases were called physicists. Nevertheless many of Bakhuis Roozeboom's students consulted van der Waals and vice versa. For example, Ph. Kohnstamm wrote an acknowledgment to Bakhuis Roozeboom in the preamble to his thesis: 'The distance separating physics and chemistry is still great, but the two fields are not so far apart as to prevent me from appealing often for your help. I am particularly grateful to you for your explanations whose value has been enhanced by the manner in which they were given' [*140*, p. x–xi]. In fact physical chemical research was carried out in both departments and it is beyond doubt that van der Waals took part in and influenced research on both sides of the line of demarcation. One direct piece of evidence of this is the beginning of van der Waals's communication to the Section of Sciences in October 1903; 'After the publication of the experiments of Dr A. Smits in the Proceedings of the September meeting, communicated by Prof. Bakhuis Roozeboom, I had a discussion with the latter chiefly on the question if, and in what way, the equilibrium liquid and gas, which can co-exist with the solid phase, may be thought to be connected by a theoretical curve at a given temperature, in consequence of the continuity between liquid and gas. It is in accord with the wish of Prof. Bakhuis Roozeboom that I communicate the following observations' [71e, p. 230].

The theory of the equilibria between solid phases and critical fluid phases (see Chapter 12) arose from discussions with Bakhuis Roozeboom and led, in particular, to the prediction of the so-called 'retrograde solidification'. As Bakhuis Roozeboom wrote in his book on *Heterogeneous Equilibria*, 'Smits was able to detect this phenomenon clearly, when he was carrying out his studies of the system ether–anthraquinone, . . . and so confirmed the shape of the vapour pressure curve predicted by van der Waals' [*33*, v. 2, p. 392]. Van der Waals had discussed these experiments with Smits and this discussion stimulated the development of the theory used by the latter, who had devoted about ten years of experimental and theoretical work to this system and had discovered a number of significant phenomena that could hardly have been found without the guidance of van der Waals's ideas[19].

When Bakhuis Roozeboom died prematurely of pneumonia in 1907 van der Waals saw to it that his research programme was continued. He contributed to the establishment of *Roozeboom's Fonds* and the vacant chair was occupied[20], with his support, by A. Smits (1870–1948), an Amsterdam pupil and co-worker of Bakhuis Roozeboom. Smits had been nominated by the faculty after Schreinemakers's refusal, but the Curators were not contented with this and proposed to invite Ostwald or Nernst. Van der Waals's reaction followed immediately; it was recorded in the minutes of the faculty (9 July 1907): 'Mr van der Waals greatly regrets that the matter has turned out this way; he attaches importance to maintaining the traditions of the vacant department, namely, that the earlier physical-chemical researches continue in the direction that has been developed in the Netherlands. This would be very desirable, and would become impossible, if a German physical chemist were invited, since physical chemistry in that country has, until now, been based upon studies of dilute solutions, which is a view that is recognised here to be obsolete. Thus Mr van der Waals is against the invitation of a foreign scientist' [251, p. 386]. After arguments which lasted for a month (for details see [251]) the Curators confirmed Smits's appointment as the professor at last. Thus the cooperation between the sub-faculty of inorganic chemistry and van der Waals continued, even after his retirement.

It was natural for Smits to use the following words in his inaugural lecture, 'The work carried out by van der Waals is so significant for chemistry that we can call him a chemist. Although van der Waals himself might cast doubt on this description, as an honorary title, let him nevertheless have the satisfaction of hearing that present-day workers in phase theory have come to believe that they must know how to use his theory' [*42*, p. 19].

Smits's name became famous at the time when the theory of allotropy was developed. This theory treated substances that have polymorphic modifications as pseudo-binary systems; it rests on the thermodynamic treatment of phase equilibria and it is hard to envisage its being developed outside the van der Waals 'school'[21].

Another of Bakhuis Roozeboom's students who worked with van der Waals was F. E. C. Scheffer (1883–1954). He discussed his theoretical work on three-phase equilibria with van der Waals, who presented it for publication in the *Proceedings of the Section of Sciences*. Many years later in a paper on van der Waals's rôle in chemistry, Scheffer wrote 'I shall always remember with gratitude the concern and encouragement shown by my great teacher' [207][22].

Van der Waals, had, of course, students who worked directly with him, of whom probably the best-known was Johannes Jacobus van Laar (1860–1938)[23], who is often remembered more as the prolific author of polemical papers than for his scientific achievements (see, for example, J. R. Partington's *History of Chemistry* [*50*]), but such a view is only part of the truth. He was also a competent physical chemist who did much to develop van der Waals's ideas. The strained and polemical tone of many of his earlier papers was probably the result of the hard times he had endured. He had a hot temper and great ambitions; his only friend was Cohen who explained that 'Tout comprendre, c'est tout pardonner' [*51*, p. 500]. Van Laar lost his mother when he was only one year old, and his father when he was thirteen. His relatives made him leave the HBS and become a naval cadet. In 1879 he graduated from the Royal Marine Institution (where he learnt his mathematics), and he took part in three voyages in ships of the Dutch navy. In 1880 he was promoted but realized that he had no real vocation for service in the navy. He secured his discharge, and returned home from the East Indies. In the autumn of 1881 he enrolled at Amsterdam University, where he took, in the next three years, the courses of lectures given by van 't Hoff and van der Waals. Because of his inadequate secondary education he was not allowed to take the examinations, so van der Waals recommended him to the Inspector of Secondary Schools, writing 'that his career promised to be a fine and glorious one' [*47*, p. 25]. Thus van Laar became a schoolteacher, first in Middelburg and then in Utrecht (1884–96), and wrote several school textbooks of mathematics. He published his first monograph on physical chemistry in 1893 – *Thermodynamics in chemistry*, with a preface by van 't Hoff – and wrote several papers for the *Zeitschrift für physikalische Chemie*. He also translated into German *The thermodynamic theory of capillarity* [37d′]. In 1896 a serious nervous disease compelled him to retire on a pension, but by 1898 he was able to lecture on mathematical chemistry as a *privaatdocent* and to collaborate with van der Waals. It was then that he translated *Binary mixtures* – the second part of the second edition of *Continuity* [61]. The task of translation was made more difficult by van der Waals's continual additions to and alterations of the Dutch manuscript. It was only just that van der Waals added a post-script expressing his gratitude to van Laar 'who has brought the translation to its conclusion with love and zeal'.

Obviously, there were other discussions between them since we have the evidence from van Laar himself who wrote to Lorentz on 18 August 1899: '. . . Van der Waals is very useful to me, he who so repeatedly returns to the old deductions, sharpens them, and constantly criticizes his own work. I am one who is in this respect often hard to please, I always see weak points. I once had quite an amusing conversation with van der Waals on this subject, who at the time was amazed that I immediately mentioned two or three weak points in his theory of Capillarity, when he asked me, or rather suggested it to me, adding that no one had ever brought that to his attention' [*47*, p. 31]

Van Laar certainly admired van der Waals, and that admiration was reflected in his important essay on van der Waals [186], already repeatedly cited, which appeared the same year as the second edition of *Continuity*. Van der Waals came to form a poor opinion of van Laar, however, and openly opposed his attempts to get a higher position in the University [*47*, p. 31–58]. That is why van Laar's articles on van der Waals's equation of state and of the theory of mixtures were presented to the Proceedings of the Section of Sciences by Lorentz, not by van der Waals. From 1903–7 van Laar was assistant to Bakhuis Roozeboom, especially in mathematical problems, and became for him 'a real co-worker whose activity he thought highly of' (Schreinemakers [*52*]). After Bakhuis Roozeboom's death van Laar lectured temporarily on the theory of phase equilibria until Smits arrived. Then he took the position of *lector* in mathematics for students in chemistry and biology since other possibilities had been closed to him. We can but regret van der Waals's strenuous opposition that hampered van Laar's scientific career and resulted even in conflict between van der Waals and Bakhuis Roozeboom in 1906 [*47*, p. 33]. In 1912 van Laar had to retire through illness and move to Switzerland. He made several unsuccessful attempts to return home, but eventually had to settle in Switzerland where he lived to an old age, losing neither his capacity for hard work nor his abrasiveness. He did not lose contact with the Netherlands; in 1914 he received a doctorate *honoria causa* from Groningen, in 1929 he received the Bakhuis Roozeboom medal, and in 1930 he became a corresponding member of the Academy[24].

During his time in Amsterdam, van Laar published many papers in physical chemistry and several well-known books (in German), *A course of mathematical chemistry* (1901), *Six lectures on the thermodynamic potential* (1906), and *A course of theoretical electrochemistry based on thermodynamics* (1907). He was one of the first to formulate exactly the concept of the complete dissociation of solutions of strong electrolytes.

More directly stimulated by van der Waals was his work on the theory of the equation of state, some of which was mentioned by van der Waals in his Nobel lecture who commented on the horrible length of the calculations. In the last year of his life van Laar wrote of these: 'My hair stands on end even today when I am reminded of that work' [167, p. 354]. This work, and some later proposals for new forms of the equation of state, were summarized in 1924 in a monograph which, he said, was founded on 'the fundamental works of genius of van der Waals, my late and highly honoured teacher' [161, p. v].

His work on binary mixtures was also based directly on the ideas of van der Waals. The equations he proposed for the partial vapour pressures of the components of a liquid mixture are still used in calculations on distillation columns. Many of his articles were devoted to combining Gibbs's analytical methods in thermodynamics with van der Waals's ideas on the form of the equation of state. In spite of their polemical style they played an important role in reinforcing the Gibbs–van der Waals method in this branch of physical chemistry. His last book of 1935, *Thermodynamics of pure substances and binary mixtures* [165], summarized his contributions to physical chemistry.

Solution of many of the problems he tackled seemed to be hopeless because of the complexity of the calculations, but these can now be tackled on modern computers and have proved to be interesting and worthwhile.

There is much evidence that the rôle of van der Waals as a teacher came mainly from the influence of his strong personality. 'For us who have the privilege of seeing you almost every day, that contact means much more than could be given by the most perfect theory or the finest experiment', wrote Kohnstamm in his acknowledgement to his promotor [*40*, p. ix]. 'The strong influence exerted by van der Waals on his pupils' was pointed out by Zeeman [198, p. 703]. That influence probably reflected van der Waals's attitude to science, for a scientific leader must know how to give of his best. For van der Waals, as also for Kamerlingh Onnes and Bakhuis Roozeboom, this truth was self-evident or may even have been held unconsciously, so readily and naturally did they present all their ideas to their students and colleagues. When van der Waals dedicated the second volume of the second edition of *Continuity* to Kamerlingh Onnes he reminded him of 'the extent to which the work of your pupils is your work', and in writing this he expressed also his own stance as a teacher. After the submission of the theses of van der Waals's pupils dozens of papers were published without his name on them. Nevertheless it is clear that his guidance of their work often included direct contributions to the content of the thesis. For example, he derived equations for the differential heat of solution of one phase in another that were incorporated in the thesis of N. J. van der Lee in 1898, who thanked his *promotor* for his readiness to allow him to use derivations not yet published [*53*]. In fact van der Waals never did publish them and it was only through Kohnstamm's initiative that they were added as an appendix to the second volume of the *Textbook of Thermodynamics* in 1912.

One should not conclude, however, that van der Waals was a nursemaid to his students. Kohnstamm, who was one of those who did his thesis with van der Waals, wrote that 'He is always ready to help, always friendly and attentive when you need his encouragement, but nevertheless he always puts first and foremost the developing of your own powers, and he teaches you to stand on your own feet and to show initiative.' So van der Waals 'with his rare experimental flair' never thrust his advice on his students and allowed them to develop their own ideas. 'But often, after several months of work that might have been fruitless in respect of achieving its goal but fruitful for you because of what it has taught you, you find that there is no better way of attaining your goal than by following the way he had recommended' [187, p. 58].

Van der Waals was proud of his students and of their many papers in which his ideas were developed. In the 1900s he spoke more and more often of the Dutch school of physics, but his pride was free from vanity and personal pretensions, as can be seen most obviously by his attitude to his students' discoveries of errors in his work. It is one thing to admit in the abstract that all research is prone to error, but another to present for publication a paper by two students, Kohnstamm and Timmermans, in which it is shown that the teacher's hypothesis is wrong [*54*].

Kohnstamm said later that, 'There was never any pettiness, or ambition, or low personal motives. He was purity itself and so was never wounded if his "prestige", that disgusting curse, was impaired. Moreover he behaved indulgently and paternally even when he had just cause for resentment. I had occasion to experience this myself. After presenting my thesis I wrote a paper on some detail in the kinetic theory of gases that had already been discussed by many others – Maxwell and Clausius as well as Korteweg and van der Waals. I suggested that it could be shown that Clausius was right and the others wrong. I thought it a good piece of work but van der Waals did not want to submit it to the *Proceedings*. Now I know why; what I had written was correct but it was a petty piece of work. "You should not try to be so logical" he said, after some discussion of the matter. I could not then understand this, and I was so irritated by his refusal that I said in haste, "I thought you wished to consider only the facts of the matter". "Well, if you put it like that, then I shall certainly publish it", was his calm and friendly response. And I never afterwards felt any trace of hostility or resentment' [206, p. 4–5]. Our impression of van der Waals as a teacher is determined not only by his skill as a pedagogue but also by his moral stature.

The Russian pianist and teacher, H. Neuhaus, once said that it is impossible to create talent, but one can create culture, and that is the soil in which talents flourish. The same is true for science, and van der Waals was, to a large degree, responsible for creating such a soil in the Netherlands. Hence his influence was felt not only by those who worked directly under his supervision but also by many others. An important channel by which this influence was propagated was the records of his lectures. These were eventually consolidated into the *Textbook of Thermodynamics* but this did not appear for more than a quarter of a century after he first started to lecture on thermodynamics in the 1880s. From the earliest days, however, notes on his lectures were much prized by intending physicists and chemists. There were then no satisfactory textbooks and for years well written-up lecture notes were passed from hand-to-hand[25]. Thus those made by G. P. Weisman in the early 1880s were used by Bakhuis Roozeboom when he first studied thermodynamics [*24*, p. 260]. At the end of the 1890s a good set of notes was compiled by two friends, D. J. Hissink (1874–1956), who was later a prominent soil scientist, and W. Reinders (1874–1951), who became a professor of physical and colloid chemistry. As Hissink recalled [*56*], they both sat in the front row and scribbled diligently through the lectures of van der Waals (on Thursdays and Fridays, from 9 to 10) and Bakhuis Roozeboom (from 10 to 11). Immediately after the lectures they wrote up a final record of what they had heard, and these records were then widely used by post-graduate students.

Van der Waals described mainly his own results at these lectures, and later he published them. Thus the second volume of the second edition of *Continuity* was composed 'mainly' of his 'lectures on thermodynamics so far as they concerned applications to mixtures' [61, p. 50]. However, as van Laar observed, 'Van der Waals never collected the results of his investigations into a complete account, so that anyone who wants to understand them must read all his papers in the order in which they were published – a long and difficult task. ... Anyone who would undertake to write a complete and integrated account of van der Waals's theories would earn the thanks of very many of us' [186d′, p. 35]. Such a man was Ph. Kohnstamm, the closest of van der Waals's students, and one to whom even now we should still be grateful, for it was he who transformed his teacher's notes into a classic textbook of chemical thermodynamics.

Van der Waals himself has related that 'My students and friends tried time and time again to persuade me to publish the contents of my lectures on thermodynamics, but I always thought it necessary to refuse. My reason was my belief that publication would be premature, and that the contents of the book would become obsolete, at least to me, as soon as they appeared. In my lectures I gave mainly the results of my own research, and, especially for the work on mixtures, it turned out every time that much still remained to be done before they were a complete account even of the main aspects. When, however, they began to ask for my consent if one of my students would prepare the lectures for the press I decided to put no more obstacles in their way, especially as one of my former students, to whom I could commit the work with absolute trust, put himself at my disposal' [91, p. v].

At first it seemed that the job would not be too difficult; his student J. J. Hallo would take down the lectures in shorthand and then read them back to Kohnstamm who would produce a written version to show to van der Waals. Unexpected problems arose from the outset when they were surprised to find grammatical errors in the text. 'Our surprise', Kohnstamm explained, 'arose from the fact that we had never noticed anything amiss when we had heard the lectures as students, although the imperfections were grave. Van der Waals never gave a smooth and complete derivation which could be listened to calmly in the deluded confidence that we now understood it all. He struggled with the material, going again through the difficulties and triumphs of discovery, and so forced his listener to work hard himself. The listener was so riveted by the account that he became totally absorbed in the work and grammatical faults totally passed him by' [206, p. 3].

The main problem was putting the material of the lectures into proper order. As van der Waals wrote, 'It is obvious that in a textbook many of the spoken words have to be deleted so that the account becomes systematic, and further smoothing occurs when the material is translated into a foreign language' [91, p. v–vi]. Thus the preparation of van der Waals's lectures for the press turned out to be a long job. Kohnstamm studied the lecture notes critically and raised queries and difficulties with van der Waals. It was no less than two years[26] before the first volume appeared early in 1908 with the title *Textbook of thermodynamics in its applications to the equilibrium of systems with gas and liquid phases*. In the foreword van der Waals gave Kohnstamm his 'heartiest thanks ... for much labour and ingenuity' in the making of the book. He assured his readers that 'in all its main features, and particularly in its general idea, the contents of the book conforms with the sense of my oral lectures' [91, p. v,vi].

The book is characterized by a clear distinction between its three basic points, the first law of thermodynamics, the second law in the form of the statement of the existence of entropy, and Gibbs's principle of equilibrium. There was also a special section on van der Waals's original work on the thermodynamic theory of capillarity [37]. Particularly interesting are the chapters on the applications of the principle of equilibrium. These cover phase equilibria for one-component systems and the chemical equilibrium of dissociation in homogeneous systems. The generalization of the principle in the presence of external forces is then considered, followed by the analysis of electrochemical and capillary systems. The originality of content and method, the consistent application of Gibbs's analytical method and its combination with apt geometrical illustrations (initially in the discussion of conditions of stability) made the publication of the *Textbook* an outstanding event in the already rich thermodynamical literature of those times. 'The reader found himself faced with a book of great originality which in many, but not all respects differs from the more familiar textbooks of thermodynamics.... The methods used are mainly those of the Dutch school and, when demonstrated by a master, they are of the greatest interest to all specialists', so wrote the German physicist K. Scheffer in his review [$91a_4$].

The second volume did not appear until 1912 [113], since work on the theory of binary mixtures was still under way and showed no sign of coming to an end. The authors were still revising the text even when reading the proofs. The first part of this volume contains a derivation and an analysis of the phase rule, an original treatment of the thermodynamics of two-phase binary systems and its application to dilute solutions. The middle sections are devoted to the theory of binary systems consisting of liquid and gaseous phases, and the coexistence of these with solids. The volume concludes with a treatment of systems of variable composition in the field of external forces (especially the effect of gravity on critical phenomena and of electrochemical equilibria), and with the analysis of the general case of equilibrium in reactive systems.

The whole *Textbook* was a landmark not only in the educational but also in the scientific field. It was, perhaps, the most original work to appear since Gibbs's *Equilibrium of heterogeneous substances* in 1876. It was in the *Textbook* that chemical thermodynamics was first expounded according to the methods of Gibbs and van der Waals, and here also was the first systematic treatment of the theory of binary mixtures. The principal impression it leaves with the reader is of its profundity. (Bancroft referred to the 'calm sanity' of the book, but this phrase describes only its style not its content [$113a_3$].) The impress of a strong and independent mind is felt in every section. The vast amount of new material that he had obtained is presented thoroughly and accurately so as to engage fully the abilities of the reader without confusing him. But there is no concession whatever to popularization. The reader is catered for by the masterly skill with which each topic is discussed several times at increasing depth and with the emphasis of new aspects. These repetitions hold the whole book together and reinforce its main theme of combining thermodynamics either with an equation of state or with experimental results to obtain the essence of the principle of equilibrium, and the equations governing phase equilibrium.

If the printed text reflects fairly the content of his lectures we may deduce that van der Waals avoided discussion of questions not directly related to the course. He probably avoided metaphysics because of his innate reserve. A passage from the book shows how briskly he dealt with one of the best-known philosophical problems of thermodynamics, the problem of the 'heat death' of the Universe. He first cites Clausius's famous dictum – the energy of the Universe is constant, the entropy of the Universe is increasing – and then remarks 'In this form the law engenders severe doubts. Even if the Universe were a proper subject for a scientific investigation, we have to face the difficulty that we have up to this point only discussed the entropy of systems at equilibrium. The Universe is, however, not at equilibrium, and we can say something of the entropy of non-equilibrium systems only in the simplest cases. Furthermore this statement overlooks the fact that the value of the entropy may depend not only on the constancy of the energy, but may be influenced by many other circumstances. . . . We may conclude that even if such a law were free from objections it would have no practical value since the world as a whole is not an isolated system' [91, p. 114–115]. Although he refused to philosophize in an abstract manner he did not, however, give up trying to understand fully and to interpret thermodynamics.

The first volume was reprinted practically without change in the year of his death, 1923, and in 1927 Kohnstamm published a new revised edition [122,123]. The principal changes were as follows: new chapters and sections were introduced to elucidate the axiomatic basis of thermodynamics, and the position in this axiomatic system of the equation of state and the principles of equilibrium, and a new chapter was added to deal with Nernst's heat theorem. Some rearrangements of the material led to the general problems of thermodynamics being covered in the first volume, while the second became a systematic exposition of van der Waals's theory of binary mixtures based on a combination of thermodynamics and his equation of state. These changes are far from negligible, but we must agree with Kohnstamm that he 'has not changed the ideas of his great teacher' [122, p. v]. Even the change from *Thermodynamics* to *Thermostatics* in the title had been proposed by van der Waals as long ago as 1900 [58].

All the aspects of van der Waals's work as a teacher – his work as a professor at the University, his collaboration with junior colleagues, and his course of lectures 'so happily edited by Kohnstamm' [198, p. 703] – let us see that he was as great a teacher as he was a scientist. His teaching was especially significant for the development of physical chemistry[27]. He himself has said of his theory of mixtures that 'It would be an impressive number were I just to list the names of the physicists and chemists who, guided by this theory, have studied and still are studying binary mixtures. It begins with Kuenen and ends with Dr Jean Timmermans of Belgium whose studies have still to be completed but which have already been published in part' [103e, p. 263]. The matter was put most clearly by van Laar: 'Van der Waals used the same material when he lectured on the theory of heat to intending physicists and intending chemists. The chemists had to know the details of thermodynamic equilibria in as great a depth as the physicists. Dutch chemists should be indebted to him for the fact that their education in physical science was and is still so solid and wide; they should be glad that they had a teacher as exacting as van der Waals, and as Lorentz at Leiden. This solid background meant that they could not be taken unawares, with an inadequate knowledge of physics and mathematics, by the continuously developing stream of science, which, in our day, has been particularly swift in theoretical and mathematical chemistry' [186d′, p. 38–39]. It was owing to van der Waals that the Dutch physical chemists were able to lead the way to the adoption of Gibbs's chemical thermodynamics throughout the world.

This view of the importance of van der Waals's teaching of Dutch chemists is reinforced when one compares the level and character of chemical papers from the Netherlands with those from neighbouring Germany. The physical and mathematical sophistication of the Dutch papers is very much higher, even in organic chemistry which was the last branch of chemistry to be treated mathematically. This difference is an indirect consequence of van der Waals's quiet influence as a teacher. This rôle was not publicly visible but van der Waals's influence on his contemporaries was silent but pervasive. This was true also of his quiet influence in the Academy where few appreciated that he was, according to Kamerlingh Onnes, 'the soul of the Academic Board' [208].

This is is an appropriate point at which to discuss the characteristic features of the Dutch school of thermodynamics, a discussion which was begun by van der Waals himself [74]. In his opinion there were two features of research in his country that allowed him to speak of a Dutch school of thermodynamics. The first 'of a somewhat formal nature' is the widespread and consistent application of geometric methods. 'We can present the results of our research either by formulae or graphically. The use of the graphical method is not, of course, restricted to Dutch scientists, but so consistently is this method used in our country to study the problems of equilibrium and the construction of equilibrium surfaces, that we may speak of a Dutch school[28]. The method flourishes here and, by following Gibbs, we have gathered abundant fruits from the surface named in his honour. To a great extent we can ascribe this flourishing to the fact that Dutch physicists and chemists study mathematics, and particularly the characteristic properties of surfaces, more thoroughly than those in other countries. We are fortunate that these mathematical studies are so organised that their useful results are not confined to the pure mathematicians. We have it almost as a fixed rule that our physicists, and even our chemists in their physical studies, must have sufficient mathematics not only to understand the interpretation of phenomena by means of geometrical surfaces, but also be able to draw further conclusions as to the particular features of these phenomena from the nature of the features of such a surface. Such mathematical ability is not so common abroad as it is here. I think this is so because of what a foreign physicist once said to me. Having paid many tributes to our investigations he could not refrain from saying to me that he had noticed only one shortcoming in our country, and that was that we taught our students too much mathematics. This remark made me rather doubtful about the value of his tributes' [74, p. 74–75].

The second characteristic feature of the Dutch school was described in the words 'In our country it has been understood more clearly than in others what thermodynamics can do and what it cannot do' [74, p. 76]. In an old-fashioned terminonology, thermodynamics comprises differential equations relating thermal quantities (A) with caloric quantities (B); symbolically $A = B$. The adjectives thermal and caloric can be associated with the two main branches of the theory of heat – thermometry and calorimetry – and the Clapeyron–Clausius equation is a typical example of a thermodynamic differential equation. Van der Waals continues, 'If I was given an equation $A = B$ then it is of no value to me unless I can determine either A or B.' Only if we know one we can benefit from our thermodynamic equations, otherwise we are trapped 'in a vicious circle' [74, p. 78–79].

'Now if we have the two quantities, A and B, we can ask for which one, or perhaps for both, can a general law be found, for the thermal A, or the caloric B? . . . We have answered this question by stating that A must be considered to be the primary quantity' [74, p. 79]. There are two arguments in favour of this choice. Firstly, theoretical calculations based on molecular kinetic ideas are feasible for the thermal quantities, for example, the equation of state of a fluid. We cannot make such calculations for caloric quantities, particularly in the case of mixtures. Secondly, we can measure only changes in caloric quantities, such as energy and entropy, whereas thermal properties such as pressure, volume and temperature, have absolute values. From this flows the principal idea of van der Waals and his followers of combining thermodynamics with the equation of state. Such a combination is the most characteristic feature of the Dutch school, and we shall meet it again later.

For an historian of science one remarkable consequence of this ideology – and this also is characteristic of the Dutch school – was the intuitive opposition in the Netherlands to Nernst's work on his Heat Theorem. They distrusted what was seen as an 'engineering' approach by the Nernst school of thermodynamics[29], that is, the search for solutions of inconsistent equations fitted to experimental results. Van der Waals's followers tackled problems from first principles and criticized the many simplifying assumptions in the papers of Nernst's school[30]. Their opposition was, however, more deeply rooted. To quote van der Waals again, 'The equations A = B are always differential equations so that one set of quantities is obtained from the other only by integration. The unknown constant of integration that is introduced in calculating caloric quantities from thermal is therefore inherent in the nature of these quantities, while the inverse is not true. I know very well that these remarks will not convince everybody, and that the matter is not so simple as I have set it out. Nevertheless there is a grain of truth in my opinion. At any rate if an alternative route turns out to be possible then we should all be glad. But this is not yet demonstrated and we have the right to say to anybody who does not accept the way we have chosen – show us another one' [74, p. 80–81].

Another way was found by Nernst in 1906, but it was too late for a seventy-year old man to be glad at the discovery, or even to accept it, although the possibility of its existence was something he thought about for a long time. Neither volume of the *Textbook of Thermodynamics* contains a word on Nernst's Heat Theorem. Many years later when passions had died down Kohnstamm included a chapter on what is now called the Third Law in the second edition of the *Textbook* in 1927, but the reserved and unenthusiastic nature of his treatment can only be understood by knowing its origin[31]. Van Laar's book on the equation of state was published 'free from the heat theorem' [161, p. iv, see also *47*].

At the same time the schools of van der Waals and Nernst shared one doctrine; both held a firm and consistent belief in the existence of atoms. This approach can be contrasted with that of Duhem who developed 'a brand of thermodynamics in which there were plenty of integrals but no atoms' [*62*, p. 54]. Van der Waals and his school always insisted on the atomic and molecular basis to their thermodynamical researches, and this viewpoint made the Dutch scientists very unreceptive to Ostwald's *Naturphilosophie*. Thus in a review of Ostwald's *Die Energie* we read that 'the author of *Algemeine Chemie* is valued much more highly than the "energetic philosopher"' [*63*][32].

We have discussed already the close interaction between mathematics, physics and chemistry in the Netherlands, both in the education of young scientists and in the development of their research. It is interesting to add Bosscha's evidence to that of van der Waals. In April 1889 he opened the Section of Physics and Chemistry of the Second Congress of the Scientists and Physicians of the Netherlands. He devoted his address to strengthening the communication between sciences in the Netherlands. 'Each science is limitless and so one would expect different branches of research to be far apart. In practice, just the reverse happens for the advance of science brings them closer together. Thus we can quote a recent series of interconnected investigations which started from purely chemical experiments, and led through regions of theoretical chemistry, physics, mechanics of molecular aggregations, to the field of geometric deductions of the properties of curved surfaces, which are the subject of study of the leading mathematicians of our day, such as Cayley. And in each of these areas we find our countrymen working to advance the subjects at a level that is as high as anywhere abroad or as in any other time. This team deliberately came together to solve a single problem of great importance' [*66*, p. 72]. This passage has not only a general significance but describes an actual occurrence – the response in 1889 to van der Waals's report two months earlier on his theory of mixtures.

The work of the Dutch school was not always entirely sound or free from error, but was always full of life, with its arguments, contradictions and the different points of view of participants of different temperament. There were disagreements, at various times, between Kuenen and van der Waals and between van Laar and Kohnstamm. In at least one case, that of unstable plait points, van der Waals characteristically acknowledged that 'the truth in the matter turned out to be mainly on Kuenen's side' [74, p. 85]. Nevertheless we can see from our point of vantage that van der Waals was quite justified in speaking of a Dutch school in science.

In conclusion it is worth discussing whether the Dutch school was somewhat closed, as one might imagine from the literature of physics and chemistry at the beginning of the century. Dutch science grew with world science, and Dutch physicists and chemists (and also mathematicians and biologists) were active participants on the world scene. They actively propagated their points of view, as, for example, in the famous and widely distributed *Communications of the Leiden Physical Laboratory* started by Kamerlingh Onnes. The Cryogenic Laboratory at Leiden was an international scientific centre that was probably more important than Ostwald's Physico-Chemical Institution at Leipzig. Nevertheless there is some truth in the belief that the Dutch school of thermodynamics was rather closed. It was not that they wished to be isolated but rather that the originality and high level of the work of the van der Waals school necessarily separated them from those working abroad. The great body of their publications was not cut off by the barrier of the Dutch language, for apart from the English *Proceedings of the Section of Sciences*, and the French *Archives néerlandaises* they published abroad also, particularly in the *Zeitschrift für physikalische Chemie*; only a small part of their work was published only in a Dutch version. The barriers were real but they were of a conceptual kind, not linguistic, and, ultimately, they were overcome.

Notes

1. There is as yet no full length biography of Kamerlingh Ohnes (1853–1926) but there are several serious biographical sketches. Here we note only that, as a seventeen-year-old student, he was awarded a gold medal for a prize essay, 'A critical study of methods for vapour density determination and of results obtained by these methods in the light of connections between the nature of chemical compounds and their vapour densities'. His doctoral thesis, brilliantly defended in 1879, is in quite another field, its title is *New proofs for the Earth's rotation*, but the interest in physical chemical measurements developed when he was young remained with him, and this was displayed in his activities at Leiden. The choice of Kamerlingh Onnes as the Professor of Experimental Physics at the University of Leiden undoubtedly owed much to the support of van der Waals who had himself been offered the chair.

2. *Through measurement to knowledge*. Here it is appropriate to quote a sentence written about a hundred years later: 'What is less widely appreciated is that important clues to a new understanding of nature can, in fact, be revealed by an improvement in the accuracy of measurement' [*6*].
3. See his Memorial Lecture on Magnus [*8*]. Kamerlingh Onnes had himself cited these words in his thesis.
4. The reference here is to Kamerlingh Onnes's work, *The General Theory of Liquids*, considered in Chapter 11, Section 4.4[92].
5. See Chapter 6, Section 2.
6. See Chapter 12, Section 2.2.
7. Van der Waals had especially emphasized the experimental skill displayed by Kuenen: 'Only Kuenen managed to observe this retrograde condensation, and the correspondence of critical phenomena with theoretical considerations presented here, because he prevented hysteresis phenomena and states where the equilibrium was yet not reached . . . Neglect of these precautions was the reason that in previous observations they could not distinguish two critical temperatures' [61, p. 130].
8. A detailed analysis of these inconsistencies has been given by J. M. H. Levelt Sengers [*14*].
9. Johannes Petrus Kuenen (1866–1922), 'a scientific grandson of van der Waals', as Kohnstamm described him [193], matriculated in 1884 in Leiden and from 1889 was the assistant to Kamerlingh Onnes; after 'promotion' he lectured as a *privaatdocent*. For a time he worked in Ramsay's laboratory and in 1895 he was appointed Professor of Physics in University College, Dundee, where he stayed until 1907. When in Scotland he worked in the field of van der Waals and Kamerlingh Onnes, and it was Kuenen who, as well as van der Waals, contributed to the *Kamerlingh Onnes Gedenkboek* (1904) by compiling a review of the thermodynamic investigations in the Physical Laboratory of Leiden. After returning to Leiden, he lectured on physics, especially thermodynamics, where he took over from Lorentz. He died suddenly a few days after his election as Rector Magnificus. Among Kuenen's works are two well-known books [158,159], and also a masterly historical work, *The contribution of the Netherlands to the development of physical science* (1919) which was referred to in Chapter 4 (Ref. *19*). For his life and personality, see [*11,16*].
10. This paper concludes with the quotation from van der Waals's paper [92] which had been published just two weeks before the liquefaction of helium and contained the assertion that it was possible to liquefy this gas also. For details see Chapter 11, Section 2.
11. Lorentz wrote in the article on the liquefaction of helium, a result 'which deserves to be recorded in golden letters', 'The Laboratory is now, after the victory has been gained, at the beginning of as new a period of development as any one dared to foresee. Anyone who can evaluate this great and brilliant achievement in the field will share our wishes that Mr Onnes may see more and more fruits of his work for the good of science and for the good of the young physicists who have the privilege of being his pupils', [*23*, p. 502]. These wishes came true in an unexpected way – everyone knows today about the startling discovery of superconductivity in 1911. That is another story, however.
12. Here it should be noted that neither van der Waals nor Kamerlingh Onnes kept their correspondence, but nevertheless more than a hundred of Kamerlingh Onnes's letters to van der Waals have been collected.
13. Zeeman commented on van der Waals's relation to Kamerlingh Onnes in the words 'Warmth appeared in his voice when he spoke about Onnes' [205, p. 111].

14. As in the case of Kamerlingh Onnes, we have as yet no full length biography of Bakhuis Roozeboom. The principal information is contained in papers [*24–27*] and in the jubilee numbers of *Chemisch Weekblad* [*28,29*].
15. To reduce his expenses and to save time, he never attended lectures at Leiden. Having married in 1879, he had to work also as a teacher in a HBS for girls from September 1881, thus doubling his earnings; his two salaries came to fl. 2000.
16. The contents of the thesis were presented in more detail in a paper [*30*]. The thesis itself is very concise (24 p.) and its most remarkable feature is the use of an installation with Cailletet tubes in which the working pressure reached 250 atm. This became possible owing to the help of Kamerlingh Onnes, as was acknowledged in the introduction.
17. It is generally accepted that the greatest discovery of Sir Humphry Davy was Michael Faraday. Perhaps it would be no exaggeration to say that the main discovery of van der Waals in the field of phase theory was Bakhuis Roozeboom.
18. As was mentioned (Chapter 4, note 14) van 't Hoff, although tired of teaching, had refused the offer from Berlin. In 1894 Max Planck arrived in Amsterdam to invite him to accept the chair of physics in place of August Kundt. This offer was refused, but the President's Address read by van 't Hoff at the Fifth Congress of Dutch Scientists and Physicians, in April 1895, concluded with words which sound like a summons for help: 'If the propagation of knowledge is of interest to the State it seems to me that its increase also has the right to be of interest. In our system of today it is this increase that is "au petit bonheur". A prize question, a grant for one year's study or an expedition; it is more or less a patchwork. It would be better if alongside our men whose duty is to teach and who, when they have inclination and time also investigate what they can, there were also others whose duty is to investigate and who, when they have inclination and time, also teach what they can' [*34*]. After a second offer from Berlin, for research work only, without any teaching, Lorentz and van Bemmelen, on behalf of all University professors of physics and chemistry, addressed to the Minister of Internal Affairs a petition to allow van 't Hoff to lecture once a week in Amsterdam and to release him from setting examinations and tests. However 'bureaucratic short-sightedness' [*35*] won the day and on 30 December 1895 there was a ceremony of farewell in the University. Van 't Hoff resigned on 1 January 1896, and in March he went to Berlin (see [*35–37*]).
19. It is interesting to examine photographs of a P–T–x-space model of this system [*41*]; its appearance readily allows us to believe that the most complicated phenomena are possible.
20. The chair was offered first to F. A. H. Schreinemakers, a pupil of Bakhuis Roozeboom in Leiden and the successor there to van Bemmelen, but he declined it. One can guess that there was some opposition between Schreinemakers and van der Waals. We know nothing about their personal relations, but there was scientific disagreement since Schreinemakers was a 'pure' thermodynamist who never combined thermodynamics with an equation of state; see also Chapter 12.
21. Smits's books on allotropy [*43,44*] contained many references to van der Waals's ideas; it is particularly interesting that Smits exploited also his notion of 'quasi-association' in connection with the formation of new phases (see his paper [*45*]). It should be remembered too that it was Smits who later described how van der Waals had helped Roozeboom, and who published van der Waals's letters to Roozeboom [220].
22. From 1917 to 1953 Scheffer was Professor of Analytical and Inorganic Chemistry in the Polytechnic at Delft. He always paid great attention to thermodynamics

and wrote an interesting textbook, *Applications of thermodynamics to chemical processes* [*46*]; his last paper, published posthumously, was on the occasion of Roozeboom's centenary.

23. A recent full-length biography of van Laar by E. P. van Emmerik [*47*] completes and, in part, changes the information on van Laar that is to be found in earlier papers [*48,49*].
24. According to its Charter, a Dutch scientist living abroad cannot be a full member of the Academy. Thus when van 't Hoff left for Berlin he became a corresponding member, although he had previously been a full member.
25. Here is a characteristic recollection by van 't Hoff's pupil, W. P. Jorissen. After his doctoral exams in February 1895 he was asked by van 't Hoff to name a good book on thermodynamics. 'Van der Waals never recommended to us books on thermodynamics; most of what he considered in his lectures was connected with his own work. I did not dare to give my notes on these lectures, however: the distance between the professor and the student seemed to be too great' [*55*].
26. This is deduced from the first sentence of Kohnstamm's paper presented in the Section of Sciences on 23 February 1907: 'Already for a considerable time I have been engaged in arranging Prof. van der Waals's thermodynamic lectures, and having arrived at the discussion of the three-phase line solid–liquid–vapour . . . I have formed on some points a different opinion from that laid down in the literature . . .' [*57*]. The topic mentioned was considered it the end of the second volume of the textbook.
27. 'One of the characteristics of his highly admired teaching was the introduction of Gibbs's great work to the chemists', Kamerlingh Onnes wrote in his obituary [208].
28. The importance of that geometrical way of consideration for chemistry and chemical technology was expressed perhaps most pithily in a book review in *Nature* [*58*] where after the assertion that 'we may fairly call the science of heterogeneous equilibria a Dutch science' the following conclusion is reached: 'The mind of the chemist, unlike that of his future colleague the engineer, really dwells in a point. He lacks the sense of space and the power of thinking spatially, or, rather, most of his work tends to produce that lack. What better corrective could there be than the study of a four-component system? If the chemist is going to rule the world – as he hopes and as he ought – he must cease to be a point-man. He must become an understander and controller of space. If not, well, then the engineer will occupy the space, and the poor old chemist will continue to sit at his point and wonder what is wrong – as he has done these many years.'
29. An early example of how this approach was proclaimed is in Nernst's review of a paper of van der Waals's [31]: 'The general thermodynamic formulae which are used have the advantage that they are valid not only for dilute solutions but also in general; they have the shortcoming, however, of being so complicated that they hardly yield any positive result' [$31c_2$].
30. See especially the discussion between Kohnstamm and Ornstein on one side and Nernst on the other [*59–61*].
31. Thus A. Rakovskii, who later edited the Russian version of the *Lehrbuch der Thermostatik*, expressed his astonishment at Kohnstamm's cool tone in this chapter; he suggested this concept was too new for Kohnstamm [$122a_2$].
32. It is worth commenting here on a rather interesting point concerning the penetration of atomistic ideas into chemical thermodynamics. It may seem to be paradoxical, but van der Waals's activity had little influence on this process. He suggested that he had combined thermodynamics and kinetic theory, but it was really a combination of thermodynamics and the equation of state, which is a fruit of the

kinetic theory, but not the theory itself, and the equation of state acquired a life of its own. According to the unfair but characteristic opinion of Ostwald in 1891, 'The kinetic hypothesis which was once received with such sympathy . . . turned out to be fruitless here also, as a matter of fact, because both of the main ideas of van der Waals's theory, which will undoubtedly govern matters in the immediate future, are detached from the kinetic hypothesis. Neither the assumption that only the volume unfilled by the substance of matter obeys Boyle's law, nor the assumption this matter has some energy of interaction implies any definite sense of kinetic ideas' [*64*]. Perhaps this example is not very convincing evidence because of Ostwald's anti-atomistic position. However, Nernst, whose atomistic convictions are beyond doubt, asserted that 'Van der Waals's work, in spite of its title "The molecular theory of a body composed of two different substances", is obviously a pure thermodynamical treatise' – a phrase which occurred in all editions of Nernst's *Theoretical Chemistry* [*65*]. One of the 'Propositions' (*Stellingen*) of Kohnstamm's thesis was an objection to this opinion; it stated that the above phrase is 'absolutely unjust', [*40*, p. 217]. This is certainly true, but if we change 'thermodynamical' to 'phenomenological', then Nernst's assertion would describe perfectly adequately the real state of affairs. Although great progress was made by atomic theory in natural sciences as a whole, chemical thermodynamics remained a purely phenomenological discipline at the turn of the century. Moreover it was a stronghold of anti-atomistics in physical chemistry, partially owing to the 'energetics' proclaimed by Ostwald, Helm, Jaumann and Duhem, and partially because of the inaccessibility of kinetic theories to chemists. It is interesting that it was Nernst's Heat Theorem that marked the real start of the movement of chemical thermodynamics to the next phase of its development which was characterized by an increasing role of atomistic ideas in chemical research. This was the result of the application of the quantum theory of heat capacities to chemical problems and, later, of the extensive contributions of theorists to the treatment of chemical equilibrium in terms of quantum and statistical theories.

References

1. Katz, J. R., The laws of surface-adsorption and the potential of molecular attraction, *Proc.*, 1912, v. 15, p. 445–454.

2. Zeeman, P., Een experiment over zoogenaamde anomale voortplanting van golven, *Versl.*, 1897, d. 6, b. 11–13.

3. Cohen, E., Kamerlingh Onnes Memorial Lecture, *J.C.S.*, 1927, May, p. 1193–1203.

4. Gerrits, G. C., *Grote Nederlanders bij die opbou der Natuurwetenschappen*, Leiden, 1948, b. 430–460.

5. Gavroglu, K. and Goudaroulis, Y., Heike Kamerlingh Onnes' researches at Leiden and their methodological implications, *Stud. Hist. Philos. Sci.*, 1988, v. 19, p. 243–274.

6. Trigg, G. L., *Landmark experiments in twentieth century physics*, New York and London, 1975, p. 121.

7. Lorentz, H. A., Het proefschrift van Prof. Kamerlingh Onnes. In *Collected papers*, Den Haag, 1939, v. 9, p. 291–307.

8. Helmholtz, H. v., Rede zum Gedachtnis an G. Magnus (Berlin, 1871). In *Vorträge und Reden*, 4.Aufl., Braunschweig, 1896, Bd 2, S. 33–51 (S. 46).

9. Kamerlingh Onnes, H., Investigations into properties of substances at low temperatures, which have led, amongst other things, to the preparation of liquid helium. In *Nobel Lectures, Physics, 1901–1921*, Amsterdam, 1967, p. 306–336.

10. Kuenen, J. P., Mésures concernant la surface de van der Waals pour les mélanges d'acide carbonique et de chlorure de méthyle, *A.N.*, 1893, t. 26, p. 354–422.

11. Kamerlingh Onnes, H., Prof J. P. Kuenen, *Nature*, 1922, v. 110, p. 673–674.

12. Kuenen, J. P., Over de retrograde condensatie, *Versl.*, 1892, d. 1, b. 15–19.

13. Kuenen, J. P., On the influence of gravitation on the critical phenomena of simple substances and of mixtures, *Leiden Comm.*, No. 17, p. 1–19; *A.N.*, 1898, t. 1, p. 342–353 (p. 349).

14. Levelt Sengers, J. M. H., Liquidons and gasons; controversies about the continuity of states, *Physica*, 1979, v. 98A, p. 363–402.

15. Kuenen, J. P., Quelques expériences relatives aux anomalies au voisinage du point critique, *A.N.*, 1898, t. 1, p. 278–296.

16. Lorentz, H. A., Johan Kuenen (1866–1922). In *Collected Papers*, Den Haag, 1939, v. 9, p. 404–406.

17. Kamerlingh Onnes, H. and Reinganum, M., Contributions to the knowledge of van der Waals' ψ-surface, 1,2, *Proc.*, 1900, v. 3, p. 275–288, 289–298.

18. Kamerlingh Onnes, H., Contribution to the knowledge of the ψ-surface of van der Waals, 11, A gas that sinks in a liquid, *Proc.* 1906, v. 9, p. 459–460.

19. Kamerlingh Onnes, H. and Keesom, W. H., Contribution to the knowledge of the ψ-surface of van der Waals, 15, *Proc.*, 1907, v. 9, p. 786–798; v. 10, p. 231–237.

20. Kamerlingh Onnes, H., The liquefaction of helium, *Proc.*, 1908, v. 11, p. 168–185; *Leiden Comm.*, No. 108.

21. Korteweg, D. J., *Versl.*, Sept. 26, 1908, d. 17, b. 182–183.

22. *Les Prix Nobel 1913*, Stockholm, 1914, p. 48.

23. Lorentz, H. A., La liquéfaction de l'Helium, *A.N.*, 1908, t. 13, p. 492–502.

24. van Bemmelen, J. M., Het leven en de werken van H. W. Bakhuis Roozeboom in zijnen Leidschen tijd, *Ch. W.*, 1907, Jg. 4, b. 249–285.

25. Jorissen, W. P. and Ringer, W. E., H. W. Bakhuis Roozeboom. In *Mannen en vrouwen van beteekenis in onze dagen*, No. 37, 1907, b. 155–218.

26. van Bemmelen, J. M., Jorissen, W. P. and Ringer, W. E., Hendrik Willem Bakhuis Roozeboom, *Ber.*, 1907, Bd 40, S. 5141–5174.

27. Stortenbeker, W., Hendrik Guillaume Bakhuis Roozeboom, *Rec.*, 1908, t. 27, p. 360–410,

28. Bakhuis Roozeboom — herdenking, 30 October 1954, *Ch. W.*, 1954, Jg. 50, N44, b. 749–763.

29. Bakhuis Roozeboom — Herdenkingssymposium, 28 October 1954, *Ch. W.*, 1955, Jg. 51, N23 (especially b. 401–413).

30. Bakhuis Roozeboom, H. W., Sur l'hydrate de l'acide sulfureux, sur l'hydrate de chlore; sur l'hydrate de brome; sur l'hydrate du gaz chlorhydrique, *Rec.*, 1884, t. 3, p. 29–58, 59–72, 73–83, 84–104.

31. Bakhuis Roozeboom, H. W., Sur les conditions d'équilibre de deux corps dans les trois états, solide, liquide et gazeux, d'après M. v.d. Waals, *Rec.*, 1886, t. 5, p. 335–350.

32. van 't Hoff, J. H., Prof Dr. Hendrik Willem Bakhuis Roozeboom, *Ch. Ztg.*, 1907, 23 Febr., S. 199.

33. Bakhuis Roozeboom, H. W., *Heterogene Gleichgewichte vom Standpunkte der Phasenlehre*, Braunschweig, Heft 1, 1901, Heft 2, Theil 1, 1904.

34. van 't Hoff, J. H., NNGC 5, Haarlem, 1895, b. 9.

35. Lorentz, H. A., Van 't Hoff's vertrek uit Nederland, *Ch. W.*, 1915, Jg. 12, b. 515–516.

36. *Jb. A.U. 1895-1896*, Amsterdam, 1897, b. 7-9, 32-34.
37. *Levensber.*, b. 66.
38. Bakhuis Roozeboom, H. W., *De wetenschappelijke beoefening der chemie en hare uitkomsten*, Rede . . ., Leiden, 1896.
39. Bakhuis Roozeboom, H. W., Erstarrungspunkte der Mischkrystalle zweier Stoffe. Umwandlungspunkte bei Mischkrystallen, *Z. ph. Ch.*, 1899, Bd 30, S. 385-412, 413-429; *A.N.*, 1900, t. 3, p. 414-444, 445-463.
40. Kohnstamm, Ph. A., *Experimenteele onderzoekingen naar aanleiding van de theorie van Van der Waals : Het p-T-X-vlak: voorstudies en methoden*, Proefschrift . . ., Amsterdam, 1901.
41. Smits, A., The *P*, *T*, *X*-spacial representation of the system ether-anthraquinone, *Proc.*, 1909-10, v. 12, p. 231-233.
42. Smits, A., *De chemie in hare oude en nieuwe banen*, Rede . . ., Delft, 1907, b. 19.
43. Smits, A., *Die Theorie der Allotropie*, Leipzig, 1921.
44. Smits, A., *Die Theorie der Komplexität und der Allotropie*, Berlin, 1938.
45. Smits, A., Das Gesetz der Umwandlungsstufen in the light of the theory of allotropy, *Proc.*, 1912, v. 14, p. 788-801 (p. 791).
46. Scheffer, F. E. C., *Toepassingen van de thermodynamica op chemische processen*, Tweede druk, Delft, 1958.
47. van Emmerik, E. P., *J. J. van Laar (1860-1938). A mathematical chemist*, Doctoral Thesis, Delft, 1991.
48. Jorissen, W. P., J. J. van Laar, *Ch. W.*, 1920, Jg. 17, b. 362-367.
Jorissen, W. P., Levensbericht van Johannis Jacobus van Laar, *Jb. KAW*, 1938-39, Amsterdam, 1939, b. 239-243
A. E. Korvezee, *Leven en werken van Dr. J. J. van Laar*, Delft, 1954.
49. van Klooster, H. S., J. J. van Laar, *J. Chem. Educ.*, 1937, v. 14, p. 459; 1962, v. 39, p. 74-76.
50. Partington, J. R., *A History of Chemistry*, v. 4, London, 1964, p. 616.
51. Cohen, E., Chemie en chemische industrie te Amsterdam in den loop der eeuwen, *Ch. W.*, 1940, Jg. 37, b. 486-502.
52. Schreinemakers, F. A. H., H. W. Bakhuis Roozeboom (1854-1907), *Ch. W.*, 1926, Jg. 23, b. 412-413.
53. van der Lee, N. J., *De invloed van den druk op de kritische mengtemperatuur*, Proefschrift . . ., Amsterdam, 1898.
54. Kohnstamm, Ph. and Timmermans, J., On vapour-pressures in binary systems with partial miscibility of the liquids, *Proc.*, 1911, v. 13, p. 865-880.
55. Jorissen, W. P., Mijn herinneringen aan Jacobus Henricus van 't Hoff en de studie der chemici aan de Universiteit van Amsterdam onder zijn leiding, *Ch. W.*, 1952, Jg. 48, N. 35, b. 647-58.
56. Hissink, D. J., Enige herinneringen uit de jaren 1896 tot Januari 1900 aan mijn leermeester Prof. Dr. H. W. Bakhuis Roozeboom, *Ch. W.*, 1954, Jg. 50, b. 756-757.
57. Kohnstamm, Ph., On the shape of the three-phase-line solid-liquid-vapour for a binary mixture, *Proc.*, 1907, v. 9, p. 639.
58. Anon, L'Espace dans la Chimie, *Nature*, 1921, v. 108, p. 171-172.
59. Kohnstamm, Ph. and Ornstein, L. S., On Nernst's theorem of heat, *Proc.*, 1911, v. 13, p. 700-715.
60. Nernst, W., On the inconsistency of my heat theorem and van der Waals' equation at very low temperatures, *Proc.* 1911, v. 14, p. 201-204; *Versl.*, 1911, d. 20, b. 64-67 (in German).
61. Kohnstamm, Ph. and Ornstein, L. S., Nernst's theorem of heat and chemical facts, *Proc.*, 1912, v. 14, p. 802-818.

62. Poincaré, H., Les conceptions nouvelles de la matière. In *Le matérialisme actuel*, Paris, 1913, p. 49–67.
63. Kruyt, H. R., Boekaankondigingen, *Ch. W.*, 1913, Jg. 10, b. 822.
64. Ostwald, W., Fortschritte der physikalischen Chemie in den letzteren Jahren (1891). In *Abhandlungen und Vorträge allgemeinen Inhaltes*, Leipzig, 1916, S. 34–52 (S. 37).
65. Nernst, W., *Theoretische Chemie vom Standpunkt der Avogadroschen Regel und der Thermodynamik*, 8–15 Aufl., Stuttgart, 1926, S. 264.
66. Bosscha, J., Voorzitter's toespraak, *NNGC 2*, Leiden, 1889, b. 71–72.

CHAPTER 9

Van der Waals and Russian Science

This chapter serves both as a 'case history' of the spread of the ideas of van der Waals to another country, and also to inform Western readers of the contributions of Russian scientists to the development of his ideas in molecular physics and physical chemistry.

The first report of van der Waals's work in Russia came from A. G. Stoletov[1], one of the few owners of the Dutch original of *Continuity*. In January 1879 he published a *Sketch of the development of our knowledge of gases* [*1*], which contained 'a new law', that is, the equation of state of van der Waals. Stoletov wrote that this 'explains exactly all the complicated phenomena if we ascribe to gas molecules only a definite volume and a definite force of cohesion. Conversely the experimental data on any gas can be fitted to the law of van der Waals, and so we can determine the size of the molecules and the strength of their mutual attraction. Thus the irregularities of gases, their deviations from 'ideal' behaviour, have a particular significance; they provide us with the means of solving problems in the world of atoms that are apparently totally inaccessible' [*1*, p. 174]. The quotation shows that even in these early days van der Waals was not alone in his ideas and Kamerlingh Onnes was right to note that 'Stoletov was, of course, among the first to understand van der Waals' [233].

After the publication of the German translation of *Continuity*, Stoletov [*2*] showed that the reduced van der Waals equation did not fit the experimental results for the limiting slope of the reduced vapour pressure curve, $(\mathrm{d}\pi/\mathrm{d}\tau)_{\mathrm{crit}}$, whereas the reduced Clausius equation gave a value of about 7 which was close to the few results available. At the same time he included a detailed discussion of van der Waals's equation in his lectures on theoretical physics in the academic year of 1881–82. The lectures were reproduced lithographically by his student D. A. Goldhammer[2] who recalled fifteen years later that 'investigations on the critical temperature were then quite new, and the account of the famous paper of van der Waals, then recently translated into German, introduced students to the most exciting and interesting problems of the day' [*3*]. In the second, and final, part of the course, the application of the equation to real gases and liquids and to critical phenomena formed about half of the lectures. They were illustrated by a model of the *P–V–T* surface for water that Stoletov constructed from the experimental data. It was Stoletov's teaching that led to his students Goldhammer and B. V. Stankevich[3] taking up problems in van der Waals's field.

In 1884 Goldhammer published his monograph on *The thermodynamic surface of water* (*after Gibbs and van der Waals*) [*4*], in which he described the ε–η–v (that is, U–S–V) surface for the three phases of water. He used van der Waals's equation ('instead of the more exact Clausius equation') because of its simplicity. Boltzmann, who knew this work, insisted that Goldhammer's graphs were displayed at the Mathematical Exhibition at Munich in 1893.

Stankevich was concerned mainly with the kinetic theory of gases and liquids. His first work, in 1884, *The mathematical treatment of the kinetic theory of gases* [*5*], was the winning entry in a prize competition based on a question posed by the Physical and Mathematical Faculty. Most of the essay was on ideal gases but several sections (Sections 48–53) were devoted to real gases composed of molecules of non-zero size that undergo elastic collisions. He attempted a more strict derivation of van der Waals's equation than that given in *Continuity* (he worked from the original Dutch) by computing the virial of the forces at collision by Lorentz's method. Several years later he wrote that van der Waals's equation was the best yet proposed but that its derivation was mathematically unsatisfactory; 'The correction of van der Waals's theory is one of the most urgent problems of the present kinetic theory of gases' [*6*]. He followed this with an important monograph, *Studies on the kinetic theory of the constitution of bodies* [7] in which he tried to improve on van der Waals's methods. He obtained the virial of attractive forces of the form $-r^{-n}$ (with $n \sim 5$ as he deduced from an analysis of experimental results) and attempted to use the virial theorem to construct a kinetic theory of liquids. He considered also the connection between the thermal equation of state, $F(P,V,T) = 0$, and the caloric equation of state, $\varepsilon = \varphi(V,T)$, where ε is the energy, and demonstrated that

$$P - T\int \frac{\partial \varepsilon}{\partial V}\cdot\frac{\mathrm{d}T}{T^2} = T\cdot\tau(V),$$

where τ is an arbitrary function. He then discussed the limitations imposed on the functions ε and τ by kinetic theories of matter, and in particular the case of van der Waals's equation of state [*8*].

N. N. Pirogov[4], a businessman who turned to science late in life, was one of the first to discuss seriously the statistical foundations of the second law of thermodynamics. He wrote also on kinetic theory[5], distinguishing between an 'ideal gas' composed of molecules of zero size and no attractive forces, a 'common gas' with molecules of non-zero size but no attractive forces, and a 'real gas' in which there were also attractive forces. He and the Polish physicist W. Natanson emphasized the importance of considering explicitly interactions between groups of three and four molecules, and he distinguished between molecular groups that can break up spontaneously and 'molecular aggregations' that can do so only after the injection of sufficient energy.

Besides these serious attempts to develop further the theory of the equation of state there were other more speculative papers. Thus K. D. Kraevich[6] deduced correctly that the force of interatomic cohesion in copper could not be Newtonian gravitation from calculations based on the density and breaking strength of the metal. He attempted also to calculate latent heats from a 'generalised' form of van der Waals's equation [*12*]. B. B. Golitsin[7] attempted to find the range of the attractive forces from measurements of vapour pressure and surface tension. He deduced that the range depended on molecular mass to the power of 0.86 [*15*], while N. P. Kasterin[8] concluded that the power was one half [*16*]. Kraevich [*17*] considered the problem of the thermodynamic limitations imposed upon any equation of state. His criteria were

$$\frac{\partial C_v}{\partial v} = T\frac{\partial^2 P}{\partial T^2}, \quad \frac{\partial C_p}{\partial P} = T\frac{\partial^2 v}{\partial T^2},$$

$$C_p = C_v + T\frac{\partial P}{\partial T}\cdot\frac{\partial v}{\partial T}.$$

From this point of view he discussed the van der Waals equation ('a generalization of high scientific importance', as he emphasized), and showed that for a body that obeyed the van der Waals equation, $\partial C_v/\partial v = 0$. The consistency of equations of state and measurements of the Joule–Thomson coefficient was examined by N. Schiller[9] who deduced that Clausius's equation of state was better than that of van der Waals [*18*].

All this work was based on a general acceptance of van der Waals's views, but there was also in Russia at least one man who was of the opposite opinion. The mathematician N. Ya. Sonin strongly criticized van der Waals's derivation [*19, 20*]. He insisted that, 'The idea of expressing physical continuity by preservation of the analytic form of the function is quite wrong', and that Lorentz's derivation [*21*] destroyed completely 'the value of van der Waals's equation as a law of nature'. He was probably the first to demonstrate the possibility of deriving a reduced equation of state of quite general form but he then went on to say that the concept of corresponding states based on van der Waals's equation 'seemed to be only an exercise in analysis having no physical meaning' [*20*]. This was the only substantial negative response in Russia, and its influence was small because it was published in a provincial journal and never appeared in a well-known one.

We may note here a distinction that is not peculiar to Russian scientists; the physicists with their better grounding in mathematics were often more critical than the chemists whose more qualitative mode of thought allowed them to appreciate more the beauty and clarity of van der Waals's thought without being disturbed by its mathematically shaky foundation. The first chemists to be influenced by his work were D. I. Mendeleev and his successor at the University of St Petersburg, D. P. Konovalov. Mendeleev obtained a copy of van der Waals's thesis in 1875 when he visited Leiden for the tercentenary celebrations of the founding of the University. The book was already hard to find and in his library catalogue Mendeleev noted in the margin that it was 'very rare'. His copy is now in the Mendeleev Archives Museum and the author's name on the title-page is underlined twice, which always denoted that he thought the work to be important. Twenty years later he bought the French translation and his library contains also a copy of van der Waals's 'Ternary systems' [65f] with the author's complimentary signature.

Mendeleev discussed the equation of state in detail in his *Principles of Chemistry*, from the fifth edition of 1889 onwards. 'It is necessary', he wrote, 'to give an idea of the results of van der Waals's considerations for they explain the gradual uninterrupted passage from a liquid into a gaseous state in the simplest form, and, although the deduction cannot be considered as complete and decisive . . . nevertheless it penetrates so deeply into the essence of the matter that its signification is not only reflected in a great number of physical investigations, but also in the province of chemistry' [*22*, p. 139]. He concluded that 'Further research will naturally give further proximity to truth, and will show the connection between composition and the constants (a and b); but a great scientific progress is seen in this form of the equation of state' [*22*, p. 141].

Van der Waals's name was honoured by an article in the Brockaus-Efron *Encyclopaedic Dictionary* edited by a group of Russian scientists which included Mendeleev. The article, entitled Валъса Формула (Waals's Formula), was written by I. Bergman, the Professor of Physics at St Petersburg, and appeared in 1891. It described departures from Boyle's and Gay-Lussac's laws in terms of van der Waals's theory and concluded that he 'had founded a kinetic theory of liquids by his memoir'.

In 1886 Konovalov[10] used van der Waals's equation [*23*] to study a crucial problem concerning the nature of the liquid state. There were at that time two different beliefs about the molecular nature of a liquid; the first was that in liquids the molecules were associated into larger groups, so that the molecular nature of a liquid was different from that of a gas; the second view was that the molecular entities were essentially the same in both phases — this was van der Waals's view. Konovalov calculated, from van der Waals's equation, the coefficients of thermal expansion and of compressibility for both gases and liquids and also the latent heats of evaporation, and found that the thermal expansions fitted well Gay-Lussac's law for gases and the expression for liquids recently obtained by Mendeleev [*24*]. He concluded that, 'There is, therefore, no reason to consider that the transformation of a gas into a liquid occurs in any other way than that envisaged by the theory of van der Waals. There is no reason to suppose that liquefaction is necessarily accompanied by the joining of many gaseous molecules into a new compound molecule in the liquid. This may occur in special cases, but these should be considered as a starting point for explaining deviations from usual behaviour, not for deriving the general laws of liquids'[11].

Another chemist who wrote extensively on van der Waals's equation was L. G. Bogaevskii[12]. In two voluminous memoirs, *On the continuity of the gaseous and liquid states* [*26*] and *On various states of matter* [*27*], he explored many of the mathematical properties of the isotherms of the equation of state. His discussion, although often naïve, reflected the great attention paid to the details of van der Waals's work. One of his conclusions was that 'just as the Boyle–Mariotte–Gay-Lussac equation has led to the establishment of a well-formed dynamical theory of gases, so the van der Waals equation can tell us much about the liquid states to which it is applicable' [*26*, p. 96]. (As Bogaevskii emphasizes, the Boyle–Mariotte–Gay-Lussac equation, although empirical in origin, expresses a perfectly definite idea, namely that of the ideal gas.)

Critical phenomena, an important aspect of van der Waals's work, attracted much attention in Russia also. As early as 1859, Mendeleev, then a post-graduate student at Heidelberg, took up problems of 'molecular mechanics' which he tried to understand through a study of capillarity. He estimated the decrease in molecular adhesion from the measurements of others, and deduced that there was an 'absolute boiling temperature' (*Die absolute Siedetemperatur* in German, later translated into English as 'the temperature of absolute boiling' [*22*]) at which the cohesion becomes zero, which he estimated to be 191 °C for ethyl ether, 279 °C for ethanol, and 543 °C for water [*28*]. Next year he studied silicon tetrachloride himself up to 40 °C and estimated the absolute boiling temperature at which the cohesion was zero, at 212 °C. His direct observations in sealed tubes gave 235–240 °C; the value accepted today is 233 °C. He concluded, 'We must consider that point to be the absolute boiling temperature at which (1) the cohesion of the liquid becomes zero, (2) the latent heat of evaporation is also zero, and at which (3) the liquid is transformed into vapour, independently of pressure and volume' [*29*]. His choice of words showed that he had an asymmetrical view of the phenomenon; liquid is transformed into vapour [183, p. 5].

After the publication of Andrews's paper in 1869 [*30*], Mendeleev responded by recalling his earlier papers and making an interesting suggestion: 'It seems to me that studies of the absolute boiling temperature will show characteristic correlations with the compositions of the substance. Kopp has increased our understanding by comparing the volumes of substances at their boiling points, and hence it is clear that a comparison of a temperature of almost zero cohesion will be still more informative' [*31*, p. 623]. We can see here a glimpse of the value of the principle of corresponding states, formulated ten years later by van der Waals.

Experimental work on critical phenomena was carried out in the Physical Laboratory of the University of Kiev by M. P. Avenarius[13] and his students from 1873 onwards. His own work [*32*] comprised the measurements of the critical temperatures of four liquids: ethyl ether (196.2 °C), carbon disulphide (276.1 °C), carbon tetrachloride (292.5 °C), and acetone (246.1 °C). These results were compared with those estimated by extrapolation from Zeuner's expressions for the change with temperature of the latent heat of evaporation. The agreement was satisfactory for the first three liquids, but poor for acetone. An interesting feature of this work was the first published account of the critical opalescence, a merit of the paper that was later noted by Kamerlingh Onnes and Keesom [160, Section 50]. (Andrews had earlier described this phenomenon in a lecture at the Royal Institution in London in June 1871.)

His theoretical ideas were still unformed and in his second paper [*33*] he revealed this very obviously by writing the Clapeyron–Clausius equation in the form

$$\lambda = P\Delta v\left[\frac{T}{P}\frac{\mathrm{d}P}{\mathrm{d}T} - 1\right],$$

and then saying that the latent heat λ could become zero at the critical point either by Δv vanishing there or if $\mathrm{d}P/\mathrm{d}T = P/T$. This error was repeated in the principal paper [*34*] of his first student and collaborator, V. I. Sayonchevskii[14]. The error did, however, lead the latter to study vapour pressures up to the critical point, and it was his results that were used by van der Waals as the basis for his discussion of the principle of corresponding states [1d$_1$, p. 130–161]. The error was soon corrected, partly by the experimental results obtained at Kiev (see [*35*]) and partly as a result of Stoletov's criticism [*2*]. In 1884 Avenarius published a review on *The critical states of substances* in which the discussion was founded on the classical concepts of equality of density and of other properties of liquid and vapour at the critical point. His second collaborator, K. N. Zhuk[15] measured many specific volumes of liquids up to the critical temperature [*36*].

This early work at Kiev was stimulated not so much by the publications of van der Waals as by those of Mendeleev and Andrews. After the publication of the German translation of *Continuity* in 1881 the new theoretical ideas were followed up, first by O. E. Strauss[16] who suggested that the critical temperature of a mixture could be represented as

$$T_\mathrm{m} = (v_1 T_1 + v_2 T_2)/(v_1 + v_2),$$

where v_1 and T_1 are the critical volume and temperature of component 1. This expression[17] was established on the basis of measurements on ethanol and ethyl ether [*37*], checked against Janssen's results for mixtures of air and carbon dioxide, and used for representing measurements of the critical temperatures of mixtures of ethanol and water that were less than 50 per cent water by volume. From these he calculated by extrapolation the critical temperature of water as 370 °C and, by the principle of corresponding states, then estimated its critical pressure as 195.5 atm [*38*] (modern values are 374.2 °C and 218.3 atm). Strauss's formula was also confirmed experimentally by Golitsin [*13*] who tried to find a theoretical expression for the decrease of critical temperature caused by an admixture of an indifferent gas; it was in the course of this work that he proposed that the cross-term in the attractive term be represented $a_{12} = (a_1 a_2)^{1/2}$ [*13*, p. 786].

A. I. Nadezhdin[18] was certainly the most talented of Avenarius's students. He esteemed van der Waals's work highly, writing, for example, in 1885 of 'the most brilliant pattern of scientific synthesis during the last decade' [*41*, p. 7]; and later, 'The theory of the changes of state of aggregation given by van der Waals . . . has emphasised the importance of the critical state and has, for the first time provided a rational way of determining the critical temperature' [*41*, p. 172]. His study of critical phenomena was wide ranging and his observations on opalescence are of interest to this day [*41*, p. 67–68].

He devised a neat method of finding the critical temperature in an opaque tube by using it as the beam of a balance which balanced only when the densities of liquid and vapour were equal. He determined the critical temperature of 35 substances, the critical pressures of 23 and the critical volumes of 20, more than a quarter of the known results at the end of his life [*45*]; some of them are the only ones available even to this day [*46*]. In 1885 he made the first direct determination of the critical temperature of water [*40*], a result that was the culmination of the work in Kiev. Little more was done after Nadezhdin's early death in 1886 and Avenarius's illness and retirement in 1890.

In the 1880s there arose some criticisms of the classical view of the critical point (see Chapter 8). In Russia these objections were raised principally by Golitsin who tried to show that the optical method could not give the true critical temperature [*47*][19]. Stoletov took up the cudgels on behalf of the original experimenters. The first of a series of papers entitled *On the critical state of substances* was published in January 1892 [*50*], and later translated into German[20]. It opened with the statement that the simple results of Andrews, and their theoretical discussion by van der Waals, Clausius and Maxwell, had been reinforced by the more recent work of Avenarius and his school, but were now being attacked by those who were raising 'objections' and introducing 'new facts'. He analysed in detail many such publications, including those of Wroblewski, Cailletet and Colardeau, Golitsin and Jamin. He showed why the temperature at which the meniscus disappeared depended on the amount of substance in the sealed tube, and suggested that many of the alleged complications were due to the admixture of impurities which were already known to show less simple behaviour than pure substances [*50*]. His next paper [*52*] criticized recent publications of Battelli, Zambiasi, Golitsin and Pellat, and suggested that partial decomposition of their materials was the cause of the apparent anomalies. In his last paper [*53*] he considered recent work of Gouy and the effect of stirring on the attainment of equilibrium. He noted that Kuenen had also, in No. 8 of the *Leiden Communications*[21], emphasized that an electromagnetically driven stirrer led to a rapid attainment of equilibrium. He quoted at length from Ramsay and Young in the *Philosophical Magazine* for February 1894, and concluded, 'Let us now hope that the clear and simple scheme propounded by their famous compatriot, Andrews, will henceforth no longer be obscured by inundations of ill-conceived experiments and irrational ideas.' The argument continued, however, in the belief that in the anomalous substances there was a difference between the gaseous and liquid molecules. Bogaevskii suggested a compromise; there could exist different clusters in the liquid and gaseous states, but the difference between them vanished at the critical point [*26*].

Let us now consider the development of the principle of corresponding states, in the early stages of which two trends can be distinguished. The first is the comparison of similar substances as exemplified[22] in de Heen's *Essai de physique comparée* of 1883, and the second the searching for 'corresponding states' which culminated in van der Waals's paper of 1880 [*12*]. Then the two trends were combined by Nadezhdin who compared similar substances under similar conditions [*44*] as, for example, when he found that the difference between critical and boiling temperatures was approximately constant for each group of organic substances [*55*]. He showed that the reduced coefficient of thermal expansion of liquids, $(T^c/V)(dV/dT)$, was constant to within about 3 per cent for 32 of a group of 35 substances, the exceptions being methyl and ethyl alcohols and chloroform, for which the deviations were about 7 per cent. Until he had made his experiments there were insufficient results to provide searching tests of the principle. He wrote that he had 'checked van der Waals's law in several ways and can now say definitely that if it is a general law then it is only a first approximation and, moreover, it is a law to which there are exceptions' [*41*, p. 106]. He studied also vapour pressures, confirming Stoletov's result that $(T^c/P^c)(dP/dT)^c \simeq 7$, but exactly the same value is found only for substances of the same class [*41*, p. 154].

After the appearance of the first papers of Sidney Young, and the freeing of the principle from any particular equation of state, an even more extensive study of deviations was made by Bogaevskii [*27*] who related them to association in the liquid state. This idea was taken up by Batschinskii[23] [*56,57*] who classified fluids as *orthomers* if the molecules of liquid and vapour were the same and as *allomers* if they were different. Van der Waals accepted this classification [73, p. 286]. By comparing acetic acid (allomer) with its isomer methyl formate (orthomer) he was able to estimate the degree of association of the liquid state of the former along the whole course of its vapour pressure curve.

By the beginning of this century, when it was becoming clear that even complicated equations of state could not wholly fit the facts, it became more common to search for relations between properties using van der Waals's equation and the principle of corresponding states as guides. Brandt[24] showed that Stefan's relation between internal pressure and latent heat was a consequence of van der Waals's equation; if it were combined with Guldberg's rule that $T^b/T^c \simeq 2/3$ then Trouton's rule for the entropy of evaporation was obtained, although his constant came out at only about half of its correct value [*58*]. Similarly, Batschinskii showed that the law of rectilinear diameters could be deduced, with slight additional assumptions, from van der Waals's equation [*59*]. He also sought further empirical correlations inspired by the equation but not directly deduced from it [*60,61*]. The physical chemist V. A. Kistyakovskii[25] similarly combined the equation with empirical principles such as Guldberg's rule in order to interpret his own measurements of surface tensions [*62,63*].

A. N. Shchukarev[26], a physical chemist and a collaborator with the thermochemist V. F. Louginin, studied calorimetrically one-component systems near their critical points [*64*]. According to van der Waals's equation, a plot of energy against density should be linear, but he found that this happened only at one temperature, which was significantly above the critical point (e.g. about 300 °C for isopentane, for which T^c is 188 °C); he called this point where the isotherm is linear the 'van der Waals temperature'. Paul Walden[27] in Riga also contributed many papers on empirical correlations of the physical properties of normal and associated liquids, the most extensive of which are the three in *Zeitschrift für physikalische Chemie* for 1909 [*65*].

Finally, the most interesting of the studies of the critical point to modern eyes are the investigations of Goldhammer [*66*, see also *67*] on what are now called the critical exponents, that is, the power of $(T^c - T)$ that governs the vanishing of some physical property (or difference of physical properties) as the critical point is approached. Van der Waals had considered this question [37] but was constrained in his thoughts by the classical exponents inherent in his equation. Goldhammer examined with less prejudice the results for a wide range of fluids. His principal results were that the orthobaric density difference $(\rho_l - \rho_v)$ vanishes as $(T^c - T)^{1/3}$, and that the surface tension vanishes as $(T^c - T)^{5/4}$, results that are close to the best modern values of these exponents of 0.325 and 1.28.

Let us now return to the 1880s and consider the contributions of the chemists to the problems of liquid mixtures[28]. Here Konovalov's work was outstanding. His classic thesis 1884, *On the vapour tension of solutions*, is known in Western Europe only through two brief papers [*68,69*], but the Russian version [*70*] is much richer. He used the vapour pressure as a means of studying problems of phase equilibria between liquid and vapour, and between two liquid phases in equilibrium with the vapour [*70*, p. 66–73][29]. This work anticipated the later studies of van der Waals, discussed in Chapter 12.

'The condition of equality of vapour pressure between coexisting solutions', Konovalov wrote, 'links the question of the mutual solubility of liquids closely to that of the change of vapour pressure with composition. If we know exactly the dependence of vapour pressure on composition in a mixture of two liquids then the problem of their mutual solubility would also be solved.' (If one substitutes 'chemical potential' for 'vapour pressure' then this sentence could appear in any modern book on thermodynamics.) He then proceeded to draw an analogy between the solution of one liquid into another and the transformation of gas into liquid. In particular he showed how the partial pressure of one component in a liquid mixture, A, would change with the addition of a second, B, according to the comparative strength of the cohesion between A–A and A–B molecular pairs. If the mixture is above its mixing temperature then the change is monotonic but if below then it could, in principle, show maxima and minima, as for the van der Waals loops for a one-component liquid–vapour system. 'The temperature of complete mixing of such liquids is a kind of critical temperature. Mixing in any proportions is possible at the temperature where maximum and minimum just described disappear in the vapour tension curve and it becomes gradually descending' [*70*, p. 68]. This work of 1884 showed how thoroughly he had mastered van der Waals's ideas and how he was able through the single concept of the critical state to unite two phenomena – condensation and miscibility – previously considered to be unrelated. His priority in this work[30] has not generally been noticed, many ascribing the idea to Orme Masson [*74*]. Later Konovalov returned to the study of critical phenomena in mixtures, and ascribed the opalescence to the fact that 'significant changes of composition can be caused by the smallest expense of work', and that 'trifling causes are sufficient to disturb the local homogeneity' [*76*].

By the early years of the twentieth century van der Waals's ideas had become well known outside the Netherlands, propagated in part by the publication of the second edition of *Continuity* [61], with its extension to mixtures, and by his report to the International Congress of Physics in Paris in 1900 [58]. His Russian followers now included E. V. Biron[31] and R. F. Hollman, the former working on the extension of the equation of state to mixtures and the latter on the thermodynamics of phase equilibria.

Most of Biron's experimental and theoretical work was summarized in his book of 1912, *The compression by mixing of normal liquids* [77], which contained also many general observations on van der Waals's theory and its value to chemistry. He stressed that van der Waals's theory of concentrated solutions was 'the most general theory so far proposed in this field; many of its conclusions have been confirmed and it throws light on many obscure and vague aspects of the field of concentrated solutions. Anyone who works in this field must take account of this theory' [*77*, p. vii].

Biron's general approach was a systematic classification of solutions, based on his experiments, into several different classes, beginning with the ideal case of no interactions between the components and in which the properties are therefore additive (he called such systems 'isofluid'), and then moving to systems of 'normal' liquids in which the interactions were 'physico-mechanical', then to systems composed of 'abnormal' or associated liquids, and finally solutions of chemically reacting components. He measured mainly the compression or change of volume on mixing since this was 'a necessary stage on the way to an equation of state for solutions' [*77*, p. xi].

For normal liquids a series of very precise measurements showed that the change of volume D was symmetric in the mole fraction x,

$$D = Kx(1 - x),$$

in which the compression constant K increased with temperature. It was zero only if 'both components of the solution belong to the same group of substances, obey the principle of corresponding states, and have similar critical pressures' [*77*, p. 78]. For components from different groups the compression constant was large even if the critical pressures were close.

He believed that the changes of volume on mixing resulted from a change of volume of the components following a change on mixing of the internal pressure, a/V^2. He therefore measured the compressibility of liquids up to 600 atm pressure, having first designed a new manometer and piezometer for this purpose. He expressed his results in the form

$$V = A + B/(C + P),$$

which he deduced from the van der Waals equation[32]. (Similar equations had been proposed earlier by Tait and by Tammann.) To describe the change of volume theoretically he followed van der Waals's use of quadratic functions of mole fraction for a_x and b_x by postulating a similar form for the internal pressure P_x. He thus obtained an equation

$$K = \left[V_1 \beta_1 \frac{P_1}{P_2} - V_2 \beta_2 \frac{P_2}{P_1} \right] (P_2 - P_1),$$

in which V_i is the molar volume, β_i the compressibility and P_i the internal pressure of component i. He verified this equation for mixtures of similar substances, such as benzene + toluene and benzene + carbon tetrachloride, but showed that it failed for mixtures of less similar components, such as those containing carbon disulphide, or for benzene with chlorobenzene and with bromobenzene.

Ideal or isofluidic solutions, for which $K = 0$, required that $P_1 = P_2$ and so $a_1 = a_2$, and he concluded that the 'rule' proposed independently by Golitsin and Daniel Berthelot that $a_{12} = (a_1 a_2)^{1/2}$ could hold only if $a_1 = a_2$, a point of view with which van der Waals would have agreed (see Chapter 12). He found later that mixtures of abnormal liquids conformed to the same equation for K if they came from a common group, for example, for mixtures of pairs of aliphatic alcohols.

His equation for the change of volume on mixing can be expressed as [*77*, p. 27]

$$V = V_1 x + V_2 (1 - x) + Kx(1 - x),$$

of which form Krichevskii was to remark many years later that 'Biron's equation was as characteristic of regular solutions as additive volumes were of ideal solutions' [*78*, p. 62]. The term 'regular solutions' was introduced by J. H. Hildebrand whose definition was, however, slightly different from that of Krichevskii and, implicitly, of Biron, whose real merit was that he distinguished experimentally the class of mixtures described by the last equation.

This 'theory of compression' was, in truth, a trivial transformation of van der Waals's equation but Biron was proud of it. Its greatest merit was perhaps the impetus it gave to the precise measurement of volumes of mixing at an early stage of the development of the thermodynamics of solutions.

Hollman[33], who remained faithful to the Dutch school of physical chemistry all his life, devoted his efforts to the thermodynamic study of heterogeneous systems. His first book, in 1907, was entitled, *On the formation and decomposition of mixed crystalline hydrates of isomorphous substances from the point of view of Gibbs's phase rule* [*79*]. It contained a detailed analysis of the conditions of stability of ternary systems of mixed crystal hydrates, and a derivation of the fifteen possible types of phase diagram. Amongst the many rules that he formulated were two generalizations of Konovalov's laws. The first was, when a mixed crystal hydrate is in equilibrium with the products of its decomposition, that is, with the liquid solution and a lower hydrate, then 'mixed crystals of the decomposing hydrate are (when compared with the products of decomposition) richer in the salt whose addition raises the temperature of decomposition' [*79*, p. 70]. The second was, 'If all the phases are formed from one or from several other phases without change of composition at some temperature of an equilibrated system, then this temperature will generally be a maximum or a minimum. If the system is monovariant then the geometrical maximum or minimum of the continuous composition–temperature curve corresponds to this equilibrium temperature. But if the system is invariant then the sharp maximum or minimum of the composition–temperature line for the states of absolute stability of the phases corresponds to the equilibrium temperature' [*79*, p. 83].

His second significant work, *A contribution to the thermodynamics of saturated solutions* [*80*] was designed to create a theoretical apparatus for handling the experimental results that followed for van der Waals's equation for the coexisting phases of a binary system (see Chapter 12, Section 2.2). Some of his many conclusions were as follows.

'If the differential change of volume and differential heat of solution of one phase (′) into another (″) are of the same sign, then a minimum of the coexistence pressure corresponds to a maximum in the coexistence temperature, and vice versa; but if the changes of volume and heat are of opposite sign then the minimum (or maximum) in the pressure corresponds to the minimum (or maximum) of the temperature' [*80*, p. 101].

'The concentration at the minimum of a x, T curve is changed by increase of pressure towards lower values of $(\partial T/\partial P)_{x'}$ and $(\partial T/\partial P)_{x''}$. The concentration at the maximum of a x, T curve is changed by increase of pressure towards higher values of the derivatives $(\partial T/\partial P)_{x'}$ and $(\partial T/\partial P)_{x''}$' [*80*, p. 158].

'The differences $(x'' - x')_T$ and $(T'' - T')_x$ increase with increasing difference between the melting points of the components, $T_1 - T_2$, or, in other words, the steeper the slope of the heterogeneous strip between the solidus and liquidus the wider it is' [*80*, p. 143].

Van der Waals's work on surface tension inspired the experimental work on binary mixtures of G. N. Antonov[34], a student of Konovalov. His best-known result, 'Antonov's rule', connected the surface tension of the interface between two immiscible liquids σ_{AB}, and the surface tension of each of the two saturated liquid phases in contact with the vapour, $\sigma_{A/B}$ and $\sigma_{B/A}$. He found that $\sigma_{AB} = \sigma_{A/B} - \sigma_{B/A}$, or 'the surface tension at the boundary of two saturated layers is equal to the difference of the tensions of the separate layers' [*82*]. The implications of the rule have been discussed by Antonov himself [*83*] and by later workers [see, e.g. *84*, *85* and 181, Chapter 8]. He found further that 'in spite of the great differences in density and composition the surface tensions become equal to each other well below the critical [solution] point' [*86*]. He notes that Laplace's treatment failed near the critical point of a single-component system but that van der Waals had found that the surface tension vanished as $(T^c - T)^{3/2}$ so that $(d\sigma/dT)^c = 0$, or σ is essentially zero near the critical point. If this held also for liquid mixtures near the critical solution temperature then his rule would require that $\sigma_{A/B}$ and $\sigma_{B/A}$ were almost the same near this temperature. 'A mixture of critical composition that passes through T^c divides into two layers of equal surface tension but of different compositions and densities' [*81*, p. 129]. It follows that if σ is independent of composition the Gibbs adsorption excess is zero, so that a 'solute must be distributed throughout the bulk of the solution and not concentrated in the surface layer' [*81*, p. 125]. He tried also to construct a theory of surface tension in which he supposed that the molecules interacted as 'doublets', that is, as electric dipoles. His essay on this topic was published first in Russian in 1914 and, after the War, in an expanded version in English [*87*].

The thermochemist Shchukarev also studied solutions near their critical points [*88*] and found that the energies and volumes of these were linear in their composition and that vapour pressure coefficients of thermal expansion and capillary rises were independent of composition, an independence that had been found earlier by Konovalov and Antonov. His explanation was in terms of Gibbs's assertion of the constancy of chemical potentials in the critical region [*71*, p. 131], but this was unfortunately omitted in the abbreviated German version of his paper.

Finally we note some work that is less directly related to that of van der Waals but which belongs in the same tradition. Gustav Tammann[35] worked at the University of Dorpat on the thermodynamic properties of pure liquids and solids. His work is a series of papers from 1896 that culminated in 1903 in his monograph on *Crystallisation and melting* [*89*]. A statement he made later could serve as an epigram for these works; 'Modern thermodynamics concerns itself exclusively and very patiently with the vapour pressure curve; it becomes much more interesting if we extend its applications to crystals' [*90*]. He suggested, wrongly, that melting lines could form a closed loop in the $P - T$ plane[36], a loop that was irreverently known to later workers as 'Tammann's egg'. A more important advance was his discovery of the polymorphism of ice, and his assertion of the impossibility of a solid–liquid critical point. This last idea appeared to be in breach of the line of thought advanced in *Continuity* but, in fact, van der Waals shared this opinion (see Chapter 12, Section 1.3). Tammann, however, rejected the theoretical route to the equation of state of liquids advocated by van der Waals in his Nobel Lecture and preferred to rely solely on experiment. His empirical equation of state for liquids under high pressures, which he derived from the observations of Amagat, was similar to those proposed earlier by Tait and by Biron.

Most of the work of the school of 'physico-chemical analysis' headed by N. Kurnakov (1860–1941) lies outside our field but one branch of it required an application of geometric methods to thermodynamics in a way that derived from the Dutch school of physical chemistry. The first 260 pages of Vittorf's[37] monograph, *The theory of alloys applied to metallic systems* [*91*] are a detailed account of the geometry of equilibrium diagrams of heterogeneous systems derived by Bakhuis Roozeboom's method of thermodynamic potential curves. Vittorf obtained about a hundred different diagrams which he classified into nine classes, only five of which were discussed by Roozeboom himself. This compendium remains to this day as the most complete set of such diagrams and it could be added to only if systems which involve liquid crystalline phases were admitted. There was a plan to publish this book in the Netherlands but this was thwarted by the start of the First World War[38]. His extensive experimental study of the system carbon + iron was published in German [*92*], but it is now only of historical interest. His work was not influenced directly by van der Waals but by the theoretical studies of Bakhuis Roozeboom[39].

A second school of physical chemistry which was also influenced indirectly was that of Paul Walden (1863–1957) at the Polytechnic University of Riga. His colleague M. G. Tsentnerszver (Centnerschwer) (1874–1944) studied the effect of non-volatile solutes on the critical temperatures of non-aqueous solvents such as ammonia and sulphur dioxide [*94*]. He made no use, however, of van der Waals's theoretical work on mixtures.

This survey shows that the problems tackled by van der Waals, in both physics and chemistry, aroused at first keen interest in Russia, and although here, as elsewhere [171,180], the field became something of a backwater between about 1910 and 1960, there was a group of physicists in Moscow – A. I. Batschinskii, K. Putilov, A. Predvoditelev and W. Sementschenko – who retained an interest in this field. This was reflected later in a substantial book [168] much of which was devoted to a discussion of van der Waals's equation and its subsequent developments. The authors commented on Mayer's derivation of the equation of state, more than sixty years after van der Waals, 'how difficult was this second step, and how glorious the first one' [168, p. 60]. This school also produced several works on 'thermodynamic similarity' [*95–97*].

In physical chemistry the dead season was less marked than in physics, both in Russia and elsewhere. Thus a two-dimensional version of van der Waals's equation of state was used by Frumkin in 1924 to derive his adsorption isotherm [*98*]. It was another Russian chemist, A. Titov, who was the first to apply van der Waals's ideas to adsorption phenomena. His important work *On adsorption of gases by coal* [*99*] had been accomplished in Lebedev's laboratory in Moscow University. He found in particular 'a remarkable fact' that the quantities of different gases adsorbed by coal under a given pressure run parallel to the values of a in van der Waals's equation [*99*, p. 674]. Arrhenius, on the basis of these results, put forward the idea that adsorption depends on van der Waals's attractions between molecules [*100*].

The thermodynamics of van der Waals and his theory of mixtures was transmitted to the next generations by A. V. Rakovskii (1879–1941), the Professor of Physical Chemistry at Moscow University. He was the author of three textbooks of thermodynamics and physical chemistry, and was the editor of the Russian version of 'van der Waals–Kohnstamm' [122,123]. In the same city I. R. Krichevskii developed his remarkable school for the study of phase equilibria under high pressures, and of the critical properties of solutions [*101–103*], while in Leningrad A. V. Storonkin and later A. I. Rusanov did important theoretical work on the thermodynamics of heterogeneous equilibria and of interfaces in which the ideas of van der Waals were extended [*104–107*] (see also Chapter 12).

It was symbolic of this strong line of research in Russia in the tradition of van der Waals and the Dutch school that in 1973 the centenary of the publication of *Continuity* should be celebrated not only in the Netherlands but also in the USSR [178]. It is appropriate, therefore, to conclude this chapter with some account of Russian participation in the various anniversaries of van der Waals's career.

The twenty-fifth anniversary (1898) of the publication of *Continuity* (see Chapter 7) was greeted by the following telegram (in French) from the physicist Nicolas A. Umov: 'Amsterdam, to Professor van der Waals. The Imperial Naturalists' Society of Moscow congratulates the author of the famous discovery that both the liquid and gaseous states are subject to the same law' [*108*]. Van der Waals replied by letter on 20 June: 'To Mr President Umoff; I received from the Naturalists' Society of Moscow the congratulatory telegram which touched me deeply. Please accept assurances of my gratitude and my good wishes for the prosperity of your Society. With devotion, yours, J. D. van der Waals' [157(b)]. Another greeting, from the University of Dorpat, was mentioned above (Chapter 7, note 7).

In 1902 friends and colleagues of van der Waals were preparing a collection of articles to mark the twenty-fifth anniversary of his professorship. Nicolas N. Schiller, the Professor of Physics at Kiev, wrote to the Moscow chemist V. A. Kistyakovskii: 'Dear Vl. Al., Let me put to you a request from Prof. Zeeman in Amsterdam. The Dutch are collecting articles and books on molecular physics to present to van der Waals.... They ask, therefore, that Russian scientists should not decline to send articles on this subject, in several languages. The most wanted are those marked in a list sent to me by Zeeman; namely, from your papers, (1) "Contributions to the van 't Hoff–Arrhenius theory of solutions", and (2) "On theories of solutions"[40]. The Russian language is no problem for them. I hope that you will respond to their appeal' [*109*][41].

At the same time Schiller sought to honour van der Waals in another way. He put a proposal to the Council of the Physical and Mathematical Faculty: 'This year the 12/21 October marks twenty-five years of activity of the famous van der Waals in the University of Amsterdam. I beg to think that the Physical and Mathematical Faculty of St Vladimir University[42], acknowledging the scientific merits of the creator of the molecular theory of the change of state of matter, will not turn down this proposal to the Council of St Vladimir University to elect van der Waals an Honorary Member of the University.' The proposal was agreed unanimously, and a copy of the diploma is kept in the University's archives. It reads as follows: 'The Council of the Imperial St Vladimir University at its meeting on 12 September, 1902, elected Van der Waals, Professor at the University of Amsterdam as an Honorary Member on the occasion of the twenty-fifth anniversary of his professorship and in recognition of his scientific merit as the creator of the molecular theory of the change of state of matter' [*111*].

The documents cited are only trivial pieces of evidence of the homage paid by Russian scientists to van der Waals; the real evidence of their indebtedness is better seen in the extent of the scientific work cited in this chapter.

Notes

1. Alexander Grigoryevich Stoletov (А. Г. Столетов, 1839–96), the best-known Russian physicist of the nineteenth century, graduated from Moscow University in 1860, then studied at Berlin, Heidelberg and Göttingen from 1862 to 1866, and lectured in physics in Moscow from 1866, becoming a full professor from 1873, when he founded the modern physical laboratory in Moscow. His collected works were published in three volumes (1939–47), and a scientific biography by M. S. Sominskii was published by Nauka in 1970.
2. Dmitrii Alexandrovich Goldhammer (Д. А. Гольдгаммер, 1860–1922) graduated from Moscow in 1884, worked in Strasbourg until 1888, and was at the University of Kazan from 1890, becoming full Professor of Physics from 1893. In this provincial university he had little opportunity for experimental work and so took up theoretical research. He anticipated Zeeman with an explanation of the magneto-optical effect in 1892.
3. Boris Vlalidimovich Stankevich (Б. В. Станкевич, 1860–1917) graduated from Moscow in 1884, took a master's degree in mathematical physics at Odessa and taught at Warsaw University from 1895 to 1902.
4. Nikolai Nikolaevich Pirogov (Н. Н. Пирогов, 1843–91) was the son of N. I. Pirogov, an eminent surgeon and educationalist, who was educated in physics and mathematics at Heidelberg, Berlin and Oxford. Until the 1880s he was in business and it was only in the last few years of his life that he was able to embark on a thorough study of the statistical foundation of the second law of thermodynamics and of the kinetic theory of imperfect gases. Some of his work was published in German [*9–11*] but it received little notice before his early death by suicide.
5. It is worth noting that Pirogov had propounded and solved what is sometimes called 'Smoluchowski's problem' (1904), namely the calculation of the fluctuation in the number of molecules in a sub-volume.
6. Konstantin Dmitrievich Kraevich (К. Д. Краевич, 1833–91) graduated from the Teachers' Training Institute at St Petersburg in 1855, the same year as Mendeleev. He taught in secondary schools and colleges but was later Professor of Physics at the Mining Institute and then, from 1883, at the Naval Academy in St Petersburg. He wrote a standard school textbook of physics and edited a popular magazine, but found time from his educational work to make measurements of the speed of sound in gases and to study deviations from ideal behaviour. He also published a few papers on the equation of state and related subjects.
7. Count Boris Borisovich Golitsin (Б. Б. Голицын, 1862–1916) as educated at the Naval Academy, where he studied under Kraevich but left the Navy to study physics in Kundt's laboratory at Strasbourg from 1887 to 1890, where he submitted his doctoral thesis *On Dalton's law* [*13*]. He taught physics in the Universities of Moscow (1891–93) and Dorpat (now Tartu) before returning to St Petersburg where he worked both in the Academy of Sciences and in the Naval Academy. From 1902 his work lay in seismology but before that he had contributed interesting papers on thermodynamics and molecular physics, including a review that gave publicity to van der Waals's theory [*14*]. His work was not entirely accurate and there was a sad occasion when Stoletov criticized Golitsin's thesis in 1893 and prevented its acceptance. A selection of his papers was published in two volumes in Moscow in 1960.

8. Nikolai Petrovich Kasterin (Н. П. Кастерин, 1869–1947) a student of Stoletov's, graduated at Moscow in 1892. His first work (1892–93) was on capillary phenomena and attracted the attention of Kamerlingh Onnes. He worked briefly in Berlin and at Leiden (1897–99) before returning to Moscow as a *Privatdocent* in theoretical physics. He received his doctorate for his main work, *On the propagation of waves in heterogeneous media*. He was Professor of Physics at Odessa from 1906 to 1922 when he returned to Moscow to work in various research laboratories. His exchange of letters with Kamerlingh Onnes [233] has been useful in preparing this biography of van der Waals. See also B. Yavelov and V. Volkov, *Janus*, 1990, v. 73, p. 39–52.
9. Nikolai Nikolaevich Schiller (Н. Н. Шиллер, 1848–1910) graduated from Moscow in 1868, worked first under Stoletov, then with Helmholtz in Berlin from 1871 to 1874, returned briefly to Moscow, and was Professor of Theoretical Physics at Kiev from 1876 to 1903. His work on the foundations of thermodynamics anticipated that of Carathéodory and for his work on phase equilibria he was elected a foreign member of the Hollandsche Maatschapij van Wetenschappen. He contributed to the commemorative volumes of the *Archives Néerlandaises* devoted to Lorentz (v. 5) and Bosscha (v. 6). For his biography, see *ЖРФО*, 1911, v. 43, p. 445–483.
10. Dmitrii Petrovich Konovalov (Д. П. Коновалов, 1856–1929) graduated from the Mining Institute in St Petersburg in 1878, and was a lecturer at the University where he also attended the lecture courses of Butlerov and Mendeleev. He then worked for a year with Kundt at Strasbourg, before returning to St Petersburg where he became Professor at the University in 1886. He did much to develop the field of physical chemistry in Russia. In 1907 he retired from the University after the student unrest, becoming first a Deputy Minister of Industry and Trade (1908–15) and, after the Revolution, the head of the Board of Weights and Measures. His principal scientific work, *On the vapour tension of solutions*, was published in 1884 and was reprinted in 1909 and 1928. A full biography by Yu. I. Soloviev and A. Ya. Kipnis was published by Nauka in 1964.
11. It is curious that this point of view was held mainly by the chemists in Russia, while the physicists, led by Nadezhdin, adhered to the concept of two kinds of molecules. In western Europe the situation was reversed, the latter view being held principally by the chemists (L. Meyer, A. Naumann) and by physicists who worked on chemical problems (O. Lehmann, P. de Heen, R. Pictet). J. M. H. Levelt Sengers has discussed in detail the concept of *liquidons* and *gasons*, particularly in its application to critical phenomena [*25*].
12. Leonid Grigoryevich Bogaevskii (Л. Г. Богаевский, 1858–1911) graduated from the Technological Institute at St Petersburg in 1885 and remained there for the rest of his life, lecturing on chemical technology and, at the same time, managing chemical factories; he was a man of many talents. Most of his work was in applied chemistry but, having encountered problems of evaporation and drying, he turned to the foundations of these subjects and wrote several papers two of which are discussed in the text [*26,27*].
13. Mikhail Petrovich Avenarius (М. П. Авенариус, 1835–95) graduated from the University of St Petersburg in 1858 and then studied physics at Berlin, under Magnus, and at Heidelberg. From 1866 to 1890 he was Professor of Physics at Kiev. He was one of the first in Russia to organize experimental research in physics and his skill, his unselfishness and his patient devotion to science made him highly esteemed by his students and co-workers. The best biography is that by Stoletov in his *Collected Works*, v. 2; see also V. P. Russakov, *ТИИЕТ*, 1955 v. 5, p. 180–215.

14. Vladimir Ivanovich Sayonchevskii (В. И. Зайончевский, 1842–1908), was referred to by van der Waals as Sayotchewsky because of a misprint in the abstract of his paper [*34*]. He graduated from Kiev in 1864 and then worked in the University as the keeper of the physical laboratory apparatus, and was also a teacher in a girls' gymnasium. His paper on *The determination of vapour tension at high temperatures* [*34*] was described as a classic by Avenarius. In 1878 he left Kiev and for the next ten years taught physics and meteorology in the Agriculture and Forestry Institute at New Alexandria near Warsaw.
15. Kassian Nikolaevich Zhuk (К. Н. Жук, 1844–1917) graduated from Kiev in 1870 and worked as a school teacher until he succeeded Sayonchevskii in the Physics Laboratory in 1878. He stayed there until the retirement of Avenarius in 1890 after which he taught and practised meteorology at the Polytechnic Institute in Kiev.
16. Oscar Emilievich Strauss (О. Э. Страус, 1856–?) was the son of a Kiev physician who graduated from the University of St Petersburg in 1880. In the summer of that year he carried out his principal experimental work in Avenarius's laboratory, *On the critical temperatures of some mixtures* [*37*]. After his return to St Petersburg he completed his measurements on the system water + alcohol and published his second important paper, *On the critical temperature and pressure of water* [*38*]. He returned to Kiev in 1896 and became responsible for the electrical section of the journal *The Engineer*. The publication of this was suspended in 1917, after which nothing further is known of Strauss's activities.
17. The same expression was obtained later by Pavlevski in Lvov [*39*].
18. Alexander Ivanovich Nadezhdin (А. И. Надеждин, 1858–86) was the son of a doctor who entered the University of Kiev in 1877 and, three years later, was awarded a gold medal for his entry in a competition set by Avenarius, *Changes in the properties of bodies near the so-called temperature of absolute boiling* (Mendeleev's term for the critical temperature). He graduated in 1882 and submitted an exceptionally able thesis, *Studies in comparative physics* in 1886. This, however, was his swan-song – he was fanatically devoted to science but worked too hard, in poor conditions and with harmful substances; his health was destroyed and he died just before his twenty-eighth birthday. His thesis and an earlier paper [*40*] were published posthumously as *Physical investigations* [*41*]. The second and third parts of his thesis were abridged by Avenarius and published in German [*42,43*]. For an account of his life and work, see Kipnis [*44*]
19. There is an interesting letter [*48*] from P. N. Lebedev to Golitsin after the latter's paper at the International Physical Congress at Paris in 1900 [*49*]. Lebedev's idea was that in the critical region 'we observe the surface of a liquid in detail for the first time. We have here a phenomenon that has been entirely ignored up to now (although Clausius and van der Waals made some references to it); we are on the verge of creating a *new science* of the boundary layer, which can clearly be observed by eye near θ' (i.e. the critical temperature).
20. Unfortunately the journal in which the German version of this paper was published survived for only two volumes in 1892, so Stoletov's work passed almost unnoticed in western Europe. It was, however, mentioned by Kuenen [*51*].
21. The relations between Stoletov and Kamerlingh Onnes have been discussed in a paper [*54*] in which it is shown that Stoletov visited the Leiden laboratory in 1895.
22. The first comparison of this kind in the natural sciences was probably Geoffroy's table of affinities in 1718.
23. Alexei Iosifovich Bachinskii (А. И. Бачинский, 1877–1944) graduated from the University of Moscow in 1899 and worked there on molecular physics and thermodynamics, becoming Professor in 1918. In 1930 he lost his sight and gave up

lecturing. His main achievement, his law of viscosity, was established in 1912. A selection of his works with a full bibliography and a biographical sketch was published in Moscow in 1960.

24. Alexander Andreevich Brandt (A. A. Брандт, 1855–1932) graduated from the Institution of Transport Engineers in St Petersburg in 1878 and worked there as a Professor in heating engineering and thermodynamics until 1918, eventually becoming the Director. He emigrated to Belgrade in 1918 where he worked in the 'Russian Scientific Institution'. He wrote a history of steam-engines in Russia (1892), a remarkable bibliography of thermodynamics and a textbook on *The foundations of thermodynamics* of which the third edition was published in two volumes in 1915–16. In this he discussed, among other matters, van der Waals's paper on quasi-association. He left to the Institute in St Petersburg a valuable collection of books and reprints on thermodynamics, including many papers of the Dutch physicists.
25. Vladimir Alexandrovich Kistyakovskii (В. А. Кистяковский, 1865–1952) graduated from the University of St Petersburg in 1889 and then studied physical chemistry under Ostwald at Leipzig. He returned briefly to the University before becoming Professor of Physical Chemistry at the Polytechnic Institute in St Petersburg. He worked mainly in electrochemistry and was one of the earliest champions of Arrhenius in Russia (see Chapter 6, Section 2). He developed several useful correlations for latent heats, boiling points and capillary constants.
26. Alexander Nikolaevich Shchukarev (А. Н. Щукарев, 1864–1936) graduated from the University of Moscow in 1889, was simultaneously a school-teacher and an assistant in Louginin's 'Thermal Laboratory' until, in 1911, he became a Professor at the Technological Institute in Kharkov. The work discussed in the text [*64*] was his Moscow thesis of 1905.
27. For Paul Walden (1863–1957) see the biography by Y. P. Stradin and Yu. I. Soloviev, published by Nauka in Moscow, 1988.
28. The mutual solubility of liquids was a traditional part of Russian chemistry (see [*67*, p. 227f]). Dmitrii Nikolaevich Abashev (Д. Н. Абашев, 1829–80) began to study the subject as early as 1857. He had graduated from the University of St Petersburg in 1851 and worked on liquids while he was a teacher in a gymnasium in Moscow (1854–58); he submitted his thesis in Moscow on mutual solubility of liquids in 1858. He injured his eye in a laboratory accident and this injury later undermined his health. He moved first to Kiev and then to Odessa in 1865 becoming Professor of Agricultural Chemistry there in 1868. His illness prevented any further research after about 1873. He studied forty-five pairs of ten different liquids, eight pairs of which showed limited solubility. He deduced that 'all liquids which dissolve each other to a certain degree at ambient temperatures will mix in all proportions at certain temperatures.' By 1873 he had tested 130 pairs of liquids in the range of temperature from $-30°C$ to $+130°C$. In eleven cases he found a transition from limited to unlimited solubility; one of these was the classic system phenol + water. He wrote, 'Plotting changes of composition of both solutions against temperature gives two curves that coincide at the temperature at which complete mixing in all proportions becomes possible. . . . It is probable that both curves of the compositions of the solutions are branches of a single curve.' His results were, however, only qualitative or, at best, semi-quantitative. Abashev's research was followed by the better known work of Vladimir Fedorovich Alekseev (В. Ф. Алексеев, 1852–1919) who graduated from the Mining Institute in St Petersburg in 1873 and worked there as a laboratory assistant until 1877 when he started his work on the mutual solubility of liquids. Like Abashev, he damaged

an eye in a laboratory accident and this injury eventually undermined his health. From 1877 he worked at the 'Coin Court' in order to earn more money to support his family. His thesis, *On the mutual solubility of liquids*, was presented at the Mining Institute in 1879. He rose to become a professor there but retired in 1899 and went to live in the country. His work on liquids came to an end in 1886 after an argument with Mendeleev and Konovalov on the nature of solutions. He wrote numerous reviews in the *Mining Journal*, one of which (1884) discussed Gibbs's principle of equilibrium. His work on mutual solubility was his major contribution to physical chemistry. He introduced the 'synthetic method' for measuring solubility and with it achieved a high level of accuracy. He noted in 1885 that 'the equation of the diameters is close to rectilinear', although he expressed it as a cubic curve. This is one year before Cailletet and Mathias established the rule of rectilinear diameters for liquid–vapour systems. In 1888 Rothmund extended this rule to liquid–liquid systems but noted that Alekseev had first drawn attention to this behaviour. He published a summary of his results in a German journal in 1886, and it is this article that is quoted by van der Waals and other western authors.

29. Such an approach stemmed from Gibbs's principle of equilibrium, expressed by him in the words, 'Hence, the conditions . . . same pressure' [*71*, p. 104–05]. Guldberg had been the first to suggest this idea in 1870; Konovalov discovered it not from the fundamental principles of thermodynamics but by generalizing experimental data. He applied it to problems of phase equilibria, osmotic pressure, the solubility of liquids and to critical phenomena in solutions [*67*, p. 188,209,229,277].
30. It has been noted [183, p. 26] that Guthrie [*72*] also used the term 'critical temperature', and, in particular, the 'critical temperature between clearness and turbidity' several months later. His aim was to discover hydrates in solution, not to study phase equilibria, and his use of the term seems to be almost casual. In 1889 the 'obvious analogy' between the turbidities in liquid–vapour and liquid–liquid systems was pointed out by Lehmann [*73*, p. 208] but the parallel only became well known to chemists after the publication of Masson's paper in 1891 [*74*]. Gibbs's discussion of the critical state in 1876 [*71*, p. 129] was generally unknown to chemists although he himself certainly realized that the separation of liquid mixtures into two layers had all the characteristics of such a state [*71*, p. 134, footnote] and he cites the observations of Duclaux published the previous year [*75*].
31. Evgenii Vladislavovich Biron (E. B. Бирон, 1874–1919) a pupil of Konovalov, graduated from the University of St Petersburg in 1897 and then taught physical chemistry there until 1910 when he became Professor of Chemistry in the Forestry Institute at St Petersburg. His main work was carried out between 1905 and 1913. It comprised experimental and theoretical studies of concentrated solutions and was marked by great care and meticulous attention to detail; even his papers were written out cleanly without corrections. In 1917 he was poisoned while investigating a German gas shell, and died two years later having been able to accomplish nothing more. His work was never published outside Russia and so remains almost unknown in western Europe.
32. Biron's interest in the problem of internal pressure can be seen from his notes in the margin of his copy of Kuenen's *Die Zustandsgleichung der Gase und Flüssigkeiten und die Kontinaitätstheorie*, which is now in the Library of the Mendeleev Chemical Society, St Petersburg. At the end of the section on molecular attraction he notes that: 'From these considerations (p. 172–74) it is seen that the smaller is v the more exact is $K = a/v^2$. This explains why van der Waals's equation is better

for a liquid than for a gas, and why the equation $v = A = B/(C + p)$ is better for measurements at higher p and lower t, that is, at smaller v.'

33. Reinhardt Friedrich Hollman (Р. Ф. Холлман, 1877–1921) was born at Dorpat (later Yuriev, now Tartu, in Estonia), and studied at the University there under Tammann, receiving a gold medal for his entry into a competition concerning the vapour pressure of mixed crystalline hydrates. He graduated in 1901, became a *Privatdocent* in 1903 and Assistant Director of the Chemical Laboratory in 1908, after he had presented his thesis at the University of Moscow [*79*]. He worked also in Bakhuis Roozeboom's laboratory in Amsterdam (1902) and in Bern, Heidelberg and Göttingen (1904–5). In his last book [*80*] he recorded van Bemmelen's recollection of meeting van der Waals in Leiden in 1886, mentioned in Chapter 8. From 1912–18 he was Professor of Chemistry in the University of Saratov and from 1918 at Yuriev (now Tartu). After one year he left for Mainz where he became head of the Scientific Laboratory of the Chemical Industries Association. He worked mainly on the thermodynamics of phase equilibria in binary and ternary systems. The contents of his thesis were published as a series of papers in the *Zeitschrift für physikalische Chemie* in 1901–6 (Bd 37, S. 193 Bd 40, S. 561,577 Bd 42, S. 597; Bd 50, S. 567 Bd 54, S. 98) but what is perhaps his most interesting work [*80*] is almost unknown.
34. Georgii Nikolaevich Antonov (Г. Н. Антонов, 1880–1959) a student of Konovalov, graduated from the University of St Petersburg in 1902, and worked there until 1908; he became an assistant also in the chemical laboratory of the Academy of Sciences from 1906, where he nominally remained until 1915. From 1908 to 1912, however, he was in England on business and worked during this time with Rutherford at Manchester. His main achievement here was the discovery of uranium-Y and this was the subject of his thesis at St Petersburg in 1914. When the War began he became involved in defence work and, in 1916, was sent to England by the Artillery Board. He never returned to Russia but taught and carried out research in England (until 1927) and then in France, the United States, and in Pakistan. The work on capillarity, for which he is best remembered, was done in St Petersburg and published there in a monograph in 1914 [*81*].
35. Gustav A. Tammann (Г. А. Тамман, 1861–1938) studied and later worked at the University of Dorpat until 1903 when he moved to Göttingen. This period of his life and his work is described by N. Ya. Ryago (Н. Я. Ряго), *ТИИЕТ*, 1956, v. 12, p. 117–20.
36. The following remark is worth quoting in this context; 'Pure thermodynamics cannot tell us if there are state diagrams of this kind; it is, indeed, so independent of the real world that it cannot tell us if state diagrams even exist' [*90*, p. 83–84]. This expression and those cited in the text are more definite and distinctly expressed than those in the earlier and more extended versions that appeared in the *Göttingen Nachrichten* for 1911.
37. Nikolai Mikhailovich Vittorf (Н. М. Витторф, 1869–1929) graduated from the Artillery Academy in 1895, and started work as an organic chemist under A. Favorskii and V. Ipatiev, who were both at the Academy. He taught in the Academy from 1900 to 1924, becoming a Professor in 1912. In 1903–4 he studied physical chemistry abroad, mainly in Tammann's laboratory at Göttingen where he met the 'never-to-be-forgotten Bakhuis Roozeboom', as he described him in the preface of his main work [*91*].
38. Some of Vittorf's manuscripts passed, by inheritance, to his daughter and thence to one of the authors of this book; it was the daughter who told us of the project mentioned in the text. Amongst the manuscripts are detailed extracts from the

papers of A. C. van Rijn van Alkemade and Bakhuis Roozeboom with thoughtful comments; their influence on Vittorf's work is clear.

39. A characteristic example is the interesting book *On solid solutions* (1904) by A. Speranskii who devoted a separate chapter to Bakhuis Roozeboom's work [*93*, p. 47–71], in which he sets out the detailed derivation of temperature–composition diagrams from thermodynamic potential–composition curves that Roozeboom had given in 1899. This was the first account of Roozeboom's work in Russia and, in his foreword, Speranskii wrote, 'I attach great importance to this chapter; in my view both the ideas and the experimental research of this Dutch school of chemists deserves more attention than is paid to them by the leaders of Russian laboratories' [*93*, p. iv].
40. These were an abstract of a preliminary communication and a full-length report of the same work, see *ЖРХО* 1897, v. 29, p. 286–87; 1898, v. 30, p. 576–85.
41. Here it is useful to give some extracts from the speech that Kistyakovskii gave to the Russian Chemical Society on 3 October 1902, to mark the twenty-fifth anniversary of van der Waals's professorship. 'His ideas on the continuity of the gaseous and liquid states underlie the whole modern doctrine of molecular weight. . . . Without "van der Waalsian continuity" we should have no theoretical basis for determining the molecular weights of liquids, and, moreover, the brilliant methods of Raoult and van 't Hoff would not have the theoretical basis that we value so highly were it not for the determination by van der Waals himself, from the continuity of the liquid and gaseous states, of the connection between the partial tension of a gas and the osmotic pressure" [*110*]. Kistyakovskii stayed, however, wholly within the German tradition of physical chemistry and praised van der Waals only in a very general way.
42. Before 1918 the University of Kiev was called St Vladimir's University, after the Grand Duke of ancient Kiev who had introducted Christianity into Russia.

References

1. Stoletov, A. G. (А. Г. Столетов), An outline development of our knowledge of gases (in Russian). In *Collected works*, v. 2, Moscow-Leningrad, 1941, p. 53–185.

2. Stoletov, A. G., Notes on critical state of bodies (in Russian), *ЖРФО*, 1882, v. 14, N5, p. 167–175; c, *J. Phys.* (*Paris*), 1882, t. 1, p. 543–548; ac, *Beibl.*, 1882, Bd 6, S. 775–777.

3. Goldhammer, D. A., (Д. А. Гольдгаммер), To the memory of Professor A. G. Stoletov (in Russian), *Izvestiya fysiko-matematicheskogo Obshchestva pri Kazanskom Universitete*, 1897, v. 7, No. 1, p. 39–54.

4. Goldhammer, D. A., The thermodynamic surface of water according to Gibbs and van der Waals (in Russian), *Uchyonie Zapiski Imperatorskogo Moskovskogo Universiteta, otdel fysiko-matematicheskii*, Issue No. 6, Moscow, 1884; ac, *Beibl.*, 1886, Bd 10, S. 333–338.

5. Stankevich, B. V. (Б. В. Станкевич), Kinetic theory of gases, mathematically presented (in Russian). *Kineticheskaya teoriya gazov v matematicheskom izlozhenii*, Moscow, 1884, 229 p.

6. Stankevich, B. V., Kinetic theory of the state of bodies and its topical problems (in Russian), *В.У.И.*, 1887, No. 3, p. 1–16.

7. Stankevich, B. V., Essays on the kinetic theory of the structure of bodies (in Russian), *В.У.И.*, 1887, N8, p. 1–32; N9, p. 33–48; 1888, No. 1, p. 49–80. The same in *Zapiski Matematicheskogo otdeleniya Novorossiiskogo Obshchestva Estestvoispitatelei*, Odessa, 1888, v. 8, p. 1–84.

8. Stankevich, B .V., A contribution to the mechanical theory of heat (in Russian), *В.У.И.*, 1887, No. 7, p. 1–16.
Stankiewitch, B. W. Zur mechanischen Wärmetheorie, *Z. Math. Phys.*, 1889, Bd 34, S. 111–116.
9. Pirogov, N. N. (Н. Н. Пирогов), On the statistical foundations of thermodynamics (in Russian), *ЖРФО*, 1886, v. 18, p. 307–326; 1887, v. 19, p. 100–120, 157–176; 1890, v. 22, p. 44–89, 173–220; ac_1, *F. Ph.*, 1886, Abth. 2, S. 237–251; ac_2, *Exner's Repert. Phys.*, 1891, Bd 27, S. 515–546.
10. Pirogov, N. N., On imperfect gases (in Russian), *ЖРФО*, 1889, v. 21, No. 4, p. 44–57; c, *F. Ph.*, 1889, Abth. 2, S. 207–209.
11. Pirogov, N. N. On the virial of forces (in Russian), *ЖРФО*, 1888, v. 20, No. 1, p. 1–19; 1889, v. 21, N9, p. 219–1248; 1891, v. 23, No. 4, p. 127–151; Ueber das Virial der Kräfte, *Z. Math. Phys.*, 1892, Bd 37, S. 257–290.
12. Kraevich, K. D. (К. Д. Краевич), On the dependence of the heat of evaporation on other observable quantities (in Russian), *ЖРФО*, 1889, v. 21, No. 6, p. 137–153; *Exner's Repert. Phys.*, 1890, Bd 26, S. 581–603.
13. Galitzine, B., Ueber das Dalton'sche Gesetz, Strassburg, 1890; or, *Nachr. Kõnigl. Ges. Wiss. Gött.*, 1890, No. 1, S. 22-29; *A. Ph.*, 1890, Bd 41, S. 588-626, 770–800.
14. Golitsin, B. B. (Б. Б. Голицын), On the gaseous and liquid state of matter (in Russian), *Vestnik opitnoi fisiki i elementarnoi matematiki*, 1889, VI semestr, p. 85–97, 133–144, 177–186, 221–231; VII semestr, p. 21–31, 62–72, 141–156; 1890, VIII semestr, p. 21–28, 41–51.
15. Galitzine, B., Über die Wirkungsweise der Molekularkräfte, *Z. ph. Ch.*, 1889, Bd 4, S. 417–426.
16. Kasterin, N. P. (Н. Р. Кастерин), On the change of the cohesion of liquids with the temperature (in Russian), *ЖРФО*, 1893, v. 25, No. 1, p. 51–72; c, *Beibl.*, 1893, Bd 17, S. 892,893.
17. Kraevich, K., A note on the van der Waals formula (in Russian), *ЖРФО*, 1887, v. 19, No. 1, p. 1–6.
18. Schiller, N., Ueber eine mögliche aus den Joule–Thomson'schen Abkühlungsversuchen herzuleitende Form der Zustandsgleichung für Gase, *A. Ph.*, 1890, Bd 40, S. 149–156; *ЖРФО*, 1890, v. 22, p. 110–115.
19. Sonin, N.Ya. (Н. Я. Сонин), On the so-called physical law of van der Waals (in Russian), *Protokoli zasedaniy otdeleniya fisiki i chimii Varshavskogo Obshchestva Estestvoispitatelei*, 1889, No. 5, p. 9–11; N6, p. 1–8.
20. Sonin, N. Ya. On the application of the virial theorem to the kinetic theory of gases (in Russian), *Protokoli zasedaniy otdeleniya fisiki i chimii Varshavskogo Obshchestva Estestvoispitatelei*, No. 7, p. 1–8; c, *F. Ph.*, 1890, Abth. 2, S. 247–248.
21. Lorentz, H. A., Über die Anwendung der Satzes vom Virial in der kinetischen Theorie der Gase, *A. Ph.*, 1881, Bd 12, S. 127–136.
22. Mendeleeff, D., *The principles of chemistry*, translated from the Russian 5th edn, London, v. 1, 1891.
23. Konowalow, D., Einige Bemerkungen zur Theorie der Flüssigkeiten, *ЖРХО*, 1886, v. 18, p. 395–402; c, *Beibl.*, 1887, Bd 11, S. 420–422; *Z. ph. Ch.*, 1887, Bd 1, S. 39–45.
24. Mendeleev, D. (Д. Менделеев), On the expansion of liquids (in Russian), *ЖРХО*, 1884, v. 16, p. 1–10; *J.C.S.*, 1884, v. 45, p. 126–135.
25. Levelt Sengers, J. M. H., Liquidons and gasons; controversies about the continuity of states, *Physica*, 1979, v. 98A, p. 363–402.

26. Bogaevskii, L. (Л. Богаевский), On the continuity of gaseous and liquid states (in Russian), *Izvestiya Sanct-Peterburgskogo Technologicheskogo Institute*, 1896, v. 11, p. 15–110. The same, as a separate book, *O neprerivnosti gasoobraznogo i zhidkogo sostoyanii*, St Petersburg, 1896, 96 p.

27. Bogaevskii, L., On the various states of matter (in Russian), *Zapiski Imp. Academii Nauk, St Petersburg*, 1897, v. 5, No. 13, p. 1–103.

28. Mendeleev, D., The molecular cohesion of some organic compounds (in Russian), *Gorny Journal* (*Mining Magazine*), 1860, Part 1, p. 365–381; *Works*, v. 5, Moscow-Leningrad, 1947, p. 8–17.

29. Mendelejeff, D., Ueber die Ausdehnung der Flüssigkeiten beim Erwärmen über ihren Siedepunkt, *Ann. Chem. Pharm.*, 1861, Bd 119, Hft. 1, S. 1–11.

30. Andrews, T., On the continuity of the gaseous and liquid states of matter. *Philos. Trans. R. Soc. London*, 1869, **159**, 575–90.

31. Mendelejeff, D., Bemerkungen zu den Untersuchungen von Andrews über die Compressibität der Kohlensäure, *A. Ph.*, 1870, Bd 141, S. 618–626.

32. Avenarius, M., Ueber innere latente Wärme, *Bull. Soc. Nat. Moscou*, 1873, t. 47, No. 3, p. 117–132; *A. Ph.*, 1874, Bd 151, S. 303–316.

33. Avenarius, M., Ueber die Ursachen welche die kritische Temperatur bedingen, *Bull. Imp. Acad. Sci. St Petersbourg*, 1876, t. 22, No. 3, p. 378–389.

34. Sayonchevskii, V. (В. Зайончевский), The determination of vapour tension for some liquids at high temperatures (in Russian), *К.У.И.*, 1878, No. 4, p. 21–49; No. 8, p. 29–39; c, *Beibl.*, 1879, Bd 3, S. 741–743.

35. Sayonchevskii, V., A few words in connection with Professor Stoletov's 'Notes on the critical state of bodies' (in Russian), *ЖРФО*, 1882, v. 14, No. 7, p. 386–388; c, *Beibl.*, 1883, Bd 7, S. 585.

36. Zhuk, K. N. (К. Н. Жук), The volume of a liquid as a function of the temperature at a constant pressure (in Russian), *ЖРФО*, 1881, v. 13, No. 5, p. 239–241; No. 9, p. 411–413; 1884, v. 16, No. 6, p. 304–306.

37. Strauss, O. (О. Страус), On the critical temperatures of some mixtures (in Russian), *ЖРФО*, 1880, v. 12, No. 8, p. 207–218; c, *Beibl.*, 1882, Bd 6, S. 282.

38. Strauss, O., On the critical temperature and the critical pressure of water (in Russian), *ЖРФО*, 1881, v. 13, p. 270; 1882; v. 14, p. 510–517; c, *Beibl.*, 1883, Bd 7, S. 676.

39. Pawlewski, B., Über die kritischen Temperaturen flüssiger Körper, *Ber.*, 1882, Bd 15, S. 460–462.

40. Nadezhdin, A. (А. Надеждин), The determination of the critical temperatures of nitrous acid, bromine, iodine and water (in Russian), *К.У.И.*, 1885, No. 6, p. 21–34; *Bull. Acad. Imp. Sci. St. Petersbourg*, 1885, t. 30, p. 327–330; c, *Beibl.*, 1885, Bd 9, S. 721–723; *Ch. Cb.*, 1886, S. 401–402.

41. Nadezhdin, A. I., *Physical Studies (Fisicheskikhie issledovaniya)*, Kiev, 1887, 172 p.

42. Nadejdine, A., Ueber die Ausdehnung der Flüssigkeiten und der Uebergang der Körper aus dem flüssigen in der gasförmigen Zustand, *Exner's Repert. Phys.*, 1887, Bd 23, S. 617–649, 685–718.

43. Nadejdine, A., Ueber die Spannkraft der gesättigten Dämpfe, *Exner's Repert. Phys.*, 1887, Bd 23, S. 759–790.

44. Kipnis, A. Ya. (А. Я. Кипнис), From the history of molecular physics: A. I. Nadezhdin (in Russian). In *Issledovaniya po istorii fisiki i mekhaniki*, Moscow, 1990, p. 5–36.

45. Hellborn, E., Zusammenstellung der bisher ermittelten kritischen Daten der Flüssigkeiten, *Z. ph. Ch.*, 1891, Bd 7, S. 601–613.

46. Landolt-Börnstein, *Zahlenwerte und Funktionen aus Naturwissenschaften und Technik*, 2 Bd, 1.Teil., Berlin, 1971 (Tab. 211161).
47. Golitsin, B. B., On critical temperatures (in Russian), *ЖРФО*, 1890, v. 22, p. 259,265–268; ac, *Beibl.*, 1891, Bd 15, S. 639.
48. Kipnis, A. Ya., P. N. Lebedev's letter on critical phenomena (in Russian), *Voprosi istorii estestvoznaniya i techniki*, 1962, Issue 13, p. 88–91.
49. Galitzine, B. and Wilip, J., L'indice critique, *Rapports présentés au Congrès Internationale de Physique*, Paris, 1900, t. 1, p. 668–681.
50. Stoletov, A. G., On the critical state of bodies, II (in Russian), *Trudi otdeleniya fisicheskikh nauk Imp. Obshchestva lyubitelei estestvoznaniya*, 1892, v. 5, issue 1, p. 1–15; *Collected Works*, v. 1, Moscow, 1939, p.276–306.
Stoletov, A. G., Ueber den kritischen Zustand der Körper, *Physik. Revue*, 1892, Bd 2, S. 44–78; *F. Ph.*, 1892 (Abt. 2), S. 190–192; c, *Beibl.*, 1894, Bd 18, S. 1006–1007.
51. Kuenen, J. P., On the anomalous phenomena near the critical point, *Leiden Commun.*, 1893, No. 8, p. 9–16.
52. Stoletov, A., On the critical state of bodies, III (in Russian), *ЖРФО*, 1893, v. 25, No. 9, p. 303–326; *Collected Works*, v. 1, p. 307–332.
53. Stoletov, A., On the critical state of bodies, IV (in Russian), *ЖРФО*, 1894, v. 26, No. 1, p. 26–30; *Collected Works*, v. 1, p. 333–338.
54. Yavelow, B. E. and Volkow, V. A., A contribution to the history of scientific relations between the Netherlands and Russia. H. Kamerlingh Onnes' correspondence with Russian physicists. I. H. Kamerlingh Onnes' letters to A. G. Stoletow, *Janus*, 1984, v. 71, p. 135–151.
55. Nadezhdin, A., Ueber den kritischen Punkt, *ЖРФО*, 1882, v. 14, p. 157–161, 536–541; 1883, v. 15, p. 25–29; c, *Beibl.*, 1883, Bd 7, S. 678–681.
56. Batschinski, A., Ueber eine Erweiterung des Begriffes der kritischen Grössen, *Z. ph. Ch.*, 1902, Bd 40, S. 629–634.
57. Batschinski, A., Über die Polymerisation der orthomeren Flüssigkeiten, insbesondere der Essigsäure, *Bull. Soc. Nat. Moscou*, 1903, t. 17, No. 2–3, p. 188–199.
58. Brandt, A. A. (А. А. Брандт), On a relation between Stefan's formula for the internal pressure of liquids and van der Waals's equation. On a relation between Trouton's formula and van der Waals's equation (in Russian), *ЖРФО*, 1903, v. 35, No. 3, p. 409–415, 417–420, 421 (Résumé).
59. Batschinski, A., Bemerkungen über das Gesetz der geraden Mittellinie, *Z. ph. Ch.*, 1902, Bd 41, S. 741–743.
60. Batschinski, A., Beziehungen für die thermischen Eigenschaften der Stoffe, *A. Ph.*, 1904, Bd 14, S. 288–308.
61. Batschinski, A., Abhandlungen über Zustandsgleichung, I, II, III, *A. Ph.*, 1906, Bd 19, S. 307–309, 310–332; Bd 21, S. 1001–1012.
62. Kistyakovskii, V. (В. Кистяковский), Relations between coefficients determining the properties of liquids at their boiling point (in Russian), *Izvestiya Sankt-Peterburgskogo Polytechnicheskogo Institute*, 1903, v. 1, Issue 3–4, p. 425–453.
63. Kistjakowsky, W., Eine der Regel von Trouton für die latente Verdampfungswärme analoge Regel für die kapillaren Erscheinungen, *Z. Elektrochem.*, 1906, Bd 12, S. 513–514.
64. Shchukarev, A. N. (А. Н. Щукарев), Studies on the internal energy of gaseous-liquid systems (in Russian), *ЖРФО*, 1904, v. 36, p. 281–365. The same as a separate book, *Issledovaniya vnutrennei energii gazoobraznozhidkikh sistem*, Moscow, 1905, 85 p.

Schükarew, A., Untersuchungen über den Zustand gasförming-flüssig, *Z. ph. Ch.*, 1903, Bd 44, S. 548–562; 1906, Bd 557 S. 99–112.

65. Walden, P., (a), Ausdehnungsmodulus, spezifische Kohäsion, Oberflächenspannung und Molekulargrösse der Lösungsmittel; (b) Über den Zusammenhang der Kapillaritätskonstanten mit der latenten Verdampfungswärme der Lösungsmittel; (c) Ueber den Zusammenhang der Oberflächenspannung mit dem Binnendruck und der van der Waalsschen Grössen *a* und *b*, *Z. ph. Ch.*, 1909, Bd 65, S. 129–225, 257–288, Bd 66, S. 385–444.

66. Goldhammer, D.A., Studien über die Theorie der übereinstimmende Zustände, *Z. ph. Ch.*, 1910, Bd 71, S. 577–624.

67. Kipnis, A. Ya. *Razvitie chimicheskoi termodinamiki v Rossii (The development of chemical thermodynamics in Russia)*, Moscow-Leningrad, 1964, 348 p.

68. Konowalow, D., Über die Dampfspannungen der Flüssigkeitsgemische, *A. Ph.*, 7 1881, Bd 14, S. 34–52, 219–226.

69. Konowalow, D. Über unzersetzt siedende Lösungen, *Ber.*, 1884, Bd 17, S. 1531–1539.

70. Konovalov, D. (Д. Коновалов), On the vapour tensions of solutions (in Russian), *ЖРХО*, 1884, v. 16, p. 11–84. The same as separate books *Ob uprugosti para rastvorov*, St Petersburg, 1884; 2nd edn St Petersburg, 1909; 3rd edn Leningrad, 1927.

71. Gibbs, J. W., *The Collected Works*, New York, 1928, v. 1.

72. Guthrie, F., (a) On salt solutions and attached water; (b) On some thermal and volume changes attending mixture, *Philos. Mag.*, 1884, v. 18, p. 22–35, 105–120, 495–517.

73. Lehmann, O., *Molekularphysik*, Leipzig, 1889, Bd 2, S. 208.

74. Masson, Orme, Eine Folgerung aus der Theorie gasähnlicher Lösungen, *Z. ph. Ch.*, 1891, Bd 7, S. 500–510.

75. Duclaux, E., Sur la séparation des liquides mélangés, et sur des nouveaux thermomètres à maxima et à minima, *C.R.*, 1875, t. 81, p. 815–817.

76. Konowalow, D., Das kritische Gebiet der Lösungen und die Erscheinungen der Opaleszenz, *ЖРХО*, 1902, v. 34, p. 738–766; *A. Ph.*, 1903, Bd 10, S. 360–392; Über die Trübung kritischer Lösungen, *A. Ph.*, 1904, Bd 12, S. 1160–1164.

77. Biron, E. V. (E. B. Бирон) *Szhatie pri smeshenii normal'nikh zhidkostei* (*Compression by mixing of normal liquids*), St Petersburg, 1912, 184 p.

78. Krichevskii, I. R. (И. P. Кричевский), *Fazovie ravnovesiya v rastvorakh pri visokikh davleniyakh* (*Phase equilibria in solutions at high pressures*), 2nd edn, Moscow-Leningrad, 1952, 168 p.

79. Hollman, R. F. (P. Ф. Холлман) *Ob obrazovanii i rashcheplenii smeshannikh kristallogidratov isomorfnikh veshchestv s tochki zreniya pravila faz Gibbsa* (*On the formation and decomposition of isomorphous substances of mixed crystals from the point of view of Gibbs's phase rule*), Yuriev, 1907, 213 p.

80. Hollman, R. F., *K termodinamike nasishchennikh rastvorov* (*A contribution to the thermodynamics of saturated solutions*), Saratov, 1917, 199 p.

81. Antonov, G. N. (Г. Н. Антонов), *O natyazhenii na granitse dvukh nesmeshivayushchichsya zhidkostei v sostoyanii ravnovesiya. Eksperimental'noe issledovanie (On the tension at the boundary of two immiscible liquids at the equilibrium state. An experimental study)*, St Petersburg, 1914, 129 p.

82. Antonoff, G., Sur la tension superficielle à la limite de deux couches, *ЖРХО*, 1907, t. 39, p. 342–353; *J. chim. phys.*, 1907, t. 5, p.372–385.

83. Antonoff, G., On the validity of Antonoff's rule, *J. Phys. Chem.*, 1942, v. 46, p. 497–499.

84. Adam, N. K., *The physics and chemistry of surfaces*, 2nd edn, Oxford, 1938, p. 214–215.
85. Adamson, A. W., *Physical chemistry of surfaces*, New York, 1976, Chap. 3.
86. Antonoff, G., Sur la tension superficielle des solutions dans la zone critique, *ЖРХО*, 1906, t. 38, p. 1258–1266; *J. chim. phys.*, 1907, t. 5, p. 364–371.
87. Antonoff, G. N., Interfacial tension and complex molecules, *Philos. Mag.*, 1918, v. 36, p. 377–396.
88. Shchukarev, A. N., *Svoistva rastvorov pri kriticheskoi temperature smesheniya* (*Solution properties at the critical temperature of mixing*), Moscow, 1908. The same in *Uchyonie zapiski Moskovskogo universiteta, otd. estestv.-ist.*, 1909, issue XXV, p. 1–67; Über die Eigenschaften der Lösungen bei ihrer kritischen Lösungstemperatur, *Z. ph. Ch.*, 1908, Bd 62, S. 601–608; 1909, Bd 71, S. 90–108.
89. Tammann, G., *Kristallisieren und Schmelzen. Ein Beitrag zur Lehre der Anderungen des Aggregatzustandes*, Leipzig, 1903, 348 S.
90. Tammann, G. A. (Г. А. Тамман), A contribution to the thermodynamics of one-component systems (in Russian), *ЖРФО*, 1912, v. 44, p. 83–101.
91. Vittorf, N. M. (Н. М. Витторф), *Teoriya splavov v primenenii k metallicheskim sistemam* (*The theory of alloys and its application to metallic systems*), St Petersburg, 1909, 433 p.
92. von Wittorf, N. M. Vorläufige Versuche über primäre Kristallisation und nachfolgende physikalisch-chemische Umwandlungen im System Eisen-Kohlenstoff mit über 4% Kohlenstoff, *ЖРХО*, 1911, v. 43, p. 1613–1690; *Z. anorg. Chem.*, 1912, Bd 79, S. 1–70.
93. Speranskii, A. (А. Сперанский), *O tvyordikh rastvorakh obrazuemikh dvumya komponentami* (*On solid solutions formed of two components*), Moscow, 1904, 154 p.
94. Centnerszwer, M., Die kritische Temperatur der Lösungen, *ЖРХО*, 1903, v. 35, p. 742–793, 897–935; *Z. ph. Ch.*, 1903, Bd 46, S. 427–501.
95. Badil'kes, I. S. (И. С. Бадылькес), *Termodinamicheskoe podobie rabochikh veshchestv i protsessov kholodil'nikh mashin (Thermodynamic similarity of working substances and the processes of refrigerating machines)*, Moscow, 1960, 64 p.
96. Putilov, K. A. (К. А. Путилов), *Termodinamika* (*Thermodynamics*), Moscow, 1971, Ch. 8.
97. Filippov, L. P. (Л. П. Филиппов), *Podobie svoistv veshchestv* (*Similarity in the properties of substances*), Moscow, 1978, 256 p.
98. Frumkin, A., Die Kapillarkurve der höheren Fettsäuren und die Zustandsgleichung der Oberflächenschicht, *Z. ph. Ch.*, 1925, Bd 116, S. 466–497.
99. Titoff, A. (А. Титов), Die Adsorption von Gasen durch Kohle, *Z. ph. Ch.*, 1910, Bd 74, S. 641–678; *Bull. Soc. Nat. Moscou*, 1910, t. 24, p. 213–331.
100. Arrhenius, S., *Theories of Solutions*, New Haven, 1912, p. 67–68.
101. Krichevskii, I. R., *Fazovie ravnovesiya v rastvorakh pri visokikh davleniyakh* (*Phase equilibria in solutions at high pressures*), Moscow, 1946, 120 p.
102. Krichevskil, I. R., *Termodinamika kriticheskikh beskonechno razbavlennikh rastvorov* (*Thermodynamics of critical infinitely dilute solutions*), Moscow, 1975, 120 p.
103. Tsiklis, D. S., Chasanova, N. E., et al. (Д. С. Циклис, Н. Е. Хазанова и др.), Papers devoted to 75th birthday of Prof. I. Krichevskii (in Russian), *ЖФХ*, 1976, v. 50, No. 6; *Russ. J. Phys. Chem.*, 1976, v. 50, No. 6.
104. Storonkin, A. V. (А. В. Сторонкин), *Ob usloviyakh termodinamicheskogo ravnovesiya mnogokomponentnikh sistem* (*On the conditions of thermodynamic equilibrium of multicomponent systems*), Leningrad, 1948, 122 p.

105. Storonkin, A. V., *Termodinamika geterogennikh sistem* (*Thermodynamics of heterogeneous systems*), Leningrad, Parts 1–2, 1967, 447 p.; Part 3, 1969, 189 p.

106. Rusanov, A. I. (А. И. Русанов), *Termodinamika poverchnostnikh yavlenii (Thermodynamics of surface phenomena)*, Leningrad, 1960, 180 p.

107. Rusanov, A. I., *Fasovie ravnovesiya i poverkhnostnie yavleniya* (*Phase equilibria and surface phenomena*), Leningrad, 1967, 388 p.

108. *The Archives of the Naturalists' Society of Moscow*, inventory 727, 1898.

109. *The Archives of the Russian Academy of Sciences*, Moscow, box 610, inventory 4, No. 48.

110. Kistyakovskii, V. A., Van der Waals's jubilee is coming (in Russian), *ЖРХО*, 1902, v. 34, No. 8, p. 735.

111. The file *On the election of Professor Van der Waals as an Honorary Member of St Vladimir University*, The City Archives of Kiev, 1902, box 116, inventory 341, No. 110.

CHAPTER 10
In retirement

November 23, 1907 was van der Waals's seventieth birthday. 'As might have been expected' [188] the day was not marked in any special way; the old scientist resisted all attempts to celebrate the anniversary. He was still full of strength and worked busily on the general theory of binary mixtures; he had only recently managed to find a new geometric method for studying the ψ-surface, 'which is easier to follow than the analytical one' and which 'throws new light' on the behaviour of binary mixtures [84e, p. 621].

Nevertheless the age of seventy is the age of retirement, the official boundary between work and rest in Dutch academic life. An academician who has reached this limit becomes a *rustend lid*; he keeps his rights but is relieved of his obligations, in particular, he is no longer required to attend the sessions of his Section, and his place in the Academy is filled at a new election. For university teachers the law requires retirement at the end of the academic year in which a professor or lector becomes seventy. On 21 March 1908 he sent to the City Council of Amsterdam a request that he might retire at the end of his current course of lectures, and expressed his thanks for the enjoyment of the Council's unfailing confidence [250, p. 278]. Van der Waals accepted that there must be this change in his life but there is no trace of his coming retirement in the last courses of lectures he gave. As in previous years, he was unfailingly in his laboratory by 9 in the morning. Once during this last winter there was an unprecedented snow storm one night, but he still was in his place in the morning to find a very thin class before him. He said, with a chuckle, 'I appreciate very much the presence of those gentlemen who were not kept away by the weather' [190, p. 328][1].

As in the previous ten years he lectured on his two favourite subjects in theoretical physics – electricity and magnetism, and thermodynamics and the theory of mixtures. In the autumn term he lectured on the mathematical theory of static electricity for two hours a week, and on the theory of binary mixtures (the course of '*p*- and *q*-lines', allowed and forbidden plait points and plait-point curves) for one hour a week. In the spring term he lectured on electricity (condensers, magnetic and electrical induction, magnetic force, electric currents) for two hours a week to the students in their final year, and for one hour a week on the mechanical theory of heat, that is, thermodynamics. Everything followed the customary pattern, only in the theory of mixtures did he venture to introduce new material with new results on the geometry of the ψ-surface. At the end of his last lecture he said, in reference to the forthcoming publication of his papers, 'I hope that you will then be able to study all this successfully.'

The ceremony to mark his retirement was on 13 July 1908 although the official decree was not issued until 1 September, after the decision on the filling of his post. Tradition was broken in a way that was typical of van der Waals; instead of a portrait in oils, he was commemorated with a tablet on the wall of the laboratory which set out his main scientific achievements[2]. About 130 people who had taken part in this ceremony with the tablet went then to the lecture hall of the Physics Laboratory, and H. W. C. E. Bückmann, one of the first of his students and at that time a gymnasium teacher, outlined van der Waals's career as a professor from 1877 to 1908. Van der Waals, who had at first no intention of speaking, responded with an interesting talk which was, in effect, his final lecture. He reviewed the state of physics, and especially the theory of heat, when he had first occupied the chair, what had been the gaps in our knowledge, and how he had developed his own theories [191].

His formal retirement was on 18 September when the Rector Magnificus again appealed to van der Waals to allow his portrait to be drawn for the benefit of historians and biologists; for historians since 'the person of the creator and the work he created are inseparable', and for the biologists so that they can study the 'comparative physiognomistics' with results that could not yet be foreseen. 'And do you really want us to be told in years to come, "What a pity that the head of van der Waals is missing from our comparative analysis"?' [*1*, p. 7]. The argument fell on deaf ears.

It was only a few days before his retirement ceremony that the news came from Kamerlingh Onnes in Leiden that helium had been liquefied, in clear confirmation of the continuing vitality of van der Waals's ideas. This triumph of both theory and experiment was quickly honoured by the award of gold medals of the Society of Natural and Medical Sciences to both van der Waals and Kamerlingh Onnes. In a similar gesture, when the first *Conférence Internationale du Froid* assembled in Paris in October 1908 under the chairmanship of Kamerlingh Onnes, van der Waals was elected its honorary chairman.

The award of the two gold medals was the first made to physicists by this Society during its existence of more than a hundred years. The presentation took place on 18 November in the University Assembly Hall with Pieter Zeeman in the chair. In his speech to van der Waals he outlined briefly the ideas in *Continuity* and of the law of corresponding states, and said 'By these works you have let us glimpse the inner harmony of the world, the harmony that is displayed best and in the most beautiful way by the laws of nature' [*2*, p. 136]. But the significance of van der Waals's works, he continued, had other aspects. In the first half of the nineteenth century one looked for scientific discoveries to Paris, or Berlin, or London, or Göttingen. But now the achievements of Dutch physicists are as good as the best of those elsewhere. 'As Dutchmen we are proud that you are one of us and that we can put your name in the front rank of physicists of all nations, convinced of the continuance of your work into the future' [*2*, p. 138].

The medal for Kamerlingh Onnes was presented by van der Waals himself to emphasize the link between their investigations. He listed, for his friend, all the great difficulties of a long piece of research, leading up to 'the memorable day of the 10th of July. During the long summer's day you had to work even longer hours than the Sun took to cross the heavens, and at the end you converted helium into a liquid[3]. By having done this you have opened up a number of scientific problems in which the effects of low temperatures are to be seen, by allowing us to get nearer to the absolute zero of temperature than heretofore. It is with admiration of your achievement, with respect for your character as a man and as an investigator, with a feeling of gratitude for this advance of Dutch science, and in the spirit of heartfelt friendship that I present you with this medal' [155].

Retirement was not an occasion for van der Waals to abandon science, nor for him to cease working with younger colleagues. At his request his vacant post was divided into two. His son, J. D. van der Waals Jr., succeeded him in the chair of physics, returning to Amsterdam from the chair in Groningen, and the second post was an *ad hominem* chair of thermodynamics which was offered to Kohnstamm with the requirement that he not only lecture but that he direct the experimental work on mixtures. The Curators did not wish to split the post and van der Waals had to visit the Burgomaster to convince him of the wisdom of this course. In a letter to him of 15 July van der Waals wrote that thermodynamics is an experimental science and that to stop teaching it would be a great blow to the University, 'and not only for students in mathematics and physics, but also and especially for those in chemistry, who would thus be deprived of the basis for further study in their field' [250, p. 279]. On 30 July the Curators agreed to adopt the Faculty propositions. Kohnstamm became also the Director of the van der Waals Fund, and it was in these years that the first work wholly supported by the Fund was undertaken by Jean Timmermans.

It is clear that relations between van der Waals and his successors remained close. He continued in office as Secretary of the Academy to which he presented the publications of his followers. His main concern at this time was, however, the systematic development of his theory of mixtures. His new geometric method opened up new possibilities and he abandoned other subjects to delve deeply into the labyrinth of the complicated relations between the stable, metastable and unstable states of mixture. A series of twenty-one communications with the general title *Contributions to the theory of binary mixtures* was published in the *Proceedings* between 1907 and October 1912.

Only one event of note disturbed the even tenor of these years; on 6 November 1910 he received a telegram which read 'The Nobel Prize for Physics has been awarded to you, details follow by letter' [244]. The citation read 'for his studies of the physical state of liquids and gases'. One may speculate as to why the Prize came so late (his compatriots van 't Hoff, Lorentz and Zeeman had been honoured in previous years) but for his pupils and colleagues it was an 'unexpected gladness' [197]. He had in fact been nominated on at least two previous occasions [*3*, p. 358,364,366].

The presentation read, 'Van der Waals's theory has also been brilliantly successful through its predictions which made it possible to calculate the conditions for converting gases to liquids. Two years ago van der Waals's most prominent pupil, Kamerlingh Onnes, in this way succeeded in compelling helium – the last previously uncondensed gas – to assume the liquid state. Yet van der Waals's studies have been of the greatest importance not only for pure research. Modern refrigeration engineering which is nowadays such a potent factor in our economy and industry, bases its vital methods mainly on van der Waals's theoretical studies' [*4*, p. 252]. Probably his theory had to bear practical fruit before its importance could not be questioned by the Nobel Committee.

In December he left for Stockholm accompanied by his son, although he had a chill. As was usual the ceremony was on 10 December, the anniversary of Nobel's death. The President of the Swedish Academy of Sciences addressed the prizewinner with the words, 'Professor van der Waals: The Royal Academy of Sciences has awarded you this year's Nobel Prize for Physics in recognition of your pioneering studies on the physical state of liquids and gases. Hammurabi's and Moses' laws are old and of great importance. The laws of nature are older still and even more important. They apply not just to certain regions on this Earth, but to the whole world. However, they are difficult to interpret. You, Professor, have succeeded in deciphering a few paragraphs of these laws. You will now receive the Nobel Prize, the highest award which our Academy can give you' [*4*, p. 252–253].

The next day, at the start of the ceremonial banquet, the President remarked to van der Waals that he was the fourth Dutchman to receive a Prize, a tribute to the ability of small countries to make a great contribution to world science. No doubt this was music to the ears of a patriotic Dutchman but his speech (he had said that he wanted to speak on the law of corresponding states) was rather more international. He started in German and then switched to French; his theme was a favourite comparison of his – physical substances, for all their diversity, belong to a single genus, just as mankind is a single genus notwithstanding racial, national and individual differences. 'The idea of a single genus embracing an infinity of variations raises many questions, and he who can answer will have done an important piece of work that you will then honour' [102].

The Nobel Lecture was the only occasion when van der Waals had to overcome his diffidence and talk in public about himself and his work [103e, p. 254]. He told in detail about his equation of state and attempts to improve it, particularly by means of the hypothesis of quasi-association, and then turned more briefly to the theory of mixtures. He concluded on a slightly sad note; 'The crowning of my studies by the esteemed Royal Swedish Academy affords me satisfaction and fills me with gratitude, a gratitude which I cannot call eternal and in my old age I cannot even promise that it will be of long duration, but, perhaps for this very reason, my gratitude is all the more intense' [103e, p. 265].

His report to his daughters in Amsterdam was modest; 'Monday 12

December 1910 was the great day. My lecture was at 2 o'clock and I omitted many things so as not to make it too long. My voice is not yet completely recovered but I think that everything went off not very well, but fairly well.' When he writes about the banquet we can see how little he was spoilt by care and attention; 'All were very kind, and when I could not at once find my place one of the high officials came up and offered me his arm to accompany me there. . . . In short, everyone tried to do all they could to see that we were happy' [231(b), p. 152].

The award of the Nobel Prize was naturally celebrated with due solemnity in his native land. On 27 November Lorentz, then President of the Academy, warmly congratulated his Secretary both personally and in the name of the Section of Sciences. After his return from Stockholm van der Waals had to give an *encore*. On 11 January 1911 a joint meeting of the Dutch Chemical Society and of the Physical Section of the Society for the Development of Natural and Medical Sciences heard with intense interest his report *On the difficulties that had to be overcome to establish the equation of state*. This Dutch version of his Nobel Lecture was published at the end of January [$103h_1'$] and in the spring the Editor of *Chemisch Weekblad* declared that requests for copies were too many for him to satisfy and recommended readers to the recently published German version of the Lecture [103_2].

One curious by-product of the award of the Nobel Prize has been much appreciated by his biographers. There was no portrait of van der Waals except for an early photograph taken in the 1870s, for he had firmly resisted all further attempts at portraiture. He had, however, to conform to the strict rules of the Nobel Committee and have his photograph taken, and this rule thus led to the production of the only well-known portrait. He then lifted his ban of forty years and some more photographs were taken in the next few years, as well as a portrait for a memorial medal[4].

In the Spring of 1912 he asked to be relieved of his post of Secretary General. The annual meeting of the Academy began as usual with a Presidential Address from Lorentz which was followed by a detailed report from the Secretary, and the reports of the Finance and Library Committees. There followed the election of the new Secretary, and Zeeman was chosen. (Van der Waals had been elected unanimously for the next period of five years in 1901, 1906 and 1911, although he himself had abstained from voting [245].) At the end of the meeting van der Waals's children were invited into the hall and heard Lorentz describe vividly the varied activities of the retiring Secretary. Then Chantepie de la Saussaye, who was taking over from Lorentz as President, spoke with great sympathy; 'Many of us', he concluded, 'esteem van der Waals not only for his science but even more for his character' [*6*].

Deeply touched, van der Waals thanked them all for everything that had been said and, his sense of humour not deserting him even on this occasion, 'for the weighty evidence of their sympathy'. They had presented him with complete sets of the Proceedings in Dutch and in English. He explained that he was resigning because 'he did not want to stay until his abilities started to decline'. Sometimes the job was hard, but sometimes, as when he had had help with the editing of the Proceedings, everything was a joy. He recalled the friends and colleagues he had worked with, van de Sande Backhuyzen, Korteweg and Chantepie de la Saussaye, and praised the permanent staff of the Academy. He concluded with his best wishes for the flourishing of the Academy in the years to come [*6*].

He continued his studies after his retirement from the post of Secretary. In the autumn of 1912 he published the twenty-first, and last *Contribution to the theory of binary mixtures*, and the substantial second volume of the *Textbook of Thermodynamics*. Having taken his study of binary mixtures as far as he could, he returned in 1913 to the equation of state of a single substance and produced a booklet entitled *Further investigations on the equations of state* [118]. He returned again to the problem of the dependence of b on density. He wrote that he 'felt the necessity, and even thought it his duty, to try again to find the cause of this dependence, and even if it proved impossible to find the form of the function $b = f(v)$, then he could perhaps show that such dependence was not the same for different substances, and so account for departures from the law of corresponding states.' This time, however, he could make no progress, and his attempts on another problem, that of the quasi-association of molecules, also resisted further attack.

In his retirement he came less and less frequently to meetings of the Section of Sciences. He gave up even his temporary posts, resigning as chairman of the Academy Committee for setting up an institution for research on the brain. He retained, however, his post as one of the five Curators of Leiden University from 1908 (after his retirement) until 1916, and it was in this capacity that he received Paul Ehrenfest who came from St Petersburg to succeed Lorentz as Professor of Theoretical Physics. This move, which was a crucial point in Ehrenfest's career, took place on 1 November 1912. The thirty-two-year-old Ehrenfest was not at first greatly impressed by van der Waals; 'a little old man (well over seventy) and, it seems, nothing more', he wrote to his closest friend Abraham Joffe [*7*, p. 98]. In his next letter he modified this opinion. It was written after his inaugural lecture *On the crisis in the light-ether hypothesis*. The new professor was heard by the Leiden students, the professors and their wives, and by the curators, among them van der Waals, who sat in the places of honour. After the lecture there was an official reception at a hotel, after which Paul and Tatiana Ehrenfest arranged a 'discussion group' at their home. Shortly afterwards Ehrenfest wrote to Joffe, not without pride, 'It all turned out very well. Van der Waals stayed with us until half-past eleven, having come specially from Amsterdam to attend my inauguration' [*7*, p. 107]. Such an hour was very late by Dutch standards. Probably the man he called the living classic turned out to be not only a relic of the physics of the past (and it is worth remembering that in 1903 Ehrenfest had carried out a study of van der Waals's equation), but had a genuine interest in the latest problems in physics, seeking in them clues as to the origin of gravitation and of the behaviour of a radiating atom.

In the summer of 1914 there started what came to be called first 'The Great War' and later 'The First World War'. At the opening of the September meeting of the Section of Sciences the Chairman, Lorentz, said 'Much has happened since the Section last met to bring us all anxiety and sorrow, but it is good that scientific work is still going ahead. Fortunately our country has so far been spared from the calamity of war and so most can continue with their research; we ardently look forward to the time when investigations abroad can be resumed also' [*8*].

Van der Waals shared the same view, but for him the ability to work now depended in the first place on his state of health. During the four years of the disastrous war he spoke only once before the Section of Sciences; in June 1916 he communicated his arguments on the increase of the parameter a at high densities [121]. He was now convinced that his attempt to ascribe this variation to quasi-association was wrong and, after many calculations, he 'became more and more convinced that the result could not be valid, even as an approximation'. He came to the idea which 'forced itself' on him that this increase in a must be a 'consequence of the ordinary regular movement of molecules which have extension' [121e, p. 359–360]. This idea is set out in an elementary way and he recognized that it limited the validity of the mean-field conception which had been the foundation of his equation of state. He felt, however, unequal to the task of resolving this problem and concluded with the hope that 'other investigators will show an interest in this research', a hope that has been fulfilled although the problem of the departures from a mean-field treatment has no exact solution to this day.

This was his last paper. He took no further part in the activities of the Academy, except, in March 1919, to present, via Lorentz, his son's paper on the theory of the viscosity of liquids.

In the autumn of 1917 he was eighty and, although the time was clearly inappropriate for celebrations, Zeeman nevertheless wrote a short but thoughtful review of his work for an issue of *Naturwissenschaften* that appeared on his birthday [198]. He summarized his main achievements – the equation of state, the law of corresponding states and the theory of binary mixtures – and observed that these were not separate ideas arrived at randomly or accidentally. 'It is only', he wrote 'after the most intensive and thoughtful work that someone arrives at such ideas. We admire not only the greatness of van der Waals's ideas but also "the earnestness that knows no fatigue"[5], the unwearying search for a cherished goal.' He ended by hoping that the great physicist might have 'many years of supportive leisure, and the opportunity to see the ripening of new fruit on that branch of physics devoted to the general properties of matter'. Old age, however, was taking its toll; the enforced leisure was no longer 'supportive'; his strength was failing and his circle of friends narrowed. 'At last we could make only short visits to show our venerated and beloved friend, whose heart was still unchanged, how much he had done for us', wrote Kamerlingh Onnes [208(a)].

In 1920 his closest friend, Chantepie de la Saussaye, passed away. Then, on 29 April 1922 there came the heavier blow of the painful death of his daughter Jacqueline; for many years she had known that she had an incurable cancer. Van der Waals's life was now one of 'patiently borne but sharply felt suffering. . . . He did not complain, he did not murmur, he was at rest although his eyes were dimmed and his powerful voice had lost its resonance. He rested and waited, sometimes sighing for death which was so late in coming' [206, p. 6].

In 1921 he had broken his leg and his grandson's memory is that of 'an old grumbling man who did not want to learn to walk again after he had broken his leg. His cigar went out regularly and was carefully relit each time by his eldest daughter' [242].

Early in 1923 Kaiser's letters of introduction (see Chapter 2) were found between some old documents. He agreed to hear his daughter read these yellowed sheets and (as told by Zeeman) 'the dying flame of his life for a moment blazed up again as his spirit returned to the beginning of his course through life' [205, p. 109].

He died on 8 March 1923 and the funeral was on the 12th. Following the wish of his children, Kohnstamm gave the graveside oration since Chantepie de la Saussaye was already dead. He was buried beside his wife and his daughter in the cemetery of Nieuwe Oosterbergraafplats. Time has almost totally erased the inscription from the plain white gravestone.

Notes

1. A play on words in Dutch, literally 'were not weatherbound by the weather'.
2. The text is as follows:

   Johannes Diderik van der Waals
   1877 – Hoogleeraar – 1908
   Toestandsvergelijking – 1873
   Wet der overeenstemmende Toestanden – 1880
   Theorie der binaire mengsels – 1891 [sic]
   Thermodynamische Theorie der Capillariteit – 1893

3. Kamerlingh Onnes reported that the experiment took from 5.45 a.m. to 9.40 p.m.
4. The artist was J. E. Jeltsema. The medal is reproduced in [197] and [215]. It is also worth note that although there was never an oil painting of van der Waals there was later a bas-relief portrait by L. Beyerman which was installed in the van der Waals Laboratory in November 1937 before the Jubilee Congress on Physics [5].
5. From Schiller's *Das Ideal und das Leben:*

   Nur dem Ernst, den keine Mühe bleichet,
   Rauscht der Wahrheit tief versteckter Born.

References

1. *Jb. AU* (1907–1908), Amsterdam, 1909.
2. Zeeman, P., Rede, *W.G.N.*, 1909, d.6, afl.2, b.135–138.
3. *Science, Technology and Society in the Time of A. Nobel*, Oxford, 1982.
4. *Les Prix Nobel en 1910*, Stockholm, 1911. Quoted from [103e].
5. *Ch.W.*, 1937, 20 Nov., Jg.34, b.765.
6. *Jb. KAW*, 1912, b.cxxxvi.
7. Ehrenfest, P.—Joffe, A.F. (П. Эренфест — А. Ф. Иоффе), *Scientific correspondence* (in Russian), ed. V.Ya. Frenkel, Leningrad, 1973.
8. *Versl.*, 1914, d.23, b.515.

CHAPTER 11

Molecular physics

11.1 The equation of state

In the autumn of 1880 Clausius referred for the first time to van der Waals's thesis describing it as 'a very interesting paper' [*1*, p. 344]. He pointed out, however, that the equation of state was of low accuracy and that the thesis had been preceded both by Andrews's papers and by the earlier publication of similar equations, such as

$PV = RT - c/TV,$	W. J. M. Rankine, 1854
$(P + l)(V - m) = RT,$	G. A. Hirn, 1865
$PV = RT(1 + f(T)/V),$	G. F. Recknagel, 1871

where $f(T)$ is a function of the temperature.

Clausius was careful to add that Hirn, his long-standing opponent who had never accepted the kinetic theory, meant m to be the 'sum of the atomic volumes' and l to be the 'sum of the internal actions' or an 'internal pressure'. He scrupulously recorded that the Munich physicist Georg Recknagel had also made a more accurate analysis of the problem in 1872, that is, in the year before the publication of van der Waals's thesis. (His paper [*2*] is dated October 1872, although it was not published until 1874.) Like van der Waals, he used the virial theorem in the derivation of his equation, although Clausius rather surprisingly makes no mention of this.

Clausius's list of equations of state should be completed by adding also

$P(V - n) = s,$	D. Bernoulli[1], 1738 [*4*]
$PV = T - t/V,$	E. Ritter[2], 1846 [*5*]
$P(V - u) = RT,$	A. Dupré[3], 1869 [*6*]

Ritter, Rankine, Joule, Dupré and James Thomson [174,183] had all, before van der Waals, pointed out the need to introduce into the equation of state of real gases a correction for the internal pressure that was proportional to the square of the density, but none of their equations was able to account for the phase transition.

11.1.1 Early discussion of the legitimacy of van der Waals's equation

The equation may be written

$$P = RT/(V - b) - a/V^2 \quad \text{or} \quad P = P_T - a/V^2. \tag{11.1}$$

In 1881, after Maxwell's criticism of 1874, and the arguments in reply from

Korteweg and van der Waals, Lorentz[4] had deduced from the virial theorem an expression for the 'thermal pressure', that is, the pressure in the absence of any mutual attraction between the molecules,

$$P_T = RT\left[\frac{1}{V} + \frac{b}{V^2}\right], \tag{11.2}$$

which is equivalent to van der Waals's result, eqn (11.1), to the second order in $1/V$; the two results differ at the next order. Lorentz failed to obtain the term in $1/V^3$ and summed up his analysis by saying that although some deviation from eqn (11.1) is possible for compressed gases 'in view of the intricacies of the problem one cannot propose a solution other than that of van der Waals' [*7*, p. 136].

In 1891 Tait attacked eqn (11.1) with his usual polemical zeal since, he said, it contradicted the expansion in powers of the density derived from the exact solution [*8*], that is, the result derived from the virial theorem, and so made no sense near the gas–liquid transition. It was 'entirely wrong in principle' [*9*] since P and V in the virial equation are the actual pressure and volume of the system, and so not subject to further correction, an objection raised by Maxwell in 1874. Rayleigh and Korteweg energetically defended van der Waals and, among other points, proved independently that eqn (11.1) is exact for a one-dimensional gas [*10–13*][5,6]. A provisional summary of these early discussions was given by Boltzmann in 1898 when, in his *Lectures on Gas Theory* he wrote that 'van der Waals has given us such a valuable tool that it would cost us much trouble to obtain by the subtlest deliberations a formula that would really be any more useful than the one that van der Waals found by inspiration, as it were' [*15* (1964), p. 355].

11.1.2 The problem of improving van der Waals's equation

It must be admitted that good luck attended the birth of *Continuity*. The 'striking concordance', as van der Waals put it [$1e_2$, p. 192], between his computed results and the measurements of Andrews was largely fortuitous since Regnault's results, which he used to calculate the parameters a and b for carbon dioxide, were not sufficiently accurate for this purpose [160, p. 61].

From the beginning van der Waals knew that his 'constant' b was not, in fact, a constant but a decreasing function of the density 'the law of which we do not yet know' [$1e_2$, p. 193]. So it was clear to him that one should only use the equation with the greatest caution in P, V, T-regions remote from those in which a and b had been determined. Nevertheless from the 1880s onwards, when his equation became well known, many physicists used it by assuming that a and b were true constants, ignoring the reservations set out in *Continuity*. This use, and the accumulation of more plentiful and more accurate experimental results, meant that there were soon many examples of disagreements with van der Waals's equation which undermined its definitive standing. There then started the unending attempts to invent new equations, some intended for special applications, some purely empirical and some based on more or less sound theoretical arguments. The first of these equations was that of Clausius who in 1880 proposed a form 'constructed', as he wrote, 'to describe in a simple manner the old and new observations of Andrews' [*1*]:

$$P = \frac{RT}{v - f} - \frac{c}{T(V + g)^2}. \tag{11.3}$$

This equation does indeed reproduce more accurately than van der Waals's equation the results for carbon dioxide, but this fact is hardly surprising since it has one more adjustable parameter. For some years it superseded van der Waals's equation, especially for the representation of experimental results[7].

Clausius's equation was followed by an avalanche of other modifications of all possible forms; a far-from-complete list published in 1948 ran to 150 different equations. Their number is, however, immaterial; what is important is that none could match the original in its combination of insight, clarity of the underlying model, simplicity of form, and convenience in application. Van der Waals himself also tried repeatedly, over a period of about thirty years, to improve his original result since, as he put it, 'That the accurate knowledge of the equation of state is of the highest importance is universally acknowledged, because in conjunction with the results of thermodynamics it will enable us to explain all phenomena relating to ponderable matter' [64, p. 185]. His efforts proved, however, no more successful than those of his rivals.

Corrections to the volume term Van der Waals devoted much effort in trying to discover the dependence of the parameter b on the density of the gas. He hoped, in this way, to be able to bring his equation into complete agreement with the facts. However it was Gustav Jäger [*16*] and his Viennese colleague Boltzmann [*17*] who first showed, independently in 1896, that the cubic term in Lorentz's expansion had the form $(5/8)(b^2/V^3)$. In the same year van der Waals [43], using his favourite method of calculating the reduction in the length of the mean-free path, arrived at $(15/32)(b^2/V^3)$. He was wrong, but acknowledged his error only in 1903 after his son had obtained the same result as Jäger and Boltzmann [70]. It is of interest that in the spring of 1899 he did not reject a challenge from Boltzmann to debate this issue but said simply that he preferred to wait for time to settle the point [56e, p. 468; *18*].

Van der Waals had, however, issued a kind of challenge himself. In the autumn of 1898 he started on the calculation of the next term in the expansion [50]. He could not, however, bring this to a successful conclusion and left the evaluation of the very difficult integrals involved to van Laar whose results were published in January 1899 [*19*]. Boltzmann's results appeared two months later, and again there was a discrepancy, van der Waals's value for the coefficient of (b^3/V^4) being about 0.17 lower than Boltzmann's. Van der Waals wrote 'As our results differ so much, we cannot have considered the same problem' [56e, p. 468]. Fifty years later Boltzmann's result was finally confirmed as the correct one. It was he who 'showed an amazing degree of insight and though he had no strict formalism to guide him, he made the proper choice at every turn' [*18*, p. 504]. Kohnstamm wrote in 1904 that the calculation of the higher terms in (b/V) led to difficulties that were at the limit of what can be done [*20*]. So van der Waals was forced to leave this systematic route, but not his search for the density dependence of the volume correction. In his Nobel Lecture of 1910 he confessed, 'I still wonder whether there is a better way [to determine this dependence]. In fact this question continually obsesses me, I can never free myself from it, it is with me even in my dreams' [103e, p. 256]. So in 1901–4 he devoted several papers [62,63,70] to his idea of 1898 of taking account of the internal degrees of freedom of the molecules as 'a better way' of getting out of the impasse. Thus he was even willing to sacrifice his favourite hard-sphere model, and consider deformable spheres of changing microscopic size as a possible route to an improved macroscopic equation of state. This approach could only be considered as an act of desperation; it proved to be a blind alley and he abandoned it after 1903.

Correction to the internal pressure From as early as 1890 van der Waals had not only made persistent attempts to improve his equation by introducing a density dependence into b but also had been brooding over how to achieve the same end by modification of the internal pressure a/V^2 [160, p. 178]. The derivation in his thesis, although based on Clausius's virial theorem, was quite qualitative. He was far from enthusiastic about tackling the extremely complicated mathematical problem of a rigorous 'microscopic' calculation of the internal pressure[8] and recognized, moreover, that in doing this one could hardly get by with some vague hypothesis for the form of law of force of the intermolecular interaction. No doubt he thought a more reasonable approach was to use the law of corresponding states to analyse the progressive accumulation of experimental results. This analysis, which revealed clearly the essential difference in behaviour of 'normal' and chemically associated substances, made him further inclined to the idea of an inhomogeneous spatial distribution of the molecules, even in normal fluids. That is, as well as simple or solitary molecules there exist in liquids (and maybe in gases also) complexes or aggregates of several molecules which take part in the thermal motion as 'a new larger unit' [103e, p. 263]. This idea was first put forward by Clausius [*21*] and was supported later by the Polish physicist Wladyslaw (Ladislaus) Natanson, by Boltzmann and his pupil Heinrich Mache, and by the Belgian physicist Pierre de Heen. It was in sharp contradiction to the fundamental notion in *Continuity* of the homogeneity of the molecular distribution, so it was not surprising that van der Waals was slow to admit it. In 1906 he made, in the Trippenhuis, a report [82] on 'quasi-association'.[9] He refrained from publishing this work in detail at that time since he 'was still doubtful about different particulars' [101e, p. 107]. It was not until 1910 that he published his results in full, when he said that one of the main reasons for this decision was the recent discovery that the kinetic behaviour of large molecular complexes is analogous to that of simple molecules. He was referring, no doubt, to Perrin's experiments of 1908 which had triumphantly confirmed Einstein's theory of Brownian motion.

His theory of quasi-association, which he considered to be only of a provisional nature, was developed with the simplifying assumption that a fluid comprised two kinds of particle, the simple molecules and the n-fold complexes. So he turned to his theory of a binary mixture in order to exploit the 'very fortunate' circumstance [103e, p. 262] of the many years he had spent studying the laws of these mixtures. He thus obtained the equation of state [101,104,105],

$$\left[P + \frac{a(1 - (1 - k)x)^2}{V^2}\right](v - b) = RT\left[1 - \left(\frac{n - 1}{n}\right)x\right], \qquad (11.4)$$

where k is a constant of about 0.5 and x is the degree of association, that is, the ratio of the number of molecules bound together in the n-fold complexes to the total number of molecules, simple and bound. This treatment of quasi-association leads to a reduction of the internal pressure, since some of the intermolecular attractions have been neutralized by being locked within the complexes, and a reduction also in the number of particles contributing to the thermal energy on the right-hand side of eqn (11.4). It is likely that van der Waals's contribution to this field was the most influential, but, however attractive it may have appeared, it scarcely lived up to his original hopes. Many others have since tried their hand at treating association in the development of the theory of the equation of state, but none has been generally accepted as of fundamental importance or even as fully reliable [168].

The idea of quasi-association has, however, reappeared in a different guise in one of the most sophisticated approaches to the statistical mechanics of real gases, namely the virial expansion developed from first principles by H. D. Ursell (1927) and J. E. Mayer (1937). The configuration integral, which is central to this treatment and which is essentially the Boltzmann factor of the free energy, can be expanded into group integrals each describing a molecular pair, a triplet, a quadruplet etc. The values of these integrals are determined by the intermolecular forces between the constituent molecules. Clusters can be regarded as 'independent entities of physical significance' and, in particular, 'every such group contributes to the pressure to the same extent as a solitary molecule' [168, p. 237]. In this way the essence of van der Waals's ideas on quasi-association have been incorporated into modern theory and so understood more clearly.

When thinking about quasi-association van der Waals had in mind principally fluids of high density and especially the liquid state. In this state, however, the concept is irrelevant since the close packing of the molecules in liquids negates any possibility of dividing them into separate complexes. It is true that the spatial distribution of molecules in a liquid is far from uniform, as is now known from determinations of liquid structure by X-ray diffraction, but the inhomogeneity, or short-range order, has nothing in common with quasi-association since it arises not from intramolecular attraction but from the geometrical effects of packing the molecules together, and so is connected rather with the repulsive forces.

11.1.3 A rigorous basis for van der Waals's equation

From 1938 until the 1980s each edition of Landau and Lifshits's *Statistical Physics* (a volume in their classic *Course of Theoretical Physics*) contained the statement: 'Van der Waals's formula is, of course, only one of an infinity of possible interpolation formulae [between the two phases] which satisfy the requirements stated, and there is no physical reason to select any one of them; van der Waals's formula is merely the simplest and most convenient' (see e.g. [*22* (1976), p. 259; (1977), p. 224])[10].

Given the remarkable success of van der Waals's equation not all physicists were, however, quite so ready to give up attempts at a rigorous derivation. As we have seen, Rayleigh and Korteweg had shown as early as 1891 that the $(V-b)^{-1}$ term is exact for a one-dimensional system[11]. But it was shown many times, and with complete rigour by van Hove [*23*] in 1950, that a one-dimensional system with interactions of finite range cannot have a phase transition. So what was the connection between Rayleigh and Korteweg's result and van der Waals's equation, the main achievement of which was its description of the condensation of a gas to a liquid?

Ornstein, in his Leiden thesis [*24*] of 1908, for which Lorentz was the 'promotor', was the first to make a sound attempt at a derivation of the equation of state from first principles, based on the powerful methods of Gibbs's *Statistical Mechanics* of 1902[12]. In 1937, at the van der Waals memorial conference in Amsterdam, Max Born remarked that, 'In principle, the methods of statistical mechanics should lead to a definite formula [i.e. equation of state] if the law of force between the molecules is given. But the mathematical difficulties are so overwhelmingly great that the majority of investigators prefer to use reasonable guesswork assisted by a knowledge of the experimental facts' [*25*, p. 1035]. This 'perfectly legitimate' way did not lead to any closed expression that was superior to that of van der Waals, in spite of much effort by many physicists [*26*], until the approximate summation of the virial series at the hands of J. K. Percus and G. J. Yevick, M. S. Wertheim and E. Thiele in 1959–63 (see e.g. [183, p. 54]).

It was a rather different route that led in the 1960s to a striking revival of van der Waals's theory. From the work of M. Kac (1959), M. Kac, G. E. Uhlenbeck and P. C. Hemmer (1963–64), N. G. van Kampen (1964), and J. L. Lebowitz and O. Penrose (1966) a clear-cut answer emerged to the question: for what physical model is the van der Waals equation the exact equation of state? The model was found to be that of hard spheres of diameter σ with an attractive potential $u(r;\gamma)$ between a pair of molecules a distance r apart given by

$$
\begin{aligned}
u(r;\gamma) &= \infty && \text{for } r < \sigma \\
&= -\alpha\gamma^d e^{-\gamma r} && \text{for } r > \sigma
\end{aligned}
\tag{11.5}
$$

where d is the dimensionality of the space considered, γ is a positive parameter whose reciprocal γ^{-1} is a measure of the range of the attractive potential, and we define

$$a = -\int_{\sigma}^{\infty} \mathrm{d}r\, u(r;\gamma). \tag{11.6}$$

If, following the route taken by Ornstein, one first takes the thermodynamic limit ($N \to \infty$, $V \to \infty$) and then the limit $\gamma \to 0$, which is sometimes called the van der Waals limit, then, in a one-dimensional system the van der Waals equation, supplemented by the Maxwell construction, emerges as the exact result. The limit $\gamma \to 0$ corresponds to an attractive potential which is everywhere weak, of order γ, but of infinite range, of order γ^{-1}. The equation of state can be written

$$P = \frac{kT}{L - l} - \frac{a}{L^2} \tag{11.7}$$

where L is the length of the system and l the sum of the lengths of the hard rods, that is, in one dimension the excluded volume is just the sum l. This result, eqn (11.7), does not violate van Hove's theorem since his prohibition of a phase transition in one dimension applies only to systems with attractive forces of finite range [*27,28*].

11.1.4 Is the question settled completely?

No one before van der Waals had put forward as convincingly what seems now to be such a natural conception of the universal existence of attractive forces between all molecules, even those of gases[13]. The existence of such forces is explicit in all his papers but it was only when he came to discuss capillarity that he committed himself to a particular form of the potential of this force, namely $u(r) = -\alpha(\beta/r)e^{-r/\beta}$, where α is a positive constant and β is a length of the order of the molecular size [37][14]. This function was chosen, however, more to illustrate his argument than as a serious attempt at obtaining a realistic estimate of the form of these forces. In his thesis he started from the old Laplacian notion of forces that were 'insensible at sensible distances'. The results he obtained then allowed him to express this more precisely. Laplace's two functions H and K, eqns (3.10) and (3.11) of Chapter 3, are related respectively to the surface tension, which can be measured experimentally, and to the internal pressure, which he could calculate from his equation of state. Their ratio is a measure of the range of the attractive forces, and this he calculated for several substances. For two of these, ether and alcohol, he had also values of the molecular diameters from the constant *b*. His results were as follows.

	Range of forces	Molecular diameter
Ether	2.9×10^{-8} cm	4.0×10^{-8} cm
Alcohol	2.5×10^{-8} cm	2.7×10^{-8} cm

Given the inaccuracies in the data and the fact that the molecular diameters depend on Maxwell's estimate of the mean-free path of molecules in air (since van der Waals did not know the value of what we now call Avogadro's constant), these results almost convinced him that the range of intermolecular attraction 'is to be considered exactly equal to the distance between the centres of the molecules during encounter. . . . By this I do not mean that there is no attraction at other distances, but the attraction at this distance is so much greater, that it is alone necessary to consider it in the calculation' [$1e_2$, p. 217].

He never wavered in his conviction that the molecular attraction was of extremely short range and disregarded the objections of a number of his colleagues, including Boltzmann. In the first chapter of the second volume of his *Lectures on Gas Theory* Boltzmann had written: 'Between the centers of two molecules there acts an attractive force (the van der Waals cohesion force), which vanishes at macroscopic distances but decreases so slowly with increasing distance that it may be considered constant within distances large compared to the average separation of two neighboring molecules' [*15* (1964), p. 220].

Van der Waals saw these words in September 1898, shortly before the publication of the second volume, when Boltzmann was his guest at a country house in Groesbeek that he had rented for his daughters [103e, p. 261; 205, p. 112; 232, p. 16]. On his return to Vienna Boltzmann was able to add a note in the fifth chapter to take account of van der Waals's comment on this statement: 'He [van der Waals] never explicitly made the assumption, explained in Section 2, that the attractive force of the molecules decreases so slowly with increasing distance that it is constant at distances large compared to the average separation of the neighboring molecules; and he believes such a force law to be improbable. Nevertheless I cannot obtain an exact foundation for his equation of state without this assumption' [*15* (1964), p. 375].

It is not without interest that Kamerlingh Onnes and Keesom, in their review *Die Zustandsgleichung*, had to distinguish between short-ranged 'van der Waals forces' and the long-ranged 'Boltzmann–van der Waals forces' [160, p. 91].

This discussion seems to settle the matter with the best of the argument again on Boltzmann's side. The matter is, however, not so clear-cut; as it was put in a report at the Van der Waals Centennial Conference in 1973: 'It is doubtful whether we would celebrate the centennial of the van der Waals equation if it were applicable only under circumstances where it has been proved to be rigorously valid. It is empirically well established that many systems whose molecules have attractive potentials that are neither long range nor weak confirm nearly quantitatively to van der Waals' model. An example is the theoretically much studied system of argon, where the attractive potential has only a "range half as large [again] as the repulsive core"' [*31*, p. 143].

11.1.5 The ubiquitous van der Waals force

Van der Waals never declared publicly his thoughts on the physical nature of the universal forces of molecular attraction that he had postulated and which soon came to bear his name. Thirty years after the publication of *Continuity* he wrote rather sadly; 'We are still perfectly in the dark as to the cause of this attraction, and opinion differs greatly as to its dependence on the distance' [64, p. 85]. It seems that it was not until 1908 that he first set out his thoughts on the subject [92, p. 144–45]. He said of the parameters *a* and *b* in his equation that they are 'introduced not on equally firm bases and with the same certitude. One infers the existence of the quantity *b* perfectly safely if one believes that matter possesses an indispensable property to occupy room. . . . It is not the same or at least not quite the same as to the quantity *a*. Why ought molecular attraction to be a necessary attribute of substance? After all it does not follow from the notion "matter is something that necessarily occupies room" that matter ought to possess attraction' [92, p. 144]. He mentioned then the first attempts by his son in 1900 and 1908, by M. Reinganum in 1901 and by W. Sutherland in 1902 to explain intermolecular forces in terms of electrostatic interactions between dipoles within the molecules. He was not enthusiastic about this approach since it seemed to be incompatible with what he saw as the universality of molecular attraction. He recalled that in his thesis of 1873 he had predicted the possibility of liquefying hydrogen (which was first accomplished by Dewar in 1898) on the basis of the somewhat metaphysical argument[15], 'we must under all circumstances allow the property of attraction to matter' [$1e_2$, p. 207][16]. So he said in Amsterdam on 27 June 1908 that there was no reason to doubt 'the existence of *a* for example for helium', and within two weeks Kamerlingh Onnes in Leiden had succeeded in liquefying it.

It was more than fifty years after the publication of his thesis before a satisfactory explanation of the origin of these forces was given. Their very universality was a stumbling block since many promising calculations were based on the existence of molecular dipole or quadrupole moments and these could not apply to all molecules. Shou Chin Wang, a Chinese graduate student at Columbia University, was, in 1927, the first to point out the long-sought origin. Three years later London worked out a rigorous quantal treatment of the force (see e.g. [*3*, p. 210–12; *29*, *34*].) The explanation is as follows. Every atom and molecule, even if electrically neutral and with a time-averaged spherical symmetry, has an instantaneous electric dipole that arises from the fact that at any given moment the electrons are not distributed symmetrically. The electric field from this fluctuating dipole couples with similar fields in other molecules and so produces an attractive dipole–dipole force. Since the electric field of a dipole falls off as r^{-3} the energy of the coupling between a pair falls off as r^{-6}. This result has been incorporated into the simplest and most widely used representation of the whole intermolecular potential, the so-called 12–6 potential,

$$u(r) = Ar^{-12} - Br^{-6}. \tag{11.8}$$

Since the 1930s, when it was first used extensively by Lennard-Jones [*29*] (whose name it now bears) this potential has been the most popular reasonably realistic representation of the potential used in statistical mechanical calculations. The first term represents the steep repulsive potential which corresponds to van der Waals's impenetrability of the molecules and was obtained only after long series of attempts to fit experimental data.

In the late 1940s there was a striking development when it was realized that the essentially short-range forces between single molecules led to long-range forces between macroscopic objects. In the 1960s these were measured successfully out to distances as large as 10^{-4} cm (see e.g. [*35*]). The theory of these forces was worked out in 1948 by H. B. G. Casimir and D. Polder who showed that the finite speed of light, and hence of the propagation of the dipole–dipole coupling, led to a 'retardation' effect that changed the energy from a dependence on r^{-6} to one of r^{-7} at these macroscopic distances. Perhaps it could be said that the theory of van der Waals reached full maturity in 1954 when E. M. Lifshits developed a new approach to the subject which ignored the existence of individual molecules and treated matter as a polarizable continuum; his results included the London (r^{-6}) and Casimir–Polder (r^{-7}) results as special cases (see e.g. [*36,37*]).

11.1.6 The van der Waals picture of a fluid[17]

As we have seen, it was ninety years after the appearance of van der Waals's equation that a model was found that rigorously led to that equation of state. It was one of freely moving hard spheres (molecules) that collide elastically and which attract each other with weak but long-range forces. This 'van der Waals' model would not enjoy its present popularity if it were only an abstract construction designed to rehabilitate an old theory; it is also the basis of modern perturbation theories of liquids.

J. G. Kirkwood in 1940 and the Russian physicist I. Z. Fisher in 1955 predicted that a system of hard spheres with no attractive forces would undergo a phase transition – crystallization – at sufficiently high densities. Their predictions were, however, far from being rigorous deductions from theory and it was only in 1957 that their conclusion was confirmed by the direct simulation on a computer of a system of elastic hard spheres. First W. W. Wood using Monte Carlo simulation, then B. J. Alder and T. E. Wainwright using molecular dynamic simulation, demonstrated the existence of the phase transition. Both methods give also $g(r)$, the radial distribution function or average density at a distance r from a reference molecule. This function, which can be determined experimentally for real fluids by X-ray or neutron diffraction, is the starting point for all calculations of the thermodynamic properties of fluids. It was found that there was remarkable agreement between $g(r)$ for liquid argon and that obtained by computer simulation for a system of hard spheres, except in the immediate neighbourhood of the hard-sphere diameters. So by the middle of the 1960s there had grown a general understanding that the structure of at least simple liquids is determined primarily by the local packing and steric effects that arise from the impenetrability of the freely moving molecules. The attractive forces provide a relatively uniform background that has little effect on the structure or dynamics but supplies the cohesive energy that makes the system stable at a particular pressure and temperature. This picture, which is in essence that of van der Waals, was advocated by J. D. Bernal from about 1936 onwards on the basis of experiments on randomly packed ball bearings, but it was not given explicit expression in statistical mechanical theory until the 1960s, probably first in the papers of H. Reiss and of H. C. Longuet-Higgins and B. Widom.

This development joined with another that had been underway since the early 1950s, the development of statistical mechanical perturbation theory. This had been used for a variety of problems in liquid state physics by J. A. Barker, J. S. Rowlinson, J. A. Pople, R. W. Zwanzig, E. B. Smith and B. J. Alder, D. A. McQuarrie and J. L. Katz and others. In some of these studies the attractive part of the intermolecular potential (usually of Lennard-Jones form) had been treated as the small parameter and the properties of the system expanded about those of the hard-core molecules. That is, the van der Waals model served as a practical and well-understood reference system, much as the ideal gas and the ideal harmonic crystal had served as reference systems for gases and solids.

The last step of combining all these efforts into a quantitatively successful theory of liquids was made in 1967–70 first by J. A. Barker and D. Henderson and then by J. D. Weeks, D. Chandler and H. C. Andersen. It consisted of finding the most effective way of specifying the radius of the hard-spheres of the reference system and of dividing the whole potential into reference and perturbation parts. Such a division is entirely in the spirit of van der Waals's ideas of separating the effects of impenetrability from those of the attractive forces[18]. The radius of the hard spheres of the reference system is found by comparing the values of the thermodynamic properties found experimentally with those calculated from a radial distribution function in which the attractive forces have been neglected. The effective size of the spheres so obtained is not a constant but is a slowly decreasing function of temperature that reflects the so-called softening of the spheres, that is, the fact that the distance of closest approach of real molecules diminishes as the energy of the collision rises.

In his review of *Continuity* Maxwell wrote that for the 'condensed state' one could hardly expect to discover 'a mode of stating these properties . . . having the simplicity and generality at all approaching to that of the gaseous laws. . . . Because in the condensed state the properties of the substance depend on the mutual action when engaged in close encounter, and this is determined by the particular properties of the encountering molecules' [*33*]. A century later the same opinion was put even more categorically in Landau and Lifshits's *Statistical Physics*; in liquids 'the interaction between molecules is strong and the specific features of their interaction (and therefore those of the liquid) depend considerably on the nature of the liquid. That is why it is impossible ... to find general formulae that describe the properties of the liquid quantitatively' [*22* (1976), p. 256; (1977), p. 222]. These pessimistic views of the impossibility of establishing a simple, analytic and universal representation of the properties of all liquids are fundamentally correct, but this does not mean that modern perturbation theory based on a van der Waals reference model is flawed. Its interpretation of the properties of simple liquids and of computer simulated results shows that it provides a full understanding of the nature of the liquid state, and it has been found that even liquids of moderate complexity such as benzene, carbon tetrachloride and acetonitrile can be treated quantitatively without a complete knowledge of their intermolecular potentials.

11.1.7 The concept of the internal field

The van der Waals (or Laplacian) internal pressure had, it was soon realized, close parallels in other branches of physics. The most notable of these was in the field of magnetism. As early as 1889 the British engineer John Hopkinson had observed that the temperatures at which nickel and iron cease to be ferromagnetic resembled formally that at which a liquid ceases to be distinguishable from a gas. He wrote that there was a 'temperature at which the magnetism disappears which we may appropriately call the critical temperature' [*40*]. Pierre Curie's experiments of 1895 [*41*] strengthened this analogy; he observed that the parallel function to the equation of state of a fluid, $f_1(\rho,P,T) = 0$ is, for magnetism, $f_2(M,H,T) = 0$, where the degree of magnetization M is the equivalent of the density ρ, and H the strength of the magnetic field is the intensive function equivalent to the pressure P.

Pierre Weiss of Zürich developed the mean-field treatment of ferromagnetism in a way that was, as he recognized, the parallel of van der Waals's theory. In a lecture to the French Physical Society in April 1907 he said that the part of H that arose from the internal magnetic moments could be given 'the name of the internal field in order to mark the analogy with the internal pressure of van der Waals' [*42*]. He and Kamerlingh Onnes [*43*] introduced the name 'Curie point' for the magnetic critical point.

From the 1930s onwards the concept of an internal field proved to be the simplest starting point for the development of the theory of a wide range of transitions, such as those of ordering in metallic alloys, a variety of electric and magnetic transitions, and the formation of liquid crystalline phases.

11.2 Capillarity

We have seen in Chapter 6 that the bold theory of capillarity that van der Waals developed between 1888 and 1893 differed fundamentally from that of Laplace which had been such an inspiration to him twenty years earlier. The clearest conceptual difference was that Laplace postulated an interface with a molecularly sharp change of density from liquid to gas while van der Waals postulated a smooth function of density. He emphasized that the most obvious physical consequence for the surface tension σ was that Laplace's theory required that $(\mathrm{d}\sigma/\mathrm{d}T)$ remained non-zero at the critical point while his theory required that it vanish. It was shown by experimental physicists of the Dutch school that van der Waals was qualitatively correct, but the matter aroused little further interest and his bold innovations in the molecular theory of non-uniform systems were forgotten by all but a few specialists. Thus when Kamerlingh Onnes wrote an obituary of van der Waals in 1923 he praised his equation of state, his theory of mixtures and his principle of corresponding states, but he did not mention his theory of capillarity [208].

For over forty years there was no real advance beyond the results of Laplace on the one hand and van der Waals on the other, and still less any attempt to reconcile their two very different approaches. In 1935, Landau and Lifshits in Kharkov [*44*] essentially recreated van der Waals's result in the context of the interface between two magnetic domains, but they did not know of the earlier work. The next advance for liquids, when it came, was modest; in 1937 R. H. Fowler in Cambridge removed from Laplace's treatment the assumption of a uniform distribution of molecules in the liquid, but retained, however, the sharp interface [*45*]. He made no mention of van der Waals. An exact treatment which removed this last assumption and so led to a result that can be regarded as the natural end of the line of theory that stems from Laplace came in 1949 in a paper by J. G. Kirkwood and F. P. Buff at Pasadena [*46*]. In this they expressed the surface tension as an integral over the virial of the intermolecular potential and the two-body distribution function.

The previous year J. Yvon of Paris had found the exact result to which van der Waals's theory is the first approximation. This is an expression of the surface tension as an integral over the product of the gradients of the density at two different heights and the direct correlation function between molecules situated at these positions. Unfortunately he only announced his result at a meeting of the International Union of Pure and Applied Physics in Brussels in 1948 and never published his derivation. His announcement was, however, remembered by Frank Buff when he and his colleagues [*47*] and, independently, D. G. Triezenberg and R. Zwanzig [*48*] obtained the same result over twenty years later. Meanwhile J. W. Cahn and J. E. Hilliard [*49*] of the General Electric Company had again rediscovered van der Waals's result, this time for the liquid–gas interface. It was only in a second paper, in 1959, that Cahn mentioned that someone had told them that their theory was essentially the same as that of van der Waals [*50*].

The edifice was finally completed in 1979 when Peter Schofield at Harwell in England [*51*] showed that the 'virial' equation of Kirkwood and Buff and the 'compressibility' equation of Yvon were equivalent, that is, one could be transformed into the other. Since that time most work has been based on density-functional theories which are approximations that are related more closely to the van der Waals–Yvon treatment than to that of Laplace–Kirkwood and Buff. As so often in statistical mechanics, the exact equations have so far resisted all attempts at rigorous solution.

11.3 Critical phenomena

The first work on what we now call critical phenomena was that of the French polymath Cagniard de la Tour who studied liquids heated in sealed tubes. He found that above a certain temperature, which was different for each substance, the liquid disappeared to give a homogeneous fluid similar to a vapour. Faraday, after some new experiments of his own, was the first to interpret the results correctly. He said in the 1840s that at what he called the 'Cagniard de la Tour point' the liquid and the vapour become one. He recognized that the so-called permanent gases could not be liquefied by pressure alone at the lowest temperatures then attainable. Mendeleev in 1861 studied the change of surface tension with temperature and, inclining more to the view of Cagniard de la Tour than to that of Faraday introduced the name of 'absolute boiling temperature' for the point at which he said the cohesion of a liquid and the latent heat of evaporation both become zero (see Chapter 9).

Thomas Andrews began his systematic study of carbon dioxide at the time Mendeleev's work was published and his most important conclusions were set out in his first Bakerian Lecture of 1869. It was he who coined the names 'critical point' and 'critical temperature' and who first established conclusively the symmetry between liquid and vapour in the critical region, in contrast to the views of his predecessors[19] (Faraday excepted) who had spoken of the transition of a liquid to a vapour [171]. He noted also the opalescence that is so characteristic of fluids near their critical points.

11.3.1 The classical theory of critical phenomena

Theoretical work on critical phenomena began in 1873 with the publication of van der Waals's thesis and Gibbs's paper on the representation of the thermodynamic properties of fluid by surfaces in a three-dimensional space [*54*]. These two works were the origin of the molecular and the thermodynamic interpretation of critical phenomena respectively; their assumptions were perfectly consistent, and the development of a fruitful interaction between the two approaches was due principally to van der Waals. He used both methods freely in his later work on the theory of mixtures (see Chapter 12). His interest in critical phenomena was, however, always subsidiary to his interest in the molecular interpretation of continuity, the equation of state, corresponding states and capillarity. Critical phenomena did not have for him quite the fascination they have for modern theoretical physicists.

His early discussion of this topic arose, in his thesis, from the form of the equation of state which enabled him to express the three critical constants in terms of the parameters a and b. Thomas Andrews and James Thomson had discussed only one constant, the critical temperature; van der Waals was the first to consider all three as of equal importance. The distinction between the stable, metastable[20] and unstable states of a fluid below its critical temperature was set out by van der Waals and by Gibbs. It was, moreover, implicit in this work and in the principle of corresponding states that the general character of critical phenomena was the same for all substances.

What is today called the classical theory of critical phenomena started as a by-product of van der Waals's work on capillarity [37]. He wanted to know how the surface tension vanishes at the critical point and so he took the equation of state and his expression for the surface tension and expanded each property about its value at the critical point in a Taylor series. He used Maxwell's equal-area rule and the two critical conditions[21] $(\partial P/\partial V)_T = 0$ and $(\partial^2 P/\partial V^2)_T = 0$ and so obtained the following results: on the critical isotherm

$$(P - P_c) \sim (V - V_c)^\delta, \quad \delta = 3;$$

on the coexistence curve

$$(\rho_l - \rho_v) \sim (T_c - T)^\beta, \quad \beta = \frac{1}{2}; \tag{11.9}$$

and for the surface tension

$$\sigma \sim (T_c - T)^\mu, \quad \mu = \frac{3}{2}.$$

(The symbols δ, β and μ are modern and are now called the critical exponents.)

He knew that these equations had a wider validity than that of his, or any other, equation of state. Experiments seemed at first to confirm these results and the related consequence that the sum of the coexistent densities $(\rho_l + \rho_v)$ is a linear function of $(T_c - T)$, which was the empirical law of rectilinear diameters discovered by Cailletet and Mathias in 1886 [*56*]. He reported that E. C. de Vries, in his Leiden thesis of 1893, had found that $\mu = 1.23$, but he was not worried by this relatively small discrepancy since he thought that the apparent value of the exponent was increasing with temperature and it would reach $\frac{3}{2}$ at the critical point.

11.3.2 Anomalies at the critical point

From the 1890s onwards evidence accumulated which showed that the classical values of the exponents were not borne out by experiment. The most assiduous worker in this field in the early years of the century was the Belgian physicist J. E. Verschaffelt who did much of his work at Leiden. He combined his own results with careful calculations from the experimental work of others, in particular Sidney Young's superb measurements on isopentane, to arrive at the set of values $\delta = 4.26$, $\beta = 0.33$–0.37 and $\mu = 1.22$–1.23. He insisted, with ever increasing confidence, that the discrepancies with the classical values were not due to the measurements being taken insufficiently close to the critical point but his evidence fell on deaf ears [*57*]. As late as 1924 van Laar was still contesting the point; he wrote that 'this $\frac{1}{3}$ power law [for β] is not a law of nature. Of course not, because theory tells us that in the vicinity of T_c, $\frac{1}{2}(\rho_l - \rho_v)$ necessarily must be proportional to $|\Delta T|^{1/2}$' [22]. It was van Laar himself who had first emphasized that the expansion of the free energy in a Taylor series leads inevitably to the classical values of the exponents and, since he had no reason to question the validity of such an expansion, he was able to make his confident assertion that β must be $\frac{1}{2}$. In 1910–12 van der Waals proposed a new function that fitted the results more accurately (and so contradicted his work of 1893) but the attempt was half-hearted and, in general, the discrepancies were left unexplained or indeed forgotten, since the interests of most physicists had moved to new fields.

11.3.3 The modern era

Interest in critical phenomena revived in the 1930s but now it was not the gas-liquid transition that was the centre of interest but the newer fields of order-disorder transitions in alloys, ferromagnetic and later anti-ferromagnetic materials, ferroelectrics and phenomena that were not at first recognized as belonging to the field such as superconductivity and the lambda-transition in liquid helium. The experimental results led to some usually rather imprecise values of the critical exponents, and of the size of the accompanying anomalies in the heat capacities, which rarely matched the simple predictions of the classical mean-field theories. The most general of these was Landau's theory of second-order phase transitions of 1937 [*59*]. It too was based on an expansion and so led inevitably to the classical values of the exponents.

The experiments became more accurate after the War and the newer and more precise values of the exponents convinced the physicists that the classical theory was seriously flawed. A breakthrough on the theoretical side was the solution by Lars Onsager of Yale of the problem of the square two-dimensional Ising model of ferromagnetism in 1942–48 [*60,61*], which led, after further work by himself and others, to the values $\delta = 15$, $\beta = \frac{1}{8}$ and $\mu = 1$. These two-dimensional values are even further from the classical ones than the experimental three-dimensional values for fluids. This work established conclusively that in two dimensions, and by implication also in three, the free energy is not an analytic function of its natural variables, volume and temperature, at a critical point. Neither van der Waals nor van Laar had ever contemplated such an extraordinary fact and although Verschaffelt had, his functions were not of the correct form.

The field soon became one of the more exciting areas of theoretical physics since it was clear that some radically new ideas were needed, the most important of which was a realization of the dominant rôle of fluctuations of density (and of composition in a mixture) in the critical region. It had long been known that when one phase is in a state that is almost at equilibrium with another then fluctuations give rise to precursors of the second phase within the bulk of the first, as for example liquid droplets in a vapour which is just below its saturation vapour pressure, or micro-domains of the 'wrong' magnetization in a paramagnetic crystal. The scale and magnitude of these fluctuations increase dramatically in the critical region and it is this growth and, in the limit, its divergence, which is responsible for the non-analytic character of the free energy.

One important characteristic of these fluctuations is that, like a fractal, they are present on all scales of length. This physical scaling is matched by a mathematical scaling of the thermodynamic functions with respect to certain groupings of their natural variables. It was only after an understanding of these two related facts that a satisfactory theory could be devised by borrowing from the theory of fundamental particles the concept of the renormalization group. It was thus shown that the exponents had their classical values in a world of four or more dimensions, and the calculated values in three and two dimensions were in reasonable accord with the known results. For this work Kenneth Wilson of Cornell received the Nobel Prize for Physics for 1982 [*62,63*].

The suppression of fluctuations or the imposition of homogeneity is inherent in van der Waals's assumption of an internal mean molecular field. It is justified in less than four dimensions only if the range of the forces is sufficiently long. There are a few ferroelectric examples where this is true but these are rare [*64*]. In most critical phenomena we see the conflict at the heart of van der Waals's work, the incompatibility of short-range forces with the assumption of a mean field. In many circumstances this conflict is qualitatively unimportant since in a three-dimensional world one can pack a large number (about 12) of molecules around a central molecule and this large number enforces a certain degree of homogeneity, but at the critical point the conflict is unresolved and the classical theory must be abandoned.

11.4 The principle of corresponding states

11.4.1 The early history

Clausius was one of the first from abroad to take note of van der Waals's principle of corresponding states. His paper of July 1881 [*21*] referred both to van der Waals's original paper of 1880 [12] and to the German edition of *Continuity* which had been published in June 1881 and which contained two new chapters, 12 and 13, that included an account of the principle. Clausius's paper dealt with the properties of the fluid in the two-phase region and he noted, no doubt with some satisfaction, that his Dutch colleague had not, as he had, been able to give detailed calculations nor an equation for the boundary curve, 'because', as van der Waals said, 'the former are too extensive and the latter too complicated.' He did, however, note that van der Waals had 'gathered a number of important consequences' that followed from the equation of state. In a second part of his paper [*65*] Clausius started to use reduced variables, but without reference to van der Waals. He says, 'It goes without saying that we can [thus] render these equations in a simpler form.' His word for *corresponding* is here *übereinstimmenden* which is the direct German equivalent of the Dutch *overeenstemmende* of van der Waals. He points out in a footnote that he had discussed this problem in an old paper of 1851 [*66*] and later, in 1864, in the first volume of his collected papers on the mechanical theory of heat [*67*]. He writes, moreover, that the notion of corresponding temperatures, that is, temperatures at which liquids had a common vapour pressure, was introduced by John Dalton as early as 1801, and was used also by J. A. Groshans in 1848–50 [*68*]. (This time Clausius writes *entsprechenden Temperaturen*, which is closer to the English *corresponding temperatures*.) Dalton said that the difference between the corresponding temperatures of two liquids was independent of the choice of value of the common vapour pressure; Groshans put this law into algebraic form. In 1851 Clausius had shown the limited validity of this law and, starting from an empirical rule of C. M. Despretz in the 1820s [*69*], had deduced thermodynamically his own law. Despretz had said that the latent heat of evaporation needed to produce a unit volume of vapour at the normal boiling point was the same for all liquids. Clausius substituted the corresponding temperature for the normal boiling point to avoid 'the accidental value of atmospheric pressure' and so was led to the following law: '. . . the same function that describes the temperature dependence of the vapour tension of one liquid is applicable to any other liquid providing that the temperature is multiplied by a certain constant which may be determined if the value of the vapour pressure is known for at least one temperature' [*67*, p.125].

A curious episode in the early history of the principle of corresponding states was a furious claim of priority from a German *Privatdozent* Eugen Dühring, who is now best remembered for the attack on him by Friedrich Engels, one of the father-figures of 'scientific communism'. Engels's book of 1878, *Anti-Dühring*, became, with Marx's *Das Kapital*, the bible of marxism. In the same year Dühring published his *New fundamental laws of rational physics and chemistry* [*70*] which claimed to bring order to all of natural science by means of five new fundamental laws. One of these was the relation

$$(T_1' - T_1)/(T_2' - T_2) = q_1/q_2, \tag{11.9}$$

where T_1' and T_2' are the normal boiling points, T_1 and T_2 are the temperatures at which the vapour pressures are equal to some arbitrary pressure, and q_1 and q_2 are constants characteristic of each liquid [*70*, p. 73].

It soon became clear after the publication of van der Waals's paper of 1880 that all earlier 'laws' for the vapour pressure were at best but small fragments of the more general law that he put forward, and that their validity was limited by special conditions. Thus, as van der Waals himself showed [1e$'_1$, p. 468], Dühring's equation holds only if the critical pressures of the two liquids are the same. Such arguments had no effect on Dühring and, aided by his son, he defended his priority for more than twenty years, accusing 'a Dutch Professor Waals' and 'a Prussian Professor Clausius' of inaccurately reproducing his law and claiming it as their own. He alleged also that van der Waals's equation of state was stolen from Bernoulli's and his own work [*71*, p. 609].

Van der Waals seemed to have been familiar with the works of his predecessors although he rarely mentions them. He did however note the contribution of Mendeleev who had, in 1870, observed that the most reasonable state in which to compare the properties of different liquids was at the 'absolute boiling points', that is, at the critical temperatures (see Chapter 9). Van der Waals was, however, the only one to realize the importance of using also the critical pressure and volume as 'reducing' parameters, and to show that this importance stemmed from their direct relation to the underlying molecular quantities. He wrote: 'In general therefore we must reject, as not possessing any rational signification, all relations from which are deduced properties for equal pressures or equal temperatures. For since the critical pressures are usually different there can never exist a general relation between the properties of different bodies at the same pressure. This will become obvious if we consider two bodies; one of which alone is at its critical pressure. In this case it is clear that the bodies are under such different conditions that their pressure relations cease to be comparable. From this it follows that most of the relations generally expressed are deprived of much of their signification' [1e$'_1$, p. 469].

It is not surprising that it was the vapour pressure curve that was used first to test the new principle. He found the principle to be almost correct and was inclined to believe that one of the causes of departures was the presence of impurities. His attempts to understand their effects was one of the factors that led him to develop his theory of mixtures [1e$'_1$, p. 473,479; see Chapter 12].

11.4.2 The testing of the principle

Levelt Sengers has written that 'The forcefulness of this theorem [i.e. the principle] was immediately obvious to van der Waals and to the scientific community' [177, p. 77]. Kamerlingh Onnes certainly responded immediately and started work on the testing of the principle in the autumn of 1880; others were less prompt. In the 1880s Nadezhdin in Kiev made some extensive studies on vapour pressure curves and showed, in 1885, that the 'law is really a general one' (see Chapter 9). Between 1882 and 1887 Ramsay and Young made a series of accurate and relevant measurements of vapour pressures, critical temperatures and molar volumes of a set of mainly organic substances. After Ramsay became involved in the work that led to the discovery of the inert gases Young continued on his own into the early years of this century. This work provided the decisive evidence for testing the principle. Gavroglu has examined in detail the relationship between the experiments of Ramsay and Young and van der

Waals's principle. It was not until 1887 that they mentioned his equation of state, not until after 1890 that Young relied frequently on the principle, which he referred to as 'van der Waals's generalisations', and it was only in 1902 that he used for the first time the term 'law of corresponding states' [184, p. 234–235]. He was then testing the law of rectilinear diameters (see Section 11.3.1) which his results confirmed, although modern theory has suggested that it may not be truly correct [177; *72*].

One of the co-discoverers of this law, the French physicist Émile Mathias, contributed most to the favourable reception of the principle of corresponding states outside the Netherlands. In his doctoral thesis of 1890 [*73*] and in later works he systematically reduced his own measurements, and those of Young and others, and concluded that 'the theorem [i.e. the principle] is particularly exact for the density of liquids', and that 'Young has demonstrated the existence of groups (the alcohols, the halo-benzenes) that confirm the beautiful theorem of van der Waals' [see 184, p. 236]. Gavroglu writes that 'Mathias was even more enthusiastic about the validity of the law of corresponding states in the epilogue he wrote for the French translation of van der Waals's thesis' [1f′]. He was impressed by the 'harmonic simplicity' of the law. '. . . According to Mathias, the law was one of the most significant results of contemporary physics for the study of heat and a method of comprehending, by a unique formula, the fundamental properties of all fluids' [184, p. 236].

Perhaps the most important conclusion from the confrontation between the experimental results and the principle was that liquids could be classed into internally conforming groups on the basis of Young's critical ratio,

$$\gamma = RT_c/P_cV_c, \tag{11.10}$$

whose value is 8/3, or 2.67, according to van der Waals's equation of state. Thus the four halo-benzenes (fluoro-, chloro-, bromo- and iodo-) all have values of γ in the narrow range 3.78 to 3.80, their reduced vapour pressure curves coincide within 1 per cent, and their orthobaric boundary curves are almost as close. A second distinctive group is formed from the highly polar or associated liquids such as water, some organic acids and the lower alcohols for which γ is above 4.3.

An elegant way of using the principle for classifying substances into groups and of determining their critical parameters was devised by E.-H. Amagat in 1896; it did not require any prior knowledge of the critical properties nor of the equation of state. He made transparencies with grids of the isotherms for the different fluids and used a simple optical projection to change the scale of the grids until they were brought into coincidence [*74*, p. 574–578]. In 1900, after ten years' experience of testing the principle, he wrote: 'As to the generality of van der Waals's law, the need of forming groups does nothing to detract from its importance. It becomes a means of classification whose usefulness will no doubt make us forget that it once had the seductive appearance of simplicity and generality that was originally foreseen and, I believe, desired of a law. The work of van der Waals retains therefore all its importance and fruitfulness' [*74*, p. p.582].

In 1945, at a time when the value of van der Waals's equation was least appreciated, E. A. Guggenheim wrote: 'The principle of corresponding states may safely be regarded as the most useful by-product of van der Waals's equation of state. Whereas this equation of state is nowadays recognised to be of little or no value, the principle of corresponding states correctly applied is extremely useful and remarkably accurate' [*75*, p. 253].

11.4.3 Attempts at improvement

The most impressive work on the principle of corresponding states was carried out by van der Waals's staunch supporter Kamerlingh Onnes who kept it at the centre of his scientific interests up to the time of his death in 1926. He wrote[23] in 1901: 'Various methods have been tried to empirically derive functions of V and T for van der Waals's a and b by means kinetic or thermodynamic considerations, but without obtaining a good agreement with the observations over whole range of the equation of state. Neither was I successful in similar attempts which were repeatedly occasioned by my continued research on the corresponding states and other investigations resulting from them at the Leiden laboratory. Whenever I seemed to have found an empirical form, I discovered after having tested it more closely that it appeared useful only within limited range to complete what had been found in a purely theoretical way by van der Waals and Boltzmann. Hence it appeared to me more and more desirable to combine systematically the entire experimental material on the isothermals of gases and liquids as independently as possible from theoretical considerations and to express them by series' [*76*, p. 3]. He was thus led to his so-called virial equation of state [*76*, *77*],

$$\frac{PV}{RT} = 1 + \frac{B(T)}{V} + \frac{C(T)}{V^2} + \frac{D(T)}{V^4} + \frac{E(T)}{V^6} + \frac{F(T)}{V^8}, \qquad (11.11)$$

in which the virial coefficients, B, C, etc. are themselves given by,

$$B(T) = b_1 + \frac{b_2}{T} + \frac{b_3}{T^2} + \frac{b_4}{T^4} + \frac{b_5}{T^6}. \qquad (11.12)$$

Kamerlingh Onnes used the expansion of van der Waals's equation in powers of the density and then reduced the expansion to a polynomial composed of a moderate number of terms each expressed as a limited set of powers of $(1/T)$. He wrote, 'The number of terms in our polynomials must be chosen so that all the isothermals of the different substances which we consider in our calculations can be represented by means of coefficients, each of which can be taken to be derived from the observations' [*77*, p. 5][24].

In the spirit of the principle of corresponding states he then turned to van der Waals's reduced variables $l = P/P_c$, $m = T/T_c$ and $n = V/V_c$ to obtain the reduced virial equation of state

$$\frac{nl}{m\gamma} = 1 + \gamma\frac{B'(m)}{n} + \gamma^2\frac{C'(m)}{n^2} + \gamma^4\frac{D'(m)}{n^4} + \gamma^6\frac{E'(m)}{n^6} + \gamma^8\frac{F'(m)}{n^8}, \tag{11.13}$$

with

$$B'(m) = b_1' + b_2'/m + b_3'/m^2 + b_4'/m^4 + b_5'/m^6, \tag{11.14}$$

and similarly for C' (m) etc.

Kamerlingh Onnes and Crommelin 'synthesised' eqn (11.13) by combining the measurements of Amagat on hydrogen, oxygen and nitrogen, of Ramsay and Young on ether and of Young on isopentane. They noted that a major problem was that the results for each substance often did not overlap, on reduced scales, with those of another. If the principle were obeyed strictly this difficulty would disappear but, in practice, it remained a problem. Nevertheless, in the end an 'equation of state has been obtained for an imaginary substance which . . . is suitable for all calculations in which the validity of the law of corresponding states is assumed. And this equation is of particular use in tracing deviations from the law . . .' [*79*, p. 3].

From the end of the last century until the mid 1920s an immense amount of work was carried out at Leiden on this reduced virial equation of state[25]. At first most of this work was on the diatomic gases, hydrogen, oxygen and nitrogen, since it was thought that their simple constitution corresponded most closely with the assumptions from which van der Waals had started. Later it was realized that the monatomic gases were even more suitable and in 1911 'helium at once usurped the place originally set apart for hydrogen' [*82*].

The sets of five virial coefficients were obtained for the various diatomic and monatomic gases but the goal of establishing a universal reduced equation of state remained unattainable. It would seem to be an act of despair to try to establish the principle on the basis of the twenty-five adjustable parameters of eqns (11.13) and (11.14)[26]. Nevertheless this approach proved to be valuable both for practical applications and because of the direct link between the virial coefficients and the intermolecular forces [*83*].

Gibbs appears to have had a particular interest in the principle of corresponding states, from a note found in his papers after his death in 1903 [*84*, p. 418] and from a letter to the American physicist Carl Barus written, most probably, in the early 1890s. Gibbs's approach to the problem in this letter is very much in character: '. . . It is worthy, I think, of much effort to find out whether, or with what degree of approximation, and within what limits, and with what exceptions, the actual laws for different substances are *similar* in the sense defined above [i.e. in the sense of the principle]. And if not similar, to find how many degrees of variation there are, in addition to the three connected with the measure of temperature, pressure and volume. In other words, to find how many independent constants there are in the general formula . . .' [*85*].

He then outlined a sequence of measurements to find the number of these independent constants needed for a sufficiently accurate representation of the thermodynamic behaviour of various groups of substances. He proposed to classify fluids first on the basis of what Kamerlingh Onnes and Keesom had called the critical pressure ratio [160, p. 139], that is, the limiting slope of the reduced vapour pressure curve[27], or $D = (\mathrm{d}\ln P/\mathrm{d}\ln T)$. He would then examine each group to see whether they had 'similar relations between P, V and T'. Gibbs's proposal for introducing an extra parameter came to maturity in the early 1950s when it was realized that such a parameter could, indeed, be used to classify systematic departures from the principle followed by the inert gases [*88–90*].

Quantum mechanical effects led to the introduction of a further parameter. As early as the spring of 1913 Kamerlingh Onnes and Keesom had suggested that at low temperatures zero-point vibrations would give rise to a contribution to the pressure, even for an ideal gas [*91*;*92*, p. 5]. Alfred Byk in Berlin tackled this problem more correctly and proposed that the parameter whose size reflects the importance of the quantal 'corrections' can be written

$$\Lambda = h(\mu k T_c)^{-1/2}(V_c)^{-1/3}, \tag{11.15}$$

where h is Planck's constant and μ the mass of the molecule [*93*, *94*]. Kamerlingh Onnes noted the 'interesting work' [*95*] of Byk, but the value of this extra parameter was not fully realized until the work of de Boer and Michels in 1938 [*96*].

11.4.4 Mechanical similarity

Van der Waals's principle of corresponding states can be written most generally as

$$F(l, m, n) = 0, \tag{11.16}$$

where the unspecified function F is identical for each substance. Kamerlingh Onnes was the first to realize that this form, or $l = f(n,m)$ as he wrote it, could be derived from a principle of mechanical similarity in the molecular interactions, without reference to any explicit form of the equation of state [*97*]. Notions of such similarity go back to Galileo and Newton, and had been used by Osborne Reynolds in his classic paper of 1883 on the criterion for laminar flow in hydrodynamics [*98*], but Kamerlingh Onnes seems to have been the first to use them in the modern sense. He argued that the corresponding states of thermodynamic systems are those that are mechanically similar, that is, they are formed of molecules whose parameters of motion and interaction are the same for each substance provided that they are measured in units specific to the molecules of that substance. The conditions needed for the principle to hold are as follows: '1. The molecules of the different substances are completely hard elastic bodies of a common shape; 2. the long-range forces that they exert emanate from homologous [corresponding] points and are proportional to the same function of the homologous separations of these; 3. the absolute temperature is proportional to the mean kinetic energy of the translational motion of the molecules' [160, p. 80].

In modern terms one might say that if the size of a molecule i is measured by a length σ_i, its mass by μ_i, and if the intermolecular force has the common form for all substances $f_i(r) = K(r/\sigma_i)$, then one can define a characteristic time for substance i by $\tau_i = (\mu_i\sigma_i/K_i)^{1/2}$ and the appropriate factors with which to scale the molar volume, the pressure and the temperature are σ_i^3, $(\mu_i/\sigma_i\tau_i^2)$ and $(\mu_i\sigma_i^2/\tau_i^2)$ respectively.

'Kamerlingh Onnes's argument was mechanical rather than statistical; nevertheless, the basic insight was there, namely that molecular interaction must be similar except for two characteristic parameters so that pressure, volume and temperature can be scaled. These very same ideas were basis for the formulation [1938–45] of corresponding states of J. de Boer and A. Michels, K. S. Pitzer, and E. A. Guggenheim . . .' [177, p. 77].

This method of dimensional analysis became the routine way of combining thermodynamic parameters into combinations that might be expected to be the same for all substances. Thus Byk arrived at the parameter Λ of eqn (11.15) by representing the unit cell of phase space of volume h^3 in terms of the critical constants. It was not until 1938 that de Boer and Michels gave it the natural interpretation of the ratio of the de Broglie wavelength of thermal motion to the characteristic spacing at the critical point $V_c^{1/3}$.

The modern form of Kamerlingh Onnes's condition is to require pairwise additive central intermolecular forces whose energy has the form

$$u_{ii}(\gamma) = \varepsilon_{ii}f(r/\sigma_{ii}), \qquad (11.17)$$

where ε_{ii} and σ_{ii} are the energy and distance that determine the strength and range of the force and f is a function common to all substances. For many years the function f was generally taken to be of the Lennard-Jones form [183] but this is now known to be inadequate even for the inert gases. Systematic departures from the principle of corresponding states can be related to changes in the form of the function f and to the presence of other terms in eqn (11.17); these changes are handled by perturbation theory in modern statistical mechanics.

The principle of corresponding states has played an important rôle as a practical tool in the history of low-temperature physics and engineering. Dewar, one of the pioneers of low temperatures, repeatedly expressed his delight at the principle and in van der Waals's works in general [184]. On 25 July 1908 he wrote to Kamerlingh Onnes: 'Van der Waals is the master of us all. He is after all the creator of all our ideas and cannot be too highly honoured' [184, p. 229]. In reality, however, his love of the principle was more platonic than practical; he made little use of it in his investigations. 'He did not mention the law of corresponding states in any of his papers until the first liquefaction of hydrogen in 1898' [184, p. 229]. In the paper in which he reported this brilliant achievement he referred to Wroblewski's fine estimate of 33 K for the critical temperature of hydrogen made ten years earlier and based on the principle. He wrote '... the accuracy of Wroblewski's deductions regarding the chief constant of liquid hydrogen following from a study of the isothermals of the gas is a signal triumph for the theory of van der Waals ...' [*99*].

It was, again, Kamerlingh Onnes who initiated a systematic use of the principle (including his form of it) as a working tool of low temperature research. His successor and close associate, W. H. Keesom, has written that when Kamerlingh Onnes was appointed Director of the Physical Laboratory at Leiden in 1882 the very choice of the main direction of the research 'was apparently determined by the appearance in 1880 of a paper by van der Waals which introduced the law of corresponding states' [*100*, p. 6]. This was confirmed by Kamerlingh Onnes himself: 'I was induced to work with condensed gases by the study of van der Waals's law of corresponding states' [*101*, p. 4]. In 1896, two years before the liquefaction of hydrogen by Dewar and ten before this was achieved in Leiden, he made a bold attempt to predict what would happen if hydrogen were substituted for oxygen in his liquefaction apparatus; he even introduced a new notion of 'thermodynamically corresponding operations' [*102*], p. 6]. In 1908, a few days after his first liquefaction of helium he sent Dewar a photograph of the apparatus with the remark, in which he refers to his 1896 paper: 'On the photograph I have put indications that will show you the way the helium circulation built [up] "corresponding" (by applying the ideas of my Comm. no 23 at least) to my hydrogen calculations.... It was a good thing to have trusted to the utmost the theory of van der Waals and to my isotherms, which could only be obtained after many years of preliminary work.... [The law of corresponding states] has been laid at foundation of the method of liquefying helium' [184, p. 211].

Keesom recalled that when Kamerlingh Onnes was contemplating his final attempt to liquefy helium there were two estimates of the critical temperature – the first by Dewar of 5–6 K and the second by Olszewski of 2 K. 'In the first case there was a prospect of success, in the latter helium must lie in the region as yet unattainable.' Kamerlingh Onnes's own estimate 'gave from the law of corresponding states a probable value of the critical temperature of about 5 K. ... This formed sufficient grounds for concentrating all energies upon the liquefaction' [*100*, p. 18].

It is interesting, although generally unknown, that the principle of corresponding states was again used – this time quite erroneously – by Kamerlingh Onnes as a guide in his second immortal achievement, the discovery of superconductivity [*103*].

Notes

1. This and other early work is discussed by Brush [179, *3*].
2. A well-argued derivation from caloric theory. The quantity t is essentially the virial of the intermolecular forces [173].
3. The parameter u, like m in Hirn's equation and n in Bernoulli's, represented the

sum of the volumes of the molecules, and was called the co-volume by Dupré and later writers [183].

4. This derivation by Lorentz [*7*] was discussed in the German translation of *Continuity* [1d_1'].
5. The day after sending his note to *Nature*, which is dated '13 October', Rayleigh wrote in a private letter: 'I am having another fling at Tait. He brings disgrace upon the country by depreciating all the work of foreigners. I am more than ever an admirer of Van der Waals' [*14*].
6. In his note Korteweg had written that 'it is extremely improbable that this question should have to be answered in a different way for linear and for three-dimensional space' [*13*].
7. Mathias wrote in 1894: 'The attention of scientists has been attached exclusively to the form into which van der Waals put his characteristic equation of fluid bodies. The modified equation proposed by Clausius as more closely related to experiment replaced it almost immediately, and many of the results van der Waals had arrived at in his famous theory of the gaseous and liquid states have lain unknown for a long time [1f′, p. 270].
8. At the turn of the century such attempts were made by W. Sutherland in Australia, by M. Reinganum in Germany, and by N. Schiller in Russia.
9. Van der Waals used the terms 'quasi-association' and 'apparent association' to describe the formation of aggregates by intermolecular (or van der Waals) forces in order to distinguish it from association due to forces of a more 'chemical' nature, such as association by means of hydrogen-bonding.
10. One of the authors (B. E. Ya.) remembers that in the mid-1960s an incautious mention by a student of 'van der Waals's theory' aroused sharp indignation in the theoretical physicists of Landau's school.
11. This, like many other early results, was forgotten and so rederived later by other physicists. Today it is often referred to as Tonks's equation [183, p. 50–61].
12. Van der Waals seems to have taken little notice of Ornstein's solid support for his equation of state, but in his Nobel Lecture he remarked that the path he had taken in molecular physics may have been 'a detour', and that, 'Perhaps there is a direct way. In search for this Gibbs' *Elementary Principles of Statistical Mechanics* will be a necessary guide' [103e, p. 263].
13. S. G. Brush has given the best account of early research on what we now call interatomic and intermolecular forces [*3*, p. 204–232; *29*].
14. Van der Waals was the first to use this potential which is now generally known as the Yukawa potential and has a wide range of uses [*30*].
15. Van der Waals thought it 'more than risky' to infer molecular attraction from Newton's omnipresent gravitational attraction. It is of interest, and perhaps of surprise, that in 1901, in his first published paper, Einstein expressed the view that 'The question whether and how there is a connection between our forces [of molecular attraction] and gravitational forces must remain therefore entirely open' [*32* (1901), S. 523; (1989), p. 20]. See also the essay *Einstein on the nature of molecular forces* [*32* (1989), p. 1–8].
16. In his review of van der Waals's thesis Maxwell had written that 'It is probable that it [hydrogen] never will be liquefied, as the attractive force is so weak' [*33*, p. 479].
17. The exposition in this section follows mainly from [183, p. 46–61; *38*,*39*].
18. The feature of van der Waals's thesis that most displeased Maxwell was the treatment of the impenetrability of molecules and their mutual attractions in entirely different ways. It is therefore paradoxical that it is just this division that underlies

much of modern theory and is the cause of much of the interest in van der Waals's work.

19. Andrews referred to all his predecessors except Mendeleev, although he was later to send to him a reprint of his second paper. Mendeleev's work was well known [*52*,*53*,160,169]; perhaps Andrews simply overlooked it.
20. This name was used first by Ostwald in 1897.
21. In his thesis van der Waals had found the position of the critical point from the point of coincidence of the three roots of his cubic equation of state. J. D. Hamilton Dickson, a Scots physicist and mathematician at Cambridge, was the first to use these simpler equations in 1880 [*55*]. They were used independently by Kamerlingh Onnes the next year [160, p. 60].
22. The translation is that of Levelt Sengers who has given full accounts of the reception of these persistent anomalies by the physicists of the day [*57*,*58*,177].
23. Kamerlingh Onnes noted the work of M. Thiesen who, in 1885, was the first to write PV as an expansion in powers of the density and who calculated the first two coefficients from Regnault's measurements of 1847 [160, p. 149].
24. In his first attempt of 1897 to establish a virial expansion Kamerlingh Onnes did not omit the odd powers of density [*78*]. Modern practice and theory include them.
25. See the discussion of Gavroglu [184] and of Gavroglu and Goudaroulis [*80*,*81*].
26. If van der Waals's equation were strictly true then there are only six adjustable parameters in Kamerlingh Onnes's virial equation; all coefficients vanish except for b'_1, b'_2, c'_1, d'_1, e'_1 and f'_1.
27. Stoletov [*86*] was the first to show that $D = 4$ for van der Waals's equation (with Maxwell's equal-area rule) by using the method with which Planck had calculated the vapour pressure curve for Clausius's equation [*87*]. Van der Waals was apparently unable to derive this result and the values he found from experiment extended to 7 and above [1e$'_1$, p. 478]. Contrary to his usual instincts he used an empirical formula

$$\log_{10} l = f(1 - m^{-1})$$

with f about 3 for 'normal' substances, although his equation of state requires $f = 1.72$.

References

1. Clausius, R., Über das Verhalten der Kohlensäure in Bezug auf Druck, Volumen und Temperatur, *A. Ph.*, 1880, Bd 9, S. 337–357.

2. Recknagel, G., Ueber Temperatur und Temperaturmaass, *A. Ph.*, 1874, Erg.Bd 6, S. 275–302.

3. Brush, S. G., *Statistical Physics and the Atomic Theory of Matter, from Boyle and Newton to Landau and Onsager*, Princeton, New Jersey, 1983.

4. Bernoulli, D., *Hydrodynamics, sive de viribus et motibus fluidorum commentarii*, Argentorati [Strasbourg], 1738; English translation of the section on kinetic theory in Brush, S. G., *Kinetic Theory*, v. 1, *The Nature of Gases and of Heat*, Oxford, 1965, p. 57–65.

5. Ritter, E., Sur la constitution physique des fluides élastiques, *Mém. Soc. Hist. Nat. Genève*, 1846, t. 11, p. 99–114.

6. Dupré, A., *Théorie mécanique de la Chaleur*, Paris, 1869.

7. Lorentz, H. A., Über die Anwendung des Satzes vom Virial in der kinetischen Theorie der Gase, *A. Ph.*, 1881, Bd 12, S. 127–136.

8. Tait, P. G. On the foundations of the kinetic theory of gases. IV, *Trans. R. Soc. Edinburgh*, 1890–91, v. 26, p. 257–272; *Collected Papers*, v. 2, p. 192–208, Cambridge, 1900.

9. Tait, P.G., On van der Waals's treatment of Laplace's pressure in the virial equation: in answer to Lord Rayleigh, *Nature*, 1891, v. 44, p. 546–547, p. 627–628.

10. Rayleigh (Lord), On van der Waals's treatment of Laplace's pressure in the virial equation: a letter to Prof. Tait, *Nature*, 1891, v. 44, p. 499.

11. Rayleigh (Lord), On van der Waals's treatment of Laplace's pressure in the virial equation: a letter to Prof. Tait, *Nature*, 1891, v. 44, p. 597.

12. Rayleigh (Lord), On the virial of a system of hard colliding bodies, *Nature*, 1891, v. 45, p. 80–82.

13. Korteweg, D.T., On van der Waals's isothermal equation, *Nature*, 1891, v. 45, p. 152–154.

14. Strutt, R. J. (4th Lord Rayleigh), *John William Strutt, Third Baron Rayleigh*, London, 1924, p. 183.

15. Boltzmann, L., *Vorlesungen über Gastheorie*, Bd 1, 2, Leipzig, 1896–1898; *Lectures on Gas Theory*, trans. S. G. Brush, Berkeley, California, 1964.

16. Jäger, G., Die Gasdruckformel unit Berücksichtigung des Molekularvolumens, *Kais. Akad. Wiss. Wien*, Sekt.IIa, 1896, Bd 105, S. 15–21.

17. Boltzmann, L., Über die Berechnung der Abweichungen der Gas vom Boyle-Charles'schen Gesetz und der Dissoziation derselben, *Kais. Akad. Wiss. Wien*, Sekt.IIa, 1896, Bd 105, S. 695–706.

18. Nairn, J. H. and Kilpatrick, J. E., Van der Waals, Boltzmann and the fourth virial coefficient of hard spheres, *Am. J. Phys.*, 1972, v. 40, p. 503–515.

19. van Laar, J. J., Berekening der tweede correctie op de grootheid *b* der toestandsvergelijking van van der Waals, *Versl.*, 1899, d. 7, b. 350–364.

20. Kohnstamm, P. A., On the equations of Clausius and van der Waals for the mean length and the number of collisions, *Versl.*, 1904, d. 12, b. 961–967.

21. Clausius, R., Über die theoretische Bestimmung des Dampfdruckes und der Volumina des Dampfes und der Flüssgkeit, *A. Ph.*, 1881, Bd 15, S. 279–290, 692–704.

22. Landau, L. D. and Lifshits, E. M., *Statistical Physics*, Part I, Moscow, 1976 (in Russian); revised 2nd English edn, Oxford, 1977.

23. van Hove, L., Sur l'intégrale de configuration pour les systèmes de particules à une dimension, *Physica*, 1950, v. 16, p. 137–143.

24. Ornstein, L. S., *Toepassing der Statistische Mechanica van Gibbs op moleculair-theoretische vraagstukken*, Leiden, 1908; augmented French translation, *A.N.*, 3a, 1918, t. 4, p. 203–303.

25. Born, M., The statistical mechanics of condensing systems, *Physica*, 1937, v. 4, p. 1034–1044.

26. de Boer, J. and Uhlenbeck, G. E., eds., *Studies in statistical mechanics*, v. 1, Amsterdam, 1962.

27. Lebowitz, J. L., Exact derivation of the van der Waals equation, *Physica*, 1974, v. 73, p. 48–60.

28. Lebowitz, J. L. and Waisman, E. M., Statistical mechanics of simple fluids: beyond van der Waals, *Physics Today*, 1980, v. 33, p. 24–30.

29. Brush, S. G., Interatomic forces and gas theory from Newton to Lennard-Jones, *Arch. Rat. Mech. Anal.*, 1970, v. 39, p. 1–29.

30. Rowlinson, J. S., The Yukawa potential, *Physica*, 1989, v. 156A, p. 15–34.

31. Alder, B. J., Alley, W. E. and Rigby, M., Corrections to the van der Waals model for mixtures and for the diffusion coefficient, *Physica*, 1974, v. 75, p. 143–155.

32. Einstein, A., Folgerungen aus den Capillaritätserscheinungen, *A. Ph.*, 1901, Bd 4, S. 513–523; reprinted with notes in *Collected Papers*, 1989, Princeton, New Jersey, v. 2, p. 10–21.
33. Maxwell, J. C., Van der Waals on the continuity of the gaseous and liquid states, *Nature*, 1874, v. 10, p. 477–480.
34. Margenau, H., Van der Waals forces, *Rev. Mod. Phys.*, 1939, v. 11, p. 1–35.
35. Israelachvili, J. N., The nature of van der Waals forces, *Contemp. Phys.*, 1974, v. 15, p. 159–177.
36. Dzyaloshinskii, I. E., Lifshits, E. M. and Pitaevskii, L. P., General theory of the van der Waals forces, (in Russian) *Uspekhi Fizitcheskikh Nauk*, 1961, t. 73, p. 381–422; e, *Adv. Phys.*, 1961, v. 10, p. 165–207.
37. Winterton, R. H. S., Van der Waals forces, *Contemp. Phys.*, 1970, v. 11, p. 559–574.
38. Barker, J. A. and Henderson, D., The fluid phases of matter, *Sci. Am.*, 1981, v. 245, p. 82–90.
39. Chandler, D., Weeks, J. D. and Andersen, H. C., Van der Waals picture of liquids, solids, and phase transformations, *Science*, 1983, v. 220, p. 787–794.
40. Hopkinson, J., Magnetic and other physical properties of iron at a high temperature, *Philos. Trans. R. Soc. London, Ser. A*, 1889, v. 180, p. 443–465; *Original Papers*, Cambridge, 1901, v. 2, p. 186–221.
41. Curie, P., Propriétés magnetiques des corps à diverses températures, *Ann. Chim.*, 1895, t. 5, p. 289–405.
42. Weiss, P., L'hypothèse du champ moléculaire et la propriété ferromagnetique, *J. Phys. (Paris)*, 1907, t. 6, p. 661–690.
43. Weiss, P. and Kamerlingh Onnes, H. Researches on magnetization at very low temperatures, *Proc.*, 1910, v. 12, p. 649–677; in French in *J. Phys. (Paris)*, 1910, t. 9, p. 555–584.
44. Landau, L. and Lifshits, E., On the theory of the dispersion of magnetic permeability in ferromagnetic bodies, *Phys. Z. Sowjetunion*, 1935, Bd 8, S.153–169; reprinted in *Collected Papers of L. D. Landau*, ed. D. ter Haar, Oxford, 1965, p. 101–114.
45. Fowler, R. H., A tentative statistical theory of MacLeod's equation for surface tension, and the parachor, *Proc. R. Soc. London, Ser. A*, 1937, v. 159, p. 229–246.
46. Kirkwood, J. G. and Buff, F. P., The statistical mechanical theory of surface tension, *J. Chem. Phys.*, 1949, v. 17, p. 338–343.
47. Lovett, R., DeHaven, P. W., Vieceli, J. J. and Buff, F. P., Generalized van der Waals theories for surface tension and interfacial width, *J. Chem. Phys.*, 1973, v. 58, p. 1880–1885.
48. Triezenberg, D. G. and Zwanzig, R., Fluctuation theory of surface tension, *Phys. Rev. Lett.*, 1972, v. 28, p. 1183–1185.
49. Cahn, J. W. and Hilliard, J. E., Free energy of a nonunifom system. I. Interfacial free energy, *J. Chem. Phys.*, 1958, v. 28, p. 258–267.
50. Cahn, J. W., Free energy of a nonuniform system. II. Thermodynamic basis, *J. Chem. Phys.*, 1959, v. 30, p. 1121–1124.
51. Schofield, P., The statistical theory of surface tension, *Chem. Phys. Lett.*, 1979, v. 62, p. 413–415.
52. Janssen, W. J., *Stickstofoxydule in den vloeibaren en gasvormigen toestand*, Leiden, 1877, ii, 56 b.
53. Meyer, O. E., *Kinetische Theorie der Gase in elementarer Darstellung*, Breslau, 1877, xv, 338 S.

54. Gibbs, J. W., A method of geometrical representation of the thermodynamic properties of substances by means of surfaces, *Trans. Connecticut Acad.*, 1873, v. 2, p. 382–404; *Collected Works*, New York, 1928, v. 1, p. 33–54.
55. Dickson, J. D. H., On the criterion by which the critical point of a gas may be determined, *Philos. Mag.*, 1880, v. 10, p. 40–43.
56. Cailletet, L. and Mathias, [E.], Recherches sur les densités des gaz liquéfiées et de leur vapeurs saturés, *C.R.*, 1880, t. 10, p. 1202–1207.
57. Levelt Sengers, J. M. H., Critical exponents at the turn of the century, *Physica*, 1976, v. 82A, p. 319–351.
58. van Laar, J.J., *Die Zustandsgleichung von Gasen und Flüssigkeiten*, Leipzig, 1924.
59. Landau, L., Zur Theorie der Phasenumwandlungen. I. . . . II., *Phys. Z. Sowjetunion*, 1937, Bd 11, S. 26–47, 545–555.
60. Onsager, L., Crystal statistics. I. A two-dimensional model with an order–disorder transition, *Phys. Rev.*, 1944, v. 65, p. 117–149.
61. Brush, S. G., History of the Lenz–Ising model, *Rev. Mod. Phys.*, 1967, v. 39, p. 883–893.
62. Wilson, K. G., Renormalization group and critical phenomena. I. . . . II., *Phys. Rev.*, 1971, v. B4, p. 3174–3183, 3184–3205.
63. Wilson, K. G., The renormalization group: Critical phenomena and the Kondo problem, *Rev. Mod. Phys.*, 1975, v. 47, p. 773–840.
64. Fleury, P. A., Phase transitions, critical phenomena, and instabilities, *Science*, 1981, v. 211, p. 125–131.
65. Clausius, R., Über die theoretische Bestimmung des Dampfdruckes und der Volumina des Dampfes und der Flüssigkeit. Zweiter Aufzatz, *A. Ph.*, 1881, Bd 14, S. 692–704.
66. Clausius, R., Über den theoretischen Zusammenhang zweier empirisch aufgestellter Gesetze über die Spannung und die latente Wärme verschiedener Dämpfe, *A. Ph.*, 1851, Bd 82, S. 273–281.
67. Clausius, R., *Abhandlungen über die mechanische Wármetheorie*, Erste Abt., Braunschweig, 1864.
68. Groshans, J. A., Bemerkungen über die Volume und die Dichtigkeiten flüssiger und gasiger Körper, *A. Ph.*, 1850, Bd 80, S. 296–302.
69. Despretz, C., Mémoire sur les chaleurs latentes de diverses vapeurs, *Ann. Chim.*, 1823, t. 24, p. 323–332, and references therein.
70. Dühring, E., *Neue Grundgesetze zur rationellen Physik und Chemie*, Leipzig, 1878.
71. Dühring, E., *Kritische Geschichte der allgemeinen Principen der Mechanik*, Leipzig, 1887.
72. Rowlinson, J.S., Are diameters rectilinear?, *Nature*, 1986, v. 319, p. 362.
73. Mathias, E., *Sur la chaleur de vaporization des gaz liquéfiés*, Thesis, 1890, Paris.
74. Amagat, E.-H., Statique expérimentale des fluides, *Rapports présentés au Congrès International de Physique réuni à Paris en 1900*, Paris, 1900, t. 1, p. 551–582.
75. Guggenheim, E. A., The principle of corresponding states, *J. Chem. Phys.*, 1945, v. 13, p. 253–261.
76. Kamerlingh Onnes, H., Expression of the equation of state of gases and liquids by means of series, *Commun. Phys. Lab. Leiden*, 1901, No. 71.
77. Kamerlingh Onnes, H., Über die Riehenentwickelung für die Zustandsgleichung der Gase und Flussigkeiten, *Commun. Phys. Lab. Leiden*, 1901, No. 4.
78. Kamerlingh Onnes, H., Théorie général de l'état fluide, *A.N.*, 1897, t. 30, p. 101–136.

79. Kamerlingh Onnes, H. and Crommelin, C. A., Isotherms of monoatomic substances and of their binary mixtures. IX. The behaviour of argon with regard to the law of corresponding states, *Commun. Phys. Lab. Leiden*, 1911, No. 120a.
80. Gavroglu, K. and Goudaroulis, Y., Heike Kamerlingh Onnes's researches at Leiden and their methodological implications, *Stud. Hist. Philos. Sci.*, 1988, v. 19, p. 243–274.
81. Gavroglu, K. and Goudaroulis, Y., *Methodological aspects of the development of low temperature physics 1881–1956: Concepts out of context*, Dordrecht, 1989, 178p.
82. Mathias, E. and Kamerlingh Onnes, H., The rectilinear diameter for argon, *Commun. Phys. Lab. Leiden*, 1911, No. 117.
83. Mason, E. A. and Spurling, T. H., *The virial equation of state*, Oxford, 1969.
84. Gibbs, J. W., *The collected works*, New York, 1928, v. 1.
85. Barus, C., Certain suggestions by J. Willard Gibbs on geophysical research, *Am. J. Sci.*, 1906, v. 21, p. 461–462.
86. Stoletov, A. G., Zametki o krititcheskom sostoyanii tel, *ЖРФО*, 1882, t. 14, part 5A, p. 167–176 (Notes on the critical state of bodies, in Russian). Summary in *J. Phys.* (*Paris*), 1882, t. 1, p. 543–548; *Beibl.*, 1882, Bd 6, S. 775–777.
87. Planck, M., Die Theorie des Sattigungsgesetzes, *A. Ph.*, 1881, Bd 13, S. 535–543.
88. Cook, D. and Rowlinson, J. S., Deviations from the principle of corresponding states, *Proc. R. Soc. London, Ser. A*, 1953, v. 219, p. 405–418.
89. Riedel, L., Untersuchungen über eine Erweiterung des Theorems der übereinstimmender Zustände, *Chem.-Ing.-Tech.*, 1954, Bd 26, S. 83–89, 259–264, 679–683.
90. Pitzer, K. S., The volumetric and thermodynamic properties of fluids, I, *J. Am. Chem. Soc.*, 1955, v. 77, p. 3427–3433 and later papers.
91. Kamerlingh Onnes, H. and Keesom, W. H., Über die Translationenergie in einatomigen Gasen beim absoluten Nullpunkt. In *Vorträge über die kinetische Theorie der Materie und der Elektrizität (gehalten in Göttingen auf Einladung der Komission der Wolfskehlstiftung)*, Leipzig und Berlin, 1914, S. 193–194.
92. Keesom, W. H., On the equation of state of an ideal monoatomic gas according to quantum theory, *Commun. Phys. Lab. Leiden*, 1913, Suppl. No. 30a.
93. Byk, A., Das Theorem der übereinstimmende Zustande und die Quantentheorie der Gase und Flüssigkeiten, *A. Ph.*, 1921, Bd 66, S. 167–216.
94. Byk, A., Zur Quantentheorie der Gase und Flüssigkeiten, *A. Ph.*, 1922, Bd 69, S. 161–201.
95. Palacios Martinez, J. and Kamerlingh Onnes, H., Isothermes de substances monoatomiques et leur mélanges binaires. XXI. Déterminations d'isothermes de l'hydrogène et de l'helium à basse température, faites en vue d'examiner si la compressibilité de ces gaz est influencée par les quanta, *Commun. Phys. Lab. Leiden*, 1923, No. 164.
96. Boer, J. de and Michels, A., Contribution to the quantum-mechanical theory of the equation of state and the law of corresponding states. Determination of the law of force of helium, *Physica*, 1938, v. 5, p. 945–957.
97. Kamerlingh Onnes, H., Algemeene theorie der vloeistoffen, *Verh.*, 1881, d. 21, b. 1–24, 1–14, 1–9; French version, Théorie général de l'état fluide, *A.N.*, 1897, t. 30, p. 101–136.
98. Reynolds, O., An experimental investigation of the circumstances which determine whether the motion of water shall be direct or sinuous, and the law of resistance in parallel channels, *Philos. Trans. R. Soc. London*, 1883, v. 174, p. 935–982.
99. Dewar, J., Liquid hydrogen, *Proc. R. Soc. London*, 1899, v. 16. p. 1–14.

100. Keesom, W. H., Prof. Dr. H. Kamerlingh Onnes. His life-work, the founding of the Cryogenic Laboratory, *Commun. Phys. Lab. Leiden*, 1926, Suppl. No. 57.

101. Kamerlingh Onnes, H., On the cryogenic laboratory at Leiden and on the production of very low temperatures, *Commun. Phys. Lab. Leiden*, 1894, No. 14.

102. Kamerlingh Onnes, H., Remarks on the liquefaction of hydrogen, on thermodynamic similarity and the use of vacuum vessels, *Commun. Phys. Lab. Leiden*, 1896, No. 23.

103. Yavelov, B. E., *Early history of superconductivity, 1911–1935*, Moscow, Institute of History of Science and Technology, 1985 (in Russian, unpublished).

CHAPTER 12
Physical chemistry

Van der Waals called himself a physicist, and in saying this he was correct, but one should remember how many there were of those who were primarily physicists who contributed also to physical chemistry. These include Helmholtz, Planck and Gibbs among the most famous, but the list is much longer and certainly includes van der Waals, whose activities are aptly delineated by Andrews's remark: 'It is often difficult to draw a distinct line of separation between the physical and chemical sciences; and it is perhaps doubtful whether the division is not really an artificial one' [*1*, p. lxxvii]. Van der Waals's equation of state, his principle of corresponding states and his theory of capillarity all belong equally to molecular physics and to physical chemistry, but there is one part of his work which nowadays would certainly fall wholly within physical chemistry, and that is his work on the thermodynamics of multicomponent non-ideal systems, that is, of real solutions and the phase equilibria between them.

12.1 General ideas and methods

12.1.1 Thermodynamics and the equation of state

No one, except perhaps Gibbs and Guldberg, realized clearly that at the end of the nineteenth century all branches of thermodynamics – general, technical and chemical – were essentially the thermodynamics of ideal systems. This restriction gave rise to the illusion that problems of equilibrium can be solved by purely thermodynamic methods. It was van der Waals who first expressed the problem clearly. 'At present thermodynamics is used in parts of physics in which such use would previously have been thought impossible, that is, in fields in which knowledge has been accumulated from other sources. It is, however, a mistake to believe, as is often done, that such information is purely thermodynamic in origin. This cannot be so by the very nature of matter. Thermodynamics teaches us how to obtain equations that contain a quantity of heat on the left and a quantity of work or energy on the right. . . . If, however, we do not know either side of the equation from other sources then the equation tells us nothing. . . . Since thermodynamics leads us naturally to a mechanical or kinetic theory of matter it must be equally possible for many of its applications to get from kinetic theory one of its most important results, namely the equation of state' [91, p. 24].

The interrelation between thermodynamics and the theory of the equation of state has also another aspect: 'It is not sufficient to know the equation of state even if one is dealing with a single substance, we need a rule to tell us when the system ceases to be uniform and to determine the properties of the coexisting phases.' It is at this point that one cannot manage without thermodynamics[1], and 'for mixtures the application of the thermodynamic rule of equilibrium is even more necessary than for a single substance' [61, p. 90,91].

12.1.2 The principle of continuity

Van der Waals based his synthesis of thermodynamics and the equation of state on the principle of continuity that he had put forward. By this he meant not only Andrews's statement of the existence of a continuous route from liquid to gas via the supercritical region, but also the generalization of this notion to include the facts that liquid and gas are subject to the same equation of state and, moreover, that it is possible to construct continuous and continuously differentiable surfaces, ψ and ζ (Gibbs's notation for the Helmholtz and Gibbs free energies), for such two-phase systems (see [65e, p. 451–52] and [169].)

A continuous transition from a liquid to a gas includes unstable states and so the corresponding values of ψ (or ζ), considered as a continuous function, cannot be realized experimentally. The real values, which describe the true phase equilibrium, are lower than the continuous ones, but are determined from them: '. . . it has been proved even for a simple substance that for the laws of coexistence the knowledge of unstable conditions is required. Think e.g. of the criterium of Maxwell for determination of the pressure of coexistence. For a binary mixture the knowledge of the plait on the ψ-surface is necessary for the deduction of critical phenomena. In all these cases the connection of what may be realized on one side of a certain limit and that which may be realized on the other side of another limit can only be fully grasped when also the conditions that cannot be realized are known — so when we assume continuity' [65e, p. 449].

It is appropriate to mention here that Planck also pointed out how important were these unrealizable states for a full thermodynamic analysis, naming van der Waals's isotherm as a prime example [*2*, p. 650–651]. For a more recent discussion of these states, see [*3*, p. 182–83].

Thus van der Waals's principle of continuity is a generalization that exceeds the bounds of pure thermodynamics; it belongs to the theory of the equation of state although it is not tied to any particular algebraic form. The idea of continuity was central to his thought, not only in the principle discussed above, but also in his need to interpret objects geometrically and so continuously. The role of geometric methods in his scientific work is essentially the same as the role of continuity. (Later N. S. Kurnakov developed the idea of continuity as the basis for a geometrical description of physicochemical systems; he acknowledged van der Waals as his predecessor, see, for example, [*4*, p. 77].)

12.1.3 Phase theory

Some preliminary remarks on definition will be useful; first, what is meant by the concept of *phase*? Van der Waals's contribution to this concept was, at first, unrecognized by research workers and writers of textbooks who were for many years content with the simplified definition of a phase as a homogeneous part of a system separated from other parts by visible boundaries. The implications of Gibbs's laconic comment that a phase is characterized by the composition and the thermodynamic state [5, p. 96] were not understood. At the end of the nineteenth and in the early years of the twentieth century the inexactness and incompleteness of the common simple definition passed unnoticed. Only when the general laws of phase equilibria, including those involving critical states, came to be studied was it realized that a more profound analysis of the concept of phase was needed; the task of providing it fell naturally to van der Waals. His definition was, as follows; a phase 'consisting of n components is to be characterised by the nature of these components and, further, by the relations, first, which establish explicitly how the thermodynamic potentials and other thermodynamic quantities of the complex of bodies belonging to this phase depend on the temperature, pressure and the composition of the complex, and, secondly, and conversely, how the fixed values of the pressure, temperature and the $(n - 1)$ thermodynamic potentials of $(n - 1)$ components determine explicitly the appropriate values of the composition and other thermodynamic quantities' [113, p. 11]. This 'law of phase construction' can be described as 'the equation of a phase'. The requirement of homogeneity is deliberately not included in this definition because different 'complexes of bodies' may belong to the same phase, and, as a rule, differ from each other only by the spatial properties, that is by 'their form, disposition and amount, and can under certain conditions become a single homogeneous complex' (as, for example, drops in an emulsion). This is, however, not necessary, and, moreover, complexes of bodies belonging to the same phase may differ not only by spatial but also by other non-thermodynamic properties, as, for example, optical enantiomers present in the same phase (see [113, p. 12]). It is important to have the two requirements, the direct and the reverse, since each alone is necessary but not sufficient for the definition of a phase (see [113, p. 336,357]).

The phase rule and thermodynamics The definition of a phase given above raises the interesting question of whether one can obtain the phase rule for univariant systems 'without using thermodynamics'. For such a derivation two conditions suffice, the first of which is that 'pressure, temperature and composition determine unequivocally, in all senses important for equilibrium, the state of a body in a given phase'. This proposition is closely linked with 'the assumption that every phase has an equation of state'. It contains in essence the definition of a phase and rests upon the facts of experience used in that definition [113, p. 21]. The second condition is that if equilibrium is established then it is not disturbed by an increase or decrease in the amount of that phase. This proposition assumes that capillary effects are without influence on the equilibrium.

Using both these propositions, van der Waals proved that the pressure of an n-component system of $(n + 1)$ phases could be changed only by a change of temperature, and not by an isothermal compression or expansion. The proof does not use Gibbs's principle of equilibrium but it assumes tacitly the existence of entropy. It indicates, therefore, a certain primacy of the idea of an equation of state over the principles of thermodynamics or, more exactly, over the principle of equilibrium. Kohnstamm took up this idea later and, according to him, 'it can be shown that a substantial part of Gibbs's results, in particularly, phase-theory, do not derive from thermodynamics, but logically precede it. That is, the basis of the theory is not founded on the first and second laws of thermodynamics but only certain assertions founded on experiment and belonging to the general theory of the equation of state. This basis is less specialised than the first and second laws, and, conversely, these laws can be founded only once a general theory of the equation of state has been formulated' [122, p. 3].

It is useful to say something here of the rôle of models in thermodynamics since van der Waals made great use of them. Maxwell defined thermodynamics as 'the investigation of the dynamical and thermal properties of bodies deduced entirely from what are called the First and Second laws of Thermodynamics, without any hypotheses as to the molecular constitution of bodies' [*6*, p. 258]. The subject soon expanded into new areas, the first of which was chemistry, and these expansions required the assumption of further information, which can, perhaps, be described as 'pre-thermodynamic' or 'extra-thermodynamic'. One is reminded here of Tammann's remark (Chapter 9, note 36) that thermodynamics alone cannot even tell us whether phase diagrams exist or not. Gibbs introduced this further information in the form of such concepts as 'phase', 'component', and so, by extension, 'critical phase', 'interphase boundary' and 'charged component'. All these are derived from the known structure of physicochemical systems, and so we can call them generalized phenomenological models derived from experience. Such models are the subject matter of thermodynamic investigations and it is this statement that is at the root of Kohnstamm's assertion quoted above.

This is, however, only part of the problem, since, even after the assumption of these 'pre-thermodynamic' models, a thermodynamic equation tells us nothing, as van der Waals observed, without the addition of information contained in the equation of state. This second aspect of the problem – the 'fertilisation of thermodynamics', as the Russian physical chemist Rakovskii called it – is the introduction of 'post-thermodynamic' models.

What is meant here by the use of the word 'model'? The data needed to make a thermodynamic equation convey information can be embodied either in experimental results for the system under consideration, or in some model which is a representation of the system. We may distinguish three kinds of model – physical, mathematical and structural. A physical model is a known system similar to that under consideration. A mathematical model, which can be analytic or geometric, is a formula such as an equation of state which is inserted into the bare thermodynamic equations or a geometric form that represents the system[2]. By a structural model is meant a set of propositions concerning the entities from which the system is constructed.

The three classes of model are clearly linked, but the structural one comes first if one is to choose adequate physical or mathematical models[3]. Van der Waals himself had begun with a structural model – small particles of non-zero volume and exerting a mutual attraction. His equations of state for pure and mixed substances derived from this model but (cf. Chapter 8, note 32) they soon acquired a life of their own that masked their structural origins.

A second structural model that van der Waals incorporated into his thermodynamics was that of an interfacial layer of non-zero thickness. He also, as we shall see below, contributed greatly to the use of geometrical models.

Critical phases A century ago thermodynamics and the molecular-kinetic theory were sharply separated, so that the position was as follows: pure thermodynamics cannot say whether or not critical phases exist, but if it is found experimentally or predicted from a molecular theory that these are states where two phases become identical, then a thermodynamic treatment of these *critical* states can be developed. The foundations of this treatment, or theory as it would then be called, were set out by Gibbs who showed, in particular, that an n-component critical phase has $(n-1)$ degrees of freedom. This development was carried further by van der Waals with the aid of his principle of continuity; 'If two coexisting phases obey the same equation of state (and a critical state is possible only if this is so) then . . . these phases are generally separated by the domain of a third *labile* [i.e. unstable] phase which obeys the same equation of state. If the two coexisting phases become identical then the domain of the labile phase must disappear. . . . In other words, if two coexisting phases become identical then they belong also to the domain of the labile phase, which separates both coexisting phases under other conditions, but which shrinks to zero in the critical state. Thus a critical phase belongs to three phase domains that meet at the critical point, and when calculating the number of phases present one must count a critical phase as three' [113, p. 38–39; 122, p. 142–143].

The following remarks are of interest both in their own right and as further evidence of his deep insight into the relations between thermodynamics, molecular theory and experiment. First, 'One may ask if there are critical phenomena of higher order, that is, cases where more than two coexisting phases become identical. . . . It is certain that this cannot happen for a two-component system, and it may not in one of three components. . . . The existence of critical points of higher order is a problem that will be solved either experimentally[4] or by an exact theory of the equation of state' [113, p. 39–40; 122, p. 143].

These speculations became scientific fact following the investigations of I. R. Krichevskii and his school in the 1960s (see [*8,9*].)[5] Critical phenomena of third order (or *tricritical* states) are now known both for gas–liquid–liquid systems and for systems with three liquid phases[6].

The second remark relates to critical states that include solid phases. Here he disagreed with Planck who had thought that such states were possible since he had used a single equation of state for both liquid and solid phases. Van der Waals wrote: 'We assume from the start that there is no difference between the gaseous and liquid states except for differences of density and composition, and of properties that depend on them. The *arrangement* of molecules in these states does not enter into the argument, but such arrangement is very important in the solid state. It is, therefore, not difficult to see why, at the same temperature, there cannot be two different arrangements with the same density, or, conversely, why different arrangements, and so different properties of matter, cannot have the same density' [113, p. 41–42; 122, p. 145][7,8].

12.1.4 Thermodynamic methods

Van der Waals held, on the whole, to Gibbs's presentation of thermodynamics but made some modifications, in part formal and in part matters of principle. In the first group there is his substitution of mass or mole fractions for the masses or numbers of molecules used by Gibbs. Such a method of reducing the number of variables by one is useful when the number of components is small, most obviously in the case of binary systems. The composition of solutions had, of course, been expressed in percentages from the beginning of the last century, and Guldberg in 1870 [*18*] and, later, Duhem [*19,20*] had used the ratios of the amounts of the components for binary systems. This method of reducing the number of variables is, however, not applicable if there are three or more components, and can give rise to problems in a binary mixture when one goes to the limit of a pure component and the ratio becomes infinite. Gibbs used occasionally the condition $m_1 + m_2 + m_3 = 1$ to represent the composition of a ternary system by a triangle, but did not use fractions generally; it was van der Waals who introduced these variables into the apparatus of chemical thermodynamics (1889–90) [29,31].

Another distinction between his work and that of Gibbs is not mathematical in nature but seems to be psychologically a natural consequence of the principle of continuity. Whereas 'Gibbs always considers a finite number of phases', van der Waals deals with the more general case in which 'this number can be infinite. This is, in fact, the case when there is an external force, for example, a gravitational field. The same is true whenever one takes capillary effects into account' [61, p. 92]. This remark can be interpreted as follows. A phase, according to Gibbs is a homogeneous body, characterized by its thermodynamic state but not by its amount or form. Hence a phase placed in the field of an external force is changed into an infinity of phases each shading into its neighbour. The same holds true for the infinity of phases in the capillary layer of non-zero thickness that separates a liquid and a vapour. Similarly what is now called a phase of variable composition, which was one of the main objects of van der Waals's studies, is not really a phase in Gibbs's sense. In the apt expression of the Russian physicist N. A. Umov it is 'a body with an infinite number of phases shading continuously into each other' (see p. 247 of Ref. *67* in Chapter 9). It was this conception that lay behind van der Waals's use of the conditions $\int(\psi/v)\mathrm{d}\tau$ = minimum, for the determination of the minimum of the free energy, since this form allows for the continuous variation of ψ from one element of volume to the next.

It is in this context that one should consider another change van der Waals introduced into chemical thermodynamics, namely the change from Gibbs's condition of equilibrium to a differential form. Gibbs wrote,

$$P^\alpha = P^\beta, \quad T^\alpha = T^\beta, \quad \mu_i^\alpha = \mu_i^\beta \quad (i = 1,2, \ldots) \tag{12.1}$$

to express the equality of pressure, temperature and the chemical potentials of each component in the two phases α and β.

An infinitely small change of state that preserves equilibrium will lead to the conditions $P^\alpha + \mathrm{d}P^\alpha = P^\beta + \mathrm{d}P^\beta$, etc., so that

$$\mathrm{d}P^\alpha = \mathrm{d}P^\beta, \quad \mathrm{d}T^\alpha = \mathrm{d}T^\beta, \quad \mathrm{d}\mu_i^\alpha = \mathrm{d}\mu_i^\beta \quad (i = 1,2, \ldots). \tag{12.2}$$

This approach to the problem of defining equilibrium introduces phases that differ only infinitesimally from each other, a sharp reminder of the ubiquity of the idea of continuity in van der Waals's thought.

The conditions (12.2) were first formulated by van der Waals in 1900 [61, p. 108] (see also [113, p. 77]), although he had used them implicitly before this. They follow so naturally from Gibbs's apparatus in chemical thermodynamics that their novelty and authorship are generally unknown.

12.2 Thermodynamics of binary and ternary systems

12.2.1 Conditions of stability

The starting point for the analysis of phase equilibria in binary and ternary systems is Gibbs's principle of equilibrium and the conditions of stability that derive from it. Following Gibbs, van der Waals considers first the stability of a phase with respect to continuous changes. A state which is unstable with respect to a change to an infinitely close state is a labile one, that is, it is incapable of continuous existence. Necessary and sufficient conditions of stability are set out successively for one-, two- and three-component phases; they are formulated as thermodynamic inequalities for different modes of isolation of the system. The most general and interesting conclusion is that, 'The conditions for stability increase in number with an increase in the number of components', [65e, p. 556–557]. The reason for this is that as systems become more complex the earlier conditions remain necessary but new ones are needed because there are then new ways of changing the composition of the phase.

The boundary between the stable and unstable states can be determined by changing the sign of inequality to one of equality in the conditions of stability. For a binary system the equation of the boundary is

$$\left(\frac{\partial^2\psi}{\partial x^2}\right)_v\left(\frac{\partial^2\psi}{\partial v^2}\right)_x - \left(\frac{\partial^2\psi}{\partial x\partial v}\right)^2 = 0 \tag{12.3}$$

and for a ternary system,

$$\begin{vmatrix} \dfrac{\partial^2\psi}{\partial v^2} & \dfrac{\partial^2\psi}{\partial x\partial v} & \dfrac{\partial^2\psi}{\partial y\partial v} \\ \dfrac{\partial^2\psi}{\partial v\partial x} & \dfrac{\partial^2\psi}{\partial x^2} & \dfrac{\partial^2\psi}{\partial y\partial x} \\ \dfrac{\partial^2\psi}{\partial v\partial y} & \dfrac{\partial^2\psi}{\partial x\partial y} & \dfrac{\partial^2\psi}{\partial y^2} \end{vmatrix} = 0. \tag{12.4}$$

Since the boundary surface defined by eqn (12.4) includes the boundary lines of the constituent binary systems, then eqn (12.3) is satisfied also.

These boundaries of absolute instability are not generally reached in a real system, but it is usually necessary to study them in order to determine the narrower regions of stable states. This is particularly important in critical states where, according to van der Waals, the stable and unstable states coincide; that is, the limits of absolute stability really are reached in such phases. This is why the equations of the boundary lines (or surfaces) have so important a role in the theory of critical phenomena.

A phase that is stable with respect to continuous (or infinitesimal) changes may not be stable with respect to discontinuous changes, that is with respect to the transformation to another phase or to the formation of a new phase from an existing one. Such states are called *metastable*; the word was introduced by Ostwald [*21*, p. 302] as a generalization of such states found on the van der Waals isotherm. 'It is not easy to determine analytically the boundary between [absolute] stable and metastable states, since whether a point lies in one region or the other depends on the peculiarities of the surface, not only at the point under consideration, but also at other points far removed from it' [91, p. 133]. (The surface is the thermodynamic ε, η, v surface for a single component system, the ψ, v, x (at constant T) for two components, or ζ, x, y (at constant P and T) for three components, where x and y are mole fractions, ε is the energy and η is the entropy.) Or again, 'The question of the stability or instability of a state depends on the properties of the system in the immediate neighbourhood of that point, while metastability is determined by the properties of the system in states quite different from that whose metastability is to be assessed' [113, p. 483].

12.2.2 Coexistence of the phases of a binary mixture

Equations of coexistence Insofar as it is possible within the framework of pure thermodynamics, the boundaries between absolutely stable and metastable states are determined by the differential equations of the coexistence of the phases. These equations connect variations of temperature, composition and pressure (or volume) of the coexisting phases, α and β. By considering an infinitesimal displacement of the equilibrium they are derived from the conditions of equilibrium in eqn (12.2). If the independent variables are P, T and x, then van der Waals's differential equation has the form

$$\begin{aligned}&[(V^{\beta} - V^{\alpha}) - (x^{\beta} - x^{\alpha})(\partial V/\partial x)^{\alpha}_{P,T}]\,\mathrm{d}P \\ &= [(\eta^{\beta} - \eta^{\alpha}) - (x^{\beta} - x^{\alpha})(\partial \eta/\partial x)^{\alpha}_{P,T}]\,\mathrm{d}T + (x^{\beta} - x^{\alpha})(\partial^2 \zeta/\partial x^2)^{\alpha}\mathrm{d}x^{\alpha}.\end{aligned} \tag{12.5}$$

This is the equation for phase α; that for phase β is obtained by exchanging the superscripts α and β. Van der Waals seldom wrote both equations or even bothered to mention that the second could be obtained by this exchange between α (liquid) and β (vapour).

If the independent variables are temperature, composition and volume (instead of pressure) then the equation for the coexistence of phases takes the form

$$\begin{aligned}&[(\eta^{\beta} - \eta^{\alpha}) - (V^{\beta} - V^{\alpha})(\partial \eta/\partial V)^{\alpha}_{x,T} - (x^{\beta} - x^{\alpha})(\partial \eta/\partial x)^{\alpha}_{V,T}]\,\mathrm{d}T \\ &+ [(V^{\beta} - V^{\alpha})(\partial^2 \psi/\partial V^2)^{\alpha}_{x,T} + (x^{\beta} - x^{\alpha})(\partial^2 \psi/\partial x \partial V)^{\alpha}_{T}]\,\mathrm{d}V^{\alpha} \\ &+ [(V^{\beta} - V^{\alpha})(\partial^2 \psi/\partial x \partial V)^{\alpha}_{T} + (x^{\beta} - x^{\alpha})(\partial^2 \psi/\partial x^2)^{\alpha}_{V,T}]\,\mathrm{d}x^{\alpha} \\ &= 0.\end{aligned} \tag{12.6}$$

Since the changes of volume of the coexistent phases produced by infinitesimal displacements of equilibrium are not equal, the pair of equations (12.6) and its complement look more complicated than eqn (12.5) and its complement. Equations (12.5) and (12.6) are, however, equivalent; either can be derived from the other. Van der Waals produced these equations bit-by-bit, starting in 1890. At first he used the free energy only and established the relation between $\mathrm{d}P$ and $\mathrm{d}x$. In 1897 he changed ψ for ζ and gave eqn (12.5) in the form quoted here [47]; eqn (12.6) followed in 1900 [61].

Let us now consider the implications of van der Waals's differential equations. The coefficients of $\mathrm{d}P$ and $\mathrm{d}T$ are the molar changes in volume, $v^{\beta\alpha}$, and entropy, $\eta^{\beta\alpha}$, that occur when one mole of one phase is dissolved in an infinite amount of the other; thus the coefficient of $\mathrm{d}P$ in eqn (12.5) is the change of volume, at fixed P and T, when one mole of phase β is dissolved in an infinite amount of phase α. The coefficient of $\mathrm{d}x^{\alpha}$ in eqn (12.5) is related to the change of thermodynamic potential ζ of the phase that arises from the change of composition; its sign determines the stability of the phase. (The coefficient of $\mathrm{d}x^{\alpha}$ in eqn (12.6) is also related to the condition of stability, but

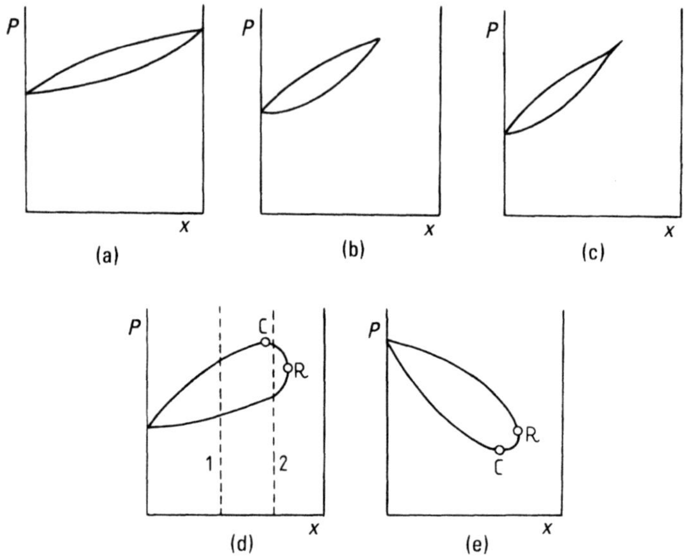

Fig. 12.1 Isotherms showing the dependence of the vapour pressure on the compositions of the liquid and of the vapour: (a), the temperature is lower than the critical temperature of the more volatile component; (b), (c), (d), (e), the temperatures are between critical temperatures of the components; (d), (e), two cases forecast by the continuity principle. The dotted lines show trajectories of a representative point of the system at the isothermal compression of the mixture for a normal (1) and for a critical (2) composition.

less simply.) The equations (12.5) and (12.6) include all the thermodynamics of a binary system.

Konovalov's laws and retrograde condensation Perhaps the best known deduction from these differential equations is a general derivation (and also a demonstration of the limits of their applicability) of Konovalov's laws on the composition of coexisting liquid and vapour phases [113, p. 114; 122, p. 221][9]. These results of van der Waals are now well known and are to be found in many textbooks of chemical thermodynamics (see e.g. [*22*]).

One important aspect is, however, usually passed over without discussion. It is generally pointed out that the two isotherms, pressure–composition for the liquid and for the vapour, do not cover the whole range of composition at temperatures above the critical temperature of the more volatile component but merge at some point where $x < 1$. This statement is correct, but it is not usually pointed out that there are three ways in which the pair of isotherms can be joined; they can intersect, they can be tangential to each other, or they can join smoothly into a common continuous curve, see Fig. 12.1(b),(c),(d) respectively. One cannot choose between these possibilities on purely thermodynamic grounds but if one invokes the principle of continuity then only the third possibility is permitted. This prediction was made in his paper *Molecular Theory*, and has now a sound experimental confirmation; no exceptions are known.

Van der Waals thus revealed peculiarities in the critical behaviour of mixtures that are not found for pure substances; the well-known features of the critical point of a pure substance are now split between two different points. At the point R we have the state of the fluid at the highest temperature at which a liquid of that composition can exist; this was the characteristic used most commonly to define a critical state, particularly by those who studied the liquefaction of gases. But at this highest temperature the liquid is in equilibrium with a vapour of a different composition. As we approach point C, on the other hand, we have two phases becoming identical in all their properties, and so this is the true critical point. This 'division' of the characteristics of a critical state arises from the fact that the coexistence of the phases is described by a continuous curve. This continuity gives rise to some remarkable phenomena such as the inversion of Konovalov's laws and retrograde condensation.

The first of these phenomena is illustrated in Fig. 12.1(d) between point C where $\mathrm{d}P/\mathrm{d}x$ is zero and point R where it is infinite. We then have the opposite of Konovalov's law since on adding to the liquid the component that increases the pressure along the coexistence curve we *increase* its concentration in the vapour.

To illustrate retrograde condensation we consider the movement of a point across the liquid–vapour diagram that represents the isothermal compression of a mixture of fixed composition. In the normal part of the diagram, at pressures well below the critical, the point moves upwards from the single-phase vapour state until it meets the line that marks the beginning of condensation – the *dew-point* curve. It crosses into the two-phase region in which, at each pressure, the ratio of liquid to vapour is determined by the well-known lever rule, until, on further compression, the vapour phase disappears on crossing the *bubble-point* curve; a further increase of pressure now merely compresses the liquid. If, however, at a pressure above the critical pressure of one of the components the composition is chosen to lie between those of points C and R, then a different course of events is observed. On compression, the vapour starts to condense, as before, when the dew-point curve is reached, but now on further compression the point leaves the two-phase region not at a bubble point but by crossing the dew-point curve a second time. This means that the liquid that is formed at first on entering the two-phase region is made to evaporate on further increasing the pressure. Kuenen predicted also *retrograde condensation of the second kind* if the points C and R are situated as shown in Fig. 12.1(e). The representative point now crosses the bubble-point curve in two places, so that on increasing the pressure vapour is first formed and then condensed again. It is clear from these figures that retrograde condensation is a consequence of the existence of a point R at which the P,x-curve has a vertical slope. This assertion can be proved rigorously with the aid of the differential equation for the coexistence of phases. This result is one of the

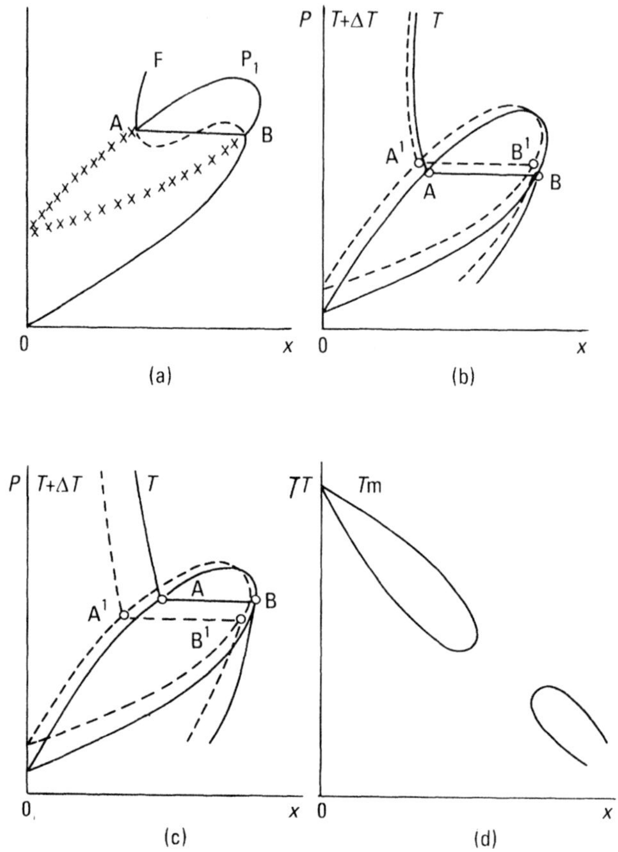

Fig. 12.2 Diagrams (as drawn by van der Waals) showing the coexistence of solid and fluid phases for a binary system in the critical region: (a), pressure–composition isotherm; (b), (c), shifts of isotherms by an increase of the temperature; (d), temperature–composition isobars.

elegant consequences of combining the principle of continuity with the thermodynamic differential equations of coexistence.

Equilibria including solid phases Van der Waals's results for the equilibria of saturated solutions of solids in the critical region are less well known but no less elegant than those for purely fluid systems. Unusual behaviour is found for this kind of system when the temperature of the melting point of the solid is above that of the critical point of the solvent. It follows that the solubility curve of the solid ends before the melting point is reached, namely at the point where the liquid state becomes identical with the vapour. One of the pressure–composition isotherms discussed by van der Waals is shown in Fig. 12.2 [71e, p. 237]. Below the curve OB is the region of vapour + solid, the horizontal line AB is at the pressure of the three-phase equilibrium, and the curve EAPBE denotes the liquid–vapour equilibrium, the part below AB being metastable and realizable only when formation of the solid phase is suppressed. He considered the two equilibrium curves AF (solid–liquid) and BO (solid–vapour) as branches of a continuous curve OBAF with an unrealizable section shown by the dotted line joining A to B. A detailed analysis using eqn (12.5) for the coexisting liquid and solid phases shows that the point at which the P,x-curve has a vertical tangent is in the stable region.

We can now see what happens when the temperature rises. From the treatment of liquid–vapour equilibrium we know that the loop EAPBE shrinks and moves upwards; only the point at which the coefficient of $\mathrm{d}T$ is zero is unchanged. The horizontal line AB that represents the three-phase region can move either way (A′B′ in Fig. 12.2(b) or (c)), but in either case the length will shrink to zero at a certain temperature. If AB moves upwards the proportion of solvent in the saturated liquid solution x^{α} diminishes at first, but not without limit. As the temperature approaches the highest value at which the three phases can coexist the x^{α} approaches the larger value x^{β} of the proportion of the solvent in the solid + vapour system, so the amount of solvent in the liquid solution passes through a minimum. Conversely if AB moves downwards with a rise of temperature then x^{α} passes through a maximum [71e].

Each of these two ways in which the three-phase equilibrium can disappear corresponds to a separate three-phase region in which solid is in equilibrium with two supercritical fluid phases. The two loops of these three-phase equilibria can be represented by an isobar in a T,x-diagram (Fig. 12.2(a), see [71e, p. 239]) in which the point T_{m} is the melting temperature of the solid at the chosen pressure. Each of these loops has a stable point at which the tangent is vertical, $(\partial T/\partial x)_P = \infty$, and so each exhibits retrograde phenomena, but now it is not condensation but retrograde solidification in the upper loop at the small mole fraction of solvent, and retrograde dissolution in the lower at the higher mole fraction. When retrograde solidification takes place a compression leads first to the separation of solid and then further compression leads to its dissolution in the dense supercritical vapour. Sometimes, depending on the composition chosen, a further rise of pressure leads to a second precipitation of solid.

Three-phase equilibrium in a binary system The two-phase equilibria described above were the main object of van der Waals's thermodynamic studies but he devoted one paper [46] to three-phase equilibria in binary systems. (An earlier paper [21] had been prompted by Bakhuis Roozeboom's work, see Chapter 8.) In it he discussed the question of whether such a system could have an 'apparent triple-point', that is, a state in which the solid, liquid and gaseous phases all have the same composition. As so often, he tackled the problem with a geometrical representation based on the ψ-surface for the liquid–vapour states but now augmented by a ψ-line of constant composition of the solid. The state of equilibrium is represented by a plane that is simultaneously tangent to the surface at two points and to the line. He derived the analytic result (obtained previously by Gibbs) that

$$\frac{\mathrm{d}P}{\mathrm{d}T} = \begin{vmatrix} x^{\mathrm{s}} & y^{\mathrm{s}} & 1 \\ x^{\mathrm{l}} & y^{\mathrm{l}} & 1 \\ x^{\mathrm{g}} & y^{\mathrm{g}} & 1 \end{vmatrix} \cdot \begin{vmatrix} x^{\mathrm{s}} & v^{\mathrm{s}} & 1 \\ x^{\mathrm{l}} & v^{\mathrm{l}} & 1 \\ x^{\mathrm{g}} & v^{\mathrm{g}} & 1 \end{vmatrix} \tag{12.7}$$

where the superscripts denote solid, liquid and gaseous phases. The equation is the natural extension of the Clapeyron–Clausius equation to a univariant system of three phases. The answer to the question that had led to this work was; no, there cannot be three phases of the same composition. If the solid component is dissociated in the liquid and vapour then it has two triple points, one of which may have the same composition as the liquid and the other as the vapour. In other words, if there are three phases present only two can have the same composition.

12.2.3 Ternary systems. Equations for the coexistence of phases

In September 1897 van der Waals first presented some results for ternary systems [47]. The principal result[10] was eqn (12.8), which is the ternary analogue of eqn (12.5):

$$\begin{aligned} &[(V^{\beta} - V^{\alpha}) - (x^{\beta} - x^{\alpha})(\partial V/\partial x)^{\alpha} - (y^{\beta} - y^{\alpha})(\partial V/\partial y)^{\alpha}]\mathrm{d}P \\ &= [(\eta^{\beta} - \eta^{\alpha}) - (x^{\beta} - x^{\alpha})(\partial \eta/\partial x)^{\alpha} - (y^{\beta} - y^{\alpha})(\partial \eta/\partial y)^{\alpha}]\mathrm{d}T \\ &+ [(x^{\beta} - x^{\alpha})(\partial^2 \zeta/\partial x^2)^{\alpha} + (y^{\beta} - y^{\alpha})(\partial^2 \zeta/\partial x \partial y)^{\alpha}]\mathrm{d}x^{\alpha} \\ &+ [(x^{\beta} - x^{\alpha})(\partial^2 \zeta/\partial x \partial y)^{\alpha} + (y^{\beta} - y^{\alpha})(\partial^2 \zeta/\partial y^2)^{\alpha}]\mathrm{d}y^{\alpha} \end{aligned} \tag{12.8}$$

Here x and y are the mole fractions of the second and third components; the equation for phase β is obtained by exchanging the superscripts α and β. Several years later, in the course of his series of papers on *Ternary systems*, he produced the second differential equation, that is, the analogue of eqn (12.6) [65e, p. 552,683]. For ternary systems and 'even for a mixture with an arbitrary number of components' [65e, p. 691], the coefficients of dP and dT again carry the meaning of the change of volume and of entropy on dissolving one mole of the first phase in an infinite amount of the second. Van der Waals had seen clearly the possibility of extending the equations to arbitrary numbers of components, but it was apparently another forty-three years before they were written down explicitly by A. V. Storonkin [*24*][11]. After the Second World War, Haase gave a more modern derivation of these equations using mean and partial molar functions [*25*] (see also [*3*, p. 183–251].) Further generalizations to include terms for both planar and curved interfaces have been obtained by Rusanov (see Ref. [*106*] and [*107*] of Chapter 9).

A detail worth noting in the evolution of van der Waals's thermodynamic studies is that, for ternary systems, the equation for the coexistence of phases expressed in terms of ζ came earlier (1897) than that in terms of ψ (1902). For binary systems it was the other way round. It follows naturally from his geometric approach to the combination of thermodynamics and the equation of state that the function ψ is more effective for the study of binary systems while ζ is more appropriate for ternary. He wrote, 'Though the geometrical representation is not necessary for the deduction of the conditions of equilibrium, and though it is not even possible to use it with increasing number of the components, yet for a binary mixture the great advantages of the graphical treatment have been sufficiently proved to continue employing it as long as we can' [65e, p. 448].

Applications of the principle of continuity In the geometrical thermodynamics of ternary systems van der Waals concentrated on the relations between stable, metastable and unstable parts of the ζ-surface which were needed in order to analyse the critical behaviour. To do this he added the principle of continuity to the thermodynamic relations. First he considered the interrelations of the different sheets of the ζ-surface (at fixed P, T) which allowed him to infer the forms of the isothermal surfaces, volume–composition and pressure–composition. The last he considered in more detail giving as his reason, 'From a theoretical point of view the relation between P, x^α and y^α at given temperature is not more important than that between V^α, x^α and y^α. But even for a simple substance the experimental investigation concerning the maximum pressure has first been executed, and only in these later years it has been followed by an investigation concerning the densities. In the same way we may expect, that also for a ternary system the experiment will occupy itself in the first place with the determination of the pressure, and that the investigation as to the densities of the phases coexisting with other phases, will follow later' [65e, p. 3]. His opinion has been confirmed by later work, particularly that of Krichevskii and his school [*26*][12].

For the volume–composition diagrams van der Waals showed that the boundary curves of binary systems become boundary surfaces in ternary systems, as for example the coexistence surface, or the surface separating unstable and metastable states. Generally speaking these surfaces are tangent along a curve and have a common point only if a ternary mixture has a minimum critical temperature. If the temperature is high enough the coexistence surface shrinks and its projection lies inside the composition triangle, while the surface itself forms a closed cavity the upper sheet of which represents the vapour and the lower the liquid. There is a curve along which the coexistence surface touches a cylindrical surface normal to the composition plane, and points on this curve are analogous to the critical points of contact (point R in Fig. 12.1) of a binary system. Plait points, that is, critical points in a mixture, can lie either on the liquid or on the vapour sheet, and to each case there corresponds a type of retrograde condensation – of the first and second kinds respectively.

In the course of studying the isothermal pressure–composition surfaces van der Waals found an interesting result for the displacement on these surfaces caused by a change of pressure. He took eqn (12.8) with $dT = 0$ and considered a change from the composition x^α, y^α, phase α, to x^β, y^β of the coexistent phase β which satisfies the following restriction

$$\frac{dx^\alpha}{x^\beta - x^\alpha} = \frac{dy^\alpha}{y^\beta - y^\alpha} = \frac{dl}{l}$$

where l is the length of the straight line that joins the representative points for these phases. He could thus find the derivative dP/dl, the sign of which determines the direction of the change of composition of phase α with pressure. If phase α is the liquid then $dP/dl > 0$, thus an increase of pressure moves the liquid composition into the heterogeneous region. 'This rule for a ternary system is equivalent to the rule of Konovalov for a binary system, if this latter rule is duly extended' [65e, p. 20–21]. Conversely, if phase α is the vapour then $dP/dl < 0$, and a decrease of pressure moves the vapour composition towards the heterogeneous region, and so it moves towards the homogeneous region when the pressure rises. Thus a change of pressure moves both branches in the same direction, or 'one of the branches retreats from the other' [65e, p. 21]. If there is a maximum in the coexistence pressure then the assertion is still correct in principle, but the picture is slightly changed; the isobars are now closed curves around the point of maximum pressure, and when the pressure rises the curve shrinks and so comes closer to the closed curve of the vapour which lies inside it; at the maximum pressure both curves collapse to the same point.

Like Konovalov's law for binary systems these rules are valid far from the critical region. In this region, however, the projection of the pressure–composition curves lie, wholly or in part, within the composition triangle, and the liquid and vapour sheets merge into each other along the curve for which $v^{\beta\alpha}$ in eqn (12.8) is zero.

Retrograde condensation and the coexistence of vapour phases Retrograde condensation of the second kind can more easily be shown experimentally in a ternary system. Van der Waals put the reasons thus: 'We may expect that it will be easier to observe this phenomenon for a ternary system, than for a binary one. In order that retrograde condensation may be easily observed a rather great distance between the two sheets of the surface of saturation [i.e. $P(x^\alpha,y^\alpha)$ and $P(x^\beta,y^\beta)$] is required; and the distance between the sheets will be more considerable in the middle than at the ends, where we have to deal with a binary mixture, because the requirements for stability and coexistence for a ternary mixture are stricter than those for a binary mixture' [65e, p. 122].

This is correct for systems with no extremum in the coexistence pressure because, if there is an extremum, as the author notes, the separation of the orthobaric sheets is not great enough for retrograde phenomena to be observed since there must be a point in the middle of the composition diagram where the sheets touch.

In this paper van der Waals considered also the interesting question of the coexistence of two vapour phases. This can occur when the temperature is above the critical temperatures of two of the components (with mole fractions x and y) but below that of the third (with a mole fraction of $1 - x - y$). There are now two possibilities for the mutual positions of the critical point C and the limiting point of vertical tangency R (where $v^{\beta\alpha} = 0$) in the composition range from $x = 0$ to $y = 0$. The first possibility is illustrated by Fig. 12.1(d) for a binary mixture and is best explained by a full quotation from the original paper; If the pressure is lower than the lowest pressure of the points R, the two branches of the curves of equal pressure are perfectly separated lines which, if the pressure is increased, will be displaced according to the rules given above. If however the pressure has risen till the pressure of a point R has been reached, then the two branches are still separated, but on the vapour branch occurs a point for which $v^{\beta\alpha} = 0$. Such a point is not displaced when the pressure increases. The locus of these points forms the limit of the mixtures which may be split up into two phases at the given temperature. From a geometrical point of view it is the envelope of the projections of the horizontal section of the surface of saturation, or the envelope of the projections of the curves of equal pressure. If the pressure has been increased till it has attained the value of the lowest of the pressures of the point C, then the two branches of the curves of equal pressure pass continuously into one another. But if we continue to call those phases represented by the lower sheet, vapour phases, and those phases represented by the higher sheet, liquid phases, then the vapour phases do not reach the point, where the connection of the two branches has taken place (the plaitpoint), but only the point where the value of $v^{\beta\alpha}$ is zero, i.e. the point where two successive curves of equal pressure intersect. For all points lying on one of the sides of that point of intersection – e.g. on the side where the plaitpoint occurs – $v^{\beta\alpha}$ is positive.

'These points will be displaced towards the conjugate point, when the pressure is increased; all points on the other side of the point of intersection will move away from the points representing coexisting phases. If we, therefore, continue to use the expressions "liquid phases" and "vapour phases" with the same meaning as we have done till now, we must say that for points between the plaitpoint and the point for which $v^{\beta\alpha} = 0$ two liquid phases coexist' [65e, p. 121–122].

The second possibility is that shown in Fig. 12.1(e), when 'the above rules would continue to hold, but in this case we find a series of *vapour* phases coexisting with *vapour* phases between the plaitpoint and the point, for which $v^{\beta\alpha} = 0$. For these points we have taken retrograde condensation of the second kind' (italics by van der Waals) [65, p. 225; 65e, p. 122; 65f, p. 422].

The series of papers on ternary mixtures concluded with the remark: 'I have discussed here, in more or less detail, only some of the properties of ternary systems. Although I intend to study other properties also I think I may consider that the work already accomplished forms a fairly complete investigation' [65f, p. 442]. His intentions had not been completely realized (but see [66]), probably because he had found ternary systems to be much more complicated than he had expected.

12.3 The theory of non-ideal multicomponent systems

12.3.1 Principles

The most characteristic part of van der Waals's legacy is his theory of mixtures based on his equation of state. It was a field that did not exist until he tackled it and, in essence, is the combination of his equation of state (generalized for a mixture) with the characteristic thermodynamic methods that he devised. As his son wrote, Gibbs's and van der Waals's work 'complemented each other in the happiest way' [*27*, p. 410].

Analytic formulation A general statement of the need to fertilize the bare thermodynamics can be set out, as follows. An expression for any of the thermodynamic potentials includes a term that depends only on the concentrations of the components and one of the form $\int P\mathrm{d}V$ which depends on their molecular nature. Thus the free energy of a binary mixture can be written

$$\psi = MRT[x\ln x + (1-x)\ln(1-x)] - \int P\mathrm{d}V, \tag{12.9}$$

and Gibbs's function ζ for a ternary mixture

$$\zeta = MRT[(1-x-y)\ln(1-x-y) + x\ln x + y\ln y] + PV - \int P\mathrm{d}V. \tag{12.10}$$

The presence of this integral is always characteristic of 'going from the theory of mixtures of low density to the theory of those of high density' [122, p. 226]. 'The theory of mixtures of low densities may be defined as that for which only the logarithmic terms need be considered. This definition makes it clear why it was that the theory of mixtures of high density was so slow in being developed since it required an investigation of the term $\int P\mathrm{d}V$, which is possible only when P is known as a function of V, x and T, that is, when the equation of state is known. However it seemed at the time, and still seems unhelpful to go down this road directly, and so is no clear way to a rational quantitative theory of mixtures of high density' [113, p. 119–120]. In 1912, when these lines were written there could have been no clearer explanation of the root of the problem posed by non-ideal mixtures in chemical thermodynamics.

The equation of state To create a theory of non-ideal solutions van der Waals needed 'the equation of state of a homogeneous mixture of any number of components' [61, p. 51]. Such an equation can be designed on the basis of the model discussed in the last chapter. He described the general idea as follows: 'The hypothesis of treating fluids as composed of a great number of mobile unalterable particles, having the property of extent and exerting an attraction which acts only at very small distances, has led to results that are generally consistent with experiment. This hypothesis explains the liquid state as the consequence of a great coming-together of the molecules and shows that it is not necessary to assume that molecules in the liquid state are more complex than those in the gaseous. . . . The same hypothesis has led naturally to the representation of mixtures, even in the liquid state, as being composed of unalterable molecules although, having adopted this hypothesis, we do not deny that there are exceptions. Mixtures in which one of the components is anomalous and, *a fortiori*, those in which both are anomalous will be exceptions. Moreover there could be chemical interactions between the two substances or one of them might dissociate electrolytically. In each case the laws that govern the behaviour of the systems will not be those for the cases in which the hypothesis holds good. Nevertheless, even in these more complicated cases we cannot arrive at the laws that govern their behaviour without knowing the laws that govern the more simple cases. One is, therefore, justified in thinking that a theory which supposes the unalterability of molecules even in the liquid state would be of great interest' [58, p. 583–584].

He therefore used his own equation of state for mixtures but with the parameters a and b as quadratic functions of the mole fractions; thus for a binary mixture,

$$a_x = a_1(1-x)^2 + 2a_{12}x(1-x) + a_2x^2, \tag{12.11}$$

$$b_x = b_1(1-x)^2 + 2b_{12}x(1-x) + b_2x^2,$$

or

$$b_x = b_1(1-x) + b_2x, \qquad \text{if } b_{12} = \tfrac{1}{2}(b_1+b_2). \tag{12.12}$$

These expressions imply that 'the number of molecules is unchanged, and that there is neither association nor dissociation, nor are chemical reactions occurring' [61, p. 52].

At first the equation of state for mixtures had no direct experimental basis, but during the 1890s some accurate P–V–T measurements were made at Leiden, and also by Amagat in Paris (see [160]). Van der Waals made systematic comparisons of the predictions of the equation of state and the experimental results of Kuenen, Hartmann and Quincke, and so convinced himself that the equation was useful [61]. He realized, however, that the short-comings that had become evident in its use for pure substances would be present also when it was used for mixtures.

The first application of this equation of state was in his paper on *Molecular Theory*; when a mixture of composition x is homogeneous it can be represented by the equation with the values $a = a_x$ and $b = b_x$, while the apparent critical constants are given by the familiar values

$$T_c = 8a_x/27Rb_x, \quad P_c = a_x/27b_x^2, \quad V_c = 3b_x. \tag{12.13}$$

Van der Waals stressed that these were only apparent values since they do not appertain to a real mixture but to 'a mixture considered as a single substance', that is, to a hypothetical substance that has the same values of a and b but which, unlike the mixture, does not separate into phases of different composition. This concept of representing a mixture by a hypothetical single substance has since proved to be a powerful technique in the calculation of phase equilibrium. In 1936 W. B. Kay generalized the principle of corresponding states by introducing 'a particular temperature and pressure for each mixture, termed the pseudo-critical temperature and pressure, for calculating the corresponding reduced quantities' [*28*, p. 1015]. Kay did not refer directly to van der Waals but the use of his ideas is clear.

The obvious problem of determining the cross-parameters a_{12} and b_{12} from those for the pure substances troubled van der Waals for many years. He realized eventually that there was no answer to this question, any more than to the problem of the relation between a and b for a single substance or, if there were answers, they were still hidden from us. Golitsin (see Chapter 9, Ref. [*13*]) and later Berthelot [*29*] suggested that

$$a_{12} = (a_1 a_2)^{1/2} \tag{12.14}$$

Van der Waals had already, in his *Molecular Theory*, considered this as the simplest relation but had rejected it; he therefore opposed Berthelot's proposal [49][13]. 'The molecular attraction', he wrote, 'does not depend simply on molecular weights but on some specific course which we do not know but which certainly exists. . . . Hence, in my view, it is not possible to calculate the density of a mixture knowing only the properties of the gases of which it is composed, . . . it is necessary to know the value of a_{12}. Maybe there are many cases for which a_{12} is close to $\sqrt{(a_1 a_2)}$ but the density calculated on this hypothesis can only be an approximation.' Berthelot agreed with him [*30*], but nevertheless the formula was widely used. For several years van der Waals tolerated this increasing use[14] but later, provoked perhaps by van Laar's many attempts to obtain a simple picture of the phase equilibria for mixtures that conformed to this rule, he spoke out against this over-simplification, saying that the fact that it made the calculation easier was no justification for its use [84e, p. 631].

Van Dalfsen, a student of van der Waals, considered in 1904 the n-component system for which

$$a = \sum_{p=1}^{n} \sum_{q=1}^{n} a_{pq} x_p x_q, \quad b = \sum_{p=1}^{n} \sum_{q=1}^{n} b_{pq} x_p x_q. \tag{12.15}$$

He showed that if $a_{pq}^2 = a_{pp} a_{qq}$ then the apparent critical temperature of the mixture, considered as a single substance, which is proportional to a/b, cannot have an extremum if $n \geq 3$, and cannot have a maximum if $n = 2$ [*31*]. Two years later van der Waals returned to the attack by showing that so strong an assumption could not lead to a minimum in the critical temperature [92][15].

Van der Waals made little use of the equation of state for one-phase systems. He did, however, write three papers on the change of volume and of pressure on mixing [51,53,55], the first of which is of some general interest and the last two of which are detailed quantitative comparisons with the experiments of Amagat, Kuenen and Verschaffelt. His aim was to test Amagat's law ($\Delta v = 0$ at constant P) and Dalton's law ($\Delta P = 0$ at constant v) and his conclusion was, 'we are induced to qualify the first rule as an approximative law. Throughout the course of an isotherm from an infinite volume down to the smallest possible volume of the substance, there may be deviations, but the deviation remains within finite limits. The second rule holds perfectly good for infinite rarefaction but it would be utterly impossible to apply it also to liquid volumes. Such a law may be qualified as a *loi limité. Considered from this point of view the law of Boyle is not an approximative law but a loi limité*' [51e, p. 191].

His main use of the equation of state for mixtures was the construction of an expression for the free energy, and hence of the ψ-surface.

Explicit expression for the free energy and its applications The calculation of the integral in eqn (12.9) from the equation of state gives ψ for a binary system as a function of its proper arguments:

$$\psi = MRT[x \ln x + (1-x)\ln(1-x)] + MRT \ln(v - b_x) - a_x/v, \quad (12.16)$$

where a_x and b_x are given by eqns (12.11) and (12.12). This expression opens up a clear route to all the thermodymaic properties and phase-equilibria of a binary mixture, although the mathematical difficulties are often severe. There are however a few results that can be obtained directly from eqn (12.16). One concerns what are now called azeotropic systems. A condition for equilibrium between phases α (liquid) and β (gas) is

$$(\partial\psi/\partial x)^{\alpha}_{v,T} = (\partial\psi/\partial x)^{\beta}_{v,T}, \quad (12.17)$$

and at an azeotrope $x^{\alpha} = x^{\beta}$ which implies also that the vapour pressure is at an extremum. Insertion of eqn (12.16) into (12.17) gives

$$\left(\frac{1}{v^{\alpha}} + \frac{1}{v^{\beta}}\right)\frac{\mathrm{d}b_x}{\mathrm{d}x} = \frac{1}{a_x}\frac{\mathrm{d}a_x}{\mathrm{d}x}. \quad (12.18)$$

At low temperatures $v^{\beta} \gg v^{\alpha} \approx b_x$, so that eqn (12.18) becomes (approximately)

$$\frac{\mathrm{d}}{\mathrm{d}x}\left(\frac{a_x}{b_x}\right) = 0. \quad (12.19)$$

We see from eqn (12.13) that this is also the condition that T_c is at an extremum, and it is not difficult to show that this must be a minimum if a_1, a_2, a_{12} etc. have reasonable values. Thus a system with a positive azeotrope (or a maximum in the vapour pressure) has also a minimum in the critical temperature. The result is not exact since approximations have been made, but all the experimental evidence is in favour of this rule [123, p. 263].

A second consequence of eqn (12.18) is, as follows. At high temperatures and pressures the volumes of both phases approach the critical limit of $3b_x$, so that eqn (12.18) becomes

$$\frac{2}{3}\frac{\mathrm{d}\ln b_x}{\mathrm{d}x} = \frac{\mathrm{d}\ln a_x}{\mathrm{d}x},$$

or, from eqn (12.13)

$$\frac{\mathrm{d}\ln T_c}{\mathrm{d}x} = -\frac{1}{3}\frac{\mathrm{d}\ln b_x}{\mathrm{d}x},$$

or the critical temperature increases when the composition shifts towards the component of smaller molecular volume [113, p. 162; 123, p.39].

A further consequence is the deduction of the existence of the phenomenon of barotropy discovered by Kamerlingh Onnes and Keesom; 'When the more volatile substance has its greatest density, M/b, then the gas phase can have a greater specific density than the liquid phase' [83e, p. 258]. Here M is the molecular weight, so that the derivation uses the specific and not the molar volume. The phases change places when $\mathrm{d}v/\mathrm{d}x$ changes sign.

Geometric formulation of plaits Rules like these can be derived purely analytically but more general problems defy explicit analysis and so van der Waals designed his theory of mixtures geometrically. His main tool, and at the same time his main subject of investigation, was the ψ-surface. It is determined by eqn (12.14) which gives the value of the free energy of a *homogeneous* mixture at a defined temperature. The surface is bounded by the vertical plane $v = 0$ and the two planes $x = 0$ and $x = 1$ which it approaches with infinite slope, that is $(\partial\psi/\partial x)_{v,T}$ becomes infinite at $x = 0$ and $x = 1$. As long as the mixture remains stable and homogeneous then the ψ-surface is convex downwards, but when the mixture is unstable with respect to separation into two phases then there appears a portion of surface which is concave downwards; this is what is known as a *plait* and represents the unstable state.

The simplest case is that of a plait that crosses the whole of the surface, from $x = 0$ to $x = 1$, which van der Waals called a transverse plait to distinguish from the longitudinal plait that runs parallel to the volume axis[16]. A transverse plait corresponds to liquid–vapour equilibrium as shown in Fig. 12.1(a). Coexisting phases are represented by the points of contact of a tangent plane with the ψ-surface. When the temperature is raised into the critical region the ψ-surface is deformed so that the plait no longer spans the whole range of composition but terminates at a point that represents the critical state at which eqn (12.3) is satisfied. Korteweg, whose results were used by van der Waals, called this point the *plait-point*, and defined it as the point 'through which a tangent can be drawn which has a higher-order point of contact with the surface than that drawn through an arbitrary point' [*33*, p. 1154].

Some particular lines on the surface are associated with the plait and plait point. The first of these is the *binodal* curve which is the locus of the pairs of points of contact of the tangent plane. Maxwell's device of rolling a greased plate over the surface generates the binodal curve. Strictly speaking, the plait is the part of the surface within and bounded by the binodal curve. The second particular line is the *spinodal curve*[17], which is the locus of points of zero curvature, that is, of the inflection points. It lies wholly within the plait and is tangential to the binodal at the plait point. Both lines have a clear thermodynamic meaning; the binodal is the boundary of absolutely stable states, and the spinodal the boundary between metastable and unstable states.

The equation of the surface $z(w,y)$ near the plait point can be represented by [*33*]

$$z = cw^2 + dwy^2 + ey^4.$$

The tangent plane at the plait point either cuts the surface or does not cut the surface, depending on the size of the difference $(4ce - d^2)$. If it is positive there is no intersection and so only a point of tangency — this is the so-called 'plait point of the first kind'. If the difference is negative there is an intersection and we have a plait point of the second kind, in which the sections are two tangent curves that lie on the convex–concave part of the surface and touch each other at the plait point.

A study of the appearance and disappearance of plait points as the surface is deformed by changes of temperature or of the van der Waals parameters a and b leads to two special cases, $d = 0$ and $d^2 - 4ce = 0$. The plait point can vanish from the region of the surface $0 < x < 1$ in one of two ways; either it moves outside the possible limits of composition or it concides with a second plait point when both become imaginary. The coincidence of two plait points of the same kind gives rise to a 'binary homogeneous plait point'; this occurs when $d = 0$. The coincidence with a plait point of the other kind gives a 'binary heterogeneous plait point'; this occurs when $d^2 = 4ce$. The appearance and disappearance of plait points of different kinds corresponds to different cases of phase equilibria in liquid–vapour, liquid–liquid, and liquid–liquid–vapour systems. The last case, which van der Waals called 'the pressure of three phases' is particularly complicated and its investigation called for great efforts.

The most important and general problem in the study of ψ-surfaces was, according to van der Waals, as follows: 'For some binary systems the plait of the ψ-surface has a simple shape. For others it is complex, or there exists a second plait. And nobody has as yet succeeded in pointing out the cause for those different forms, not even in bringing them in connection with other properties of the special groups of mixtures' [84e,a p. 621]. First one must ascertain 'whether for a given binary mixture a three-phase-pressure is to be expected or not. And everything that can contribute to elucidate a matter of such practical importance, must necessarily be considered important' [95e, p. 317].

These problems had already been raised in his *Molecular Theory* where the way to solve them was also set out, namely to study the course of the spinodal line or lines. A knowledge of the course of these lines allows us to map out the 'plait-point line' (the term is due to Kuenen), or what is now called the critical line. This line is one of the most important geometric determinants of the phase behaviour of the system [*33*], and it is easily enough obtained to allow for an experimental test of the theoretical conclusions and for a comparison of different classes of substances. It was for these reasons that for many years 'the plait-point line was a particular target for detailed investigation' [113, p. 290]. His strategy is based on his conception of a mixture as a hypothetical single substance. If, for example, points C and R in Fig. 12.1(d) or 12.1(e) were to coincide, then they coincide also with a third point, namely the point on the isotherm at which $(\partial P/\partial v)$ and $(\partial^2 P/\partial v^2)$ vanish. At this last point eqn (12.13) holds good and so 'in all cases where the three points are not far from each other we can plot approximately the plait-point line' [48f, p. 82]. This is quite effective for systems such as $C_2H_6 + N_2O$, studied by Kuenen, but more general methods are clearly needed to handle other systems.

12.3.2 Networks of p and q lines

An analytic study of the course of the plait-point line is difficult. It is readily seen how cumbersome the spinodal eqn (12.3) becomes if ψ is substituted from eqn (12.15) and the derivatives db_x/dx and db_x/dx inserted. Analytic attempts at a classification of ψ-surfaces have similar problems, and so a search was made for other ways. 'I have tried to find a method of treatment of the theory, which is easier to follow than the analytical one, and which, as the result proved, enables us to point out a cause for the different shape of the plaits, and which in general throws new light upon other already more or less known phenomena' [84e, p. 62].

The method chosen was the formation and subsequent analysis of two systems of 'isolines', the p-lines and the q-lines. At a given temperature coexistent phases have equal values of three quantities. The first is the pressure or $-(\partial\psi/\partial v)_{x,T}$, the second is mathematically analogous, $q - (\partial\psi/\partial x)_{v,T}$, where $q = M_2\mu_2 - M_1\mu_1$, or the difference of 'molecular, or chemical potentials, and the third is the potential $M_1\mu_1$ expressed in the form $\psi - v(\partial\psi/\partial v)_{x,T} - x(\partial\psi/\partial x)_{v,T}$. The elegance of van der Waals's method stems partly from the fact that a qualitative analysis of phase equilibria can dispense with the third quantity.

What can we obtain from the networks of p- and q-lines? From the equation of the spinodal, eqn (12.3), it follows that

$$(\partial v/\partial x)_p = (\partial v/\partial x)_q$$

and so the spinodal is the locus of the tangent points to the p- and q-lines. He summed up: 'So if we are able to derive from the properties of the components of a mixture what the course of the p- and of the q-lines is, we can derive much, if not everything, about the shape of the spinodal curve. And even when the course of these lines can only be predicted qualitatively, and the quantitatively accurate knowledge is wanting, the quantitatively accurate shape of the spinodal curve will, indeed, not be known, but yet in large traits the reasons may be stated, why in many cases the shape of the plait is so simple as we are used to consider as the normal course, whereas in other cases the plait is more complex, and there are even cases that there is a second plait' [84e, p. 622–623].

Van der Waals built the network of p-lines directly from the equation of state since this gives explicitly $p = f(v,x)$, and so the network, that is the projection of the isobars, p = constant, on the v–x plane can readily be made. For different ranges of temperature and for different ratios of a_{12}, a_1 and a_2, six different kinds of network can be obtained.

Similarly networks of q-lines, that is the projection of q = constant onto the v–x plane, can be built up from the expression for q,

$$q = MRT\ln\frac{x}{1-x} + \int_{v_0}^{\infty}\left(\frac{\partial P}{\partial x}\right)_v \mathrm{d}v.$$

This equation follows from differentiation of the expression for the free energy (12.9); the derivative $(\partial P/\partial x)_v$ is obtained from the equation of state. Again there are six different kinds of network.

12.3.3 The classification of the different kinds of phase separation: phase separation in gaseous mixtures

Further studies concerned the various ways in which the p- and q-lines were disposed and, in particular, how these dispositions depended on the existence, or not, of lines on the ψ-surface on which $(\partial^2\psi/\partial v^2)_x = 0$ (that is $(\partial P/\partial v)_x = 0$) and $(\partial^2\psi/\partial x^2)_v = 0$. It was found that the presence or absence of one or both of these lines, and their mutual disposition if both were present, were important factors in determining which forms of critical lines occurred and were used by van der Waals and Kohnstamm for classifying the phase separation in binary mixtures. While this work was in progress van der Waals wrote, 'The phenomena are really not complex in the case of a single substance and it is an easy matter to obtain a general picture of them. It is therefore very surprising that in a mere binary mixture they become so complex that they have often been compared with a labyrinth. This is particularly so where a three-phase pressure can exist' [103e, p. 263]. Nevertheless they managed to establish the six main types of phase diagram, a summary of which is given in Table 12.1.

The most unusual type is the last one which depicts a paradoxical situation that for many years had been thought to be impossible. This is the now well-known phenomenon of the separation of a gas into two phases in the supercritical region. Such a possibility was foreseen by van der Waals as early as 1894, although in an obscure way, as he had similarly anticipated Kuenen's discovery of retrograde condensation. In a little-quoted paper entitled *On the*

Table 12.1 A review of main types of phase separation in binary mixtures (after [123] and [163])

Type	The characteristics of the plait-point line (PPL)	Explanatory notes
I	PPL has two heterogeneous double plait points, that is, it has two turning points. PPL for the liquid–vapour equilibrium is separate from liquid–liquid.	The simplest case of three-phase equilibria liquid–liquid–vapour.
II	PPL with two turning points forms the single complicated curve connecting the critical points of the components.	One of the two liquid layers merges with the vapour in the critical region.
III	PPL breaks into two parts, the first of which has two heterogeneous plait points and the second has a double homogeneous plait point.	The case is analogous to type I, but differs from it by the existence of the maximum in the pressure and of the minimum in the temperature at the coexistence of the phases.
IV	PPL is divided into two parts, one of which has two heterogeneous and two homogeneous plait points.	Two liquid phases coexist at pressures high enough to suppress the vapour phase.
V	PPL has two turning points and a minimum in temperature.	A complication of type II for the case where the three-phase pressure is higher than the vapour pressure of pure components.
VIa	PPL has two turning points and a maximum in temperature.	Separation into layers of gaseous mixtures at supercritical pressures and temperatures.
VIb	PPL has two heterogeneous and two homogeneous plait points and connects the critical points of the components.	

absence of or the change in the critical phenomena of a mixture owing to the presence of a longitudinal plait in the ψ-surface at higher temperatures [38], he discussed Kuenen's results on liquid–liquid equilibria and their possible alterations on raising the temperature. Apart from the usual case in which the longitudinal plait disappears leading to complete miscibility, he considered also the case where the plait persisted: 'Such a case, perhaps the only such case, is where the liquid phase disappears as a consequence of the presence of the longitudinal plait on the ψ-surface at temperatures which exceed the critical temperatures of the components. This does not occur for substances that mix with the evolution of heat since for these there is no longitudinal plait at any temperature. It does not occur either for substances that mix with a small absorption of heat, for then the longitudinal plait exists only at low temperatures. But when mixing occurs with a great absorption of heat such a case becomes possible' [38, p. 134–135].

Van der Waals did not call this the separation of a gaseous mixture, possibly because he did not like sensational phraseology, or possibly because he did not distinguish between liquid and gas in the supercritical region. But it is clear to those familiar with the field that the existence of a critical curve and of phase equilibrium above the critical temperatures of both components implies that the coexistent phases are not liquids in the usual sense of the word, but perhaps 'fluids' if one shies away from the word 'gas'. The first to draw attention to this phenomenon were Kamerlingh Onnes and Keesom who did not think it necessary to avoid the expression 'gas–gas equilibrium'[19]. Later van der Waals adopted it also in the *Textbook of Thermodynamics* [113, p. 401]. He deduced that two kinds of separation of a gaseous mixture could occur.

Nearly thirty years later Krichevskii and Bolshakov, in the autumn of 1940, submitted for publication the results of experiments that showed that the system ammonia–nitrogen exhibited the barotropic effect, that is, that the gas phase was denser than the liquid and so went to the bottom of the vessel and, furthermore, that the critical line passed through a minimum at between 85 and 90 °C and then moved to higher temperatures. They followed it to 125 °C and 5800 atm. The arguments brought forward by van der Waals led them to believe that the separation of the system into two gaseous phases could therefore be achieved. 'Experiments which we hope to carry out soon,' they wrote, 'will show how the system ammonia–nitrogen will behave at temperatures above 125 °C, and so we shall see if separation into two gaseous phases is possible above the critical temperature of ammonia' [*35*, p. 171]. In September 1941 a short paper by Krichevskii and Tsiklis was published: 'We have been able to follow the course of the critical line up to 148 °C where the critical pressure is 9500 kg cm^{-2}. Thus the phenomenon of limited mutual miscibility has been observed in the system ammonia–nitrogen at temperatures above that of the critical point of ammonia (133.1 °C). We have studied also the system methane–ammonia, methane–nitrogen–ammonia, and nitrogen–hydrogen–ammonia which also show limited mutual miscibility. Further experiments are under way' [*36*][20]. Thirty-five years later Tsiklis wrote '. . . in 1941 Krichevskii published the first results of experiments in which it was shown that gas–gas equilibrium did occur. Thus it is to him that the credit is due for reviving the forgotten ideas on this subject, for testing them experimentally, and for making further developments of the theory of this kind of phase equilibria' [*37*, p. 1366].

In 1943 Tsiklis confirmed van der Waals's prediction that there would be two types of gas–gas equilibrium; in one case the critical line on leaving the critical point of the less volatile component moves at once to higher temperatures, and in the other it moves first to lower temperatures, passes through a minimum, and then goes to higher temperatures. Ammonia–helium is of the first kind and ammonia–nitrogen of the second, in which the gas–gas equilibrium occurs at temperatures below the critical point of ammonia, but above the minimum of the critical line, which can be called a double homogeneous plait point[21]. On the shapes of the predicted *P–T* and *P–x* diagrams of the first kind Tsiklis later wrote, 'We can only be astonished at how closely they are confirmed by the experiments' [*38*, p. 89]. At present supercritical (or gas–gas) phase equilibria are known for several dozen systems[22].

Van der Waals spent much effort in trying to establish the relations between his six main types of phase equilibrium and the ratios of the characteristic parameters of the binary system,

$$n = b_2/b_1, \quad k = (a_2/a_1)^{1/2}, \; l = a_{12}/(a_1 a_2)^{1/2}.$$

Later Kohnstamm who took part in this work confessed that 'even with various simplifying assumptions we have not been able to obtain a clear picture' [123, p. 285]. Only with the advent of powerful computers could the problem be tackled properly. The first results were those of P. H. van Konynenburg in his thesis of 1968, which were later published with some additions and corrections in two papers written with his supervisor R. L. Scott [*40,41*]. These led to the classification of phase diagrams that is now most commonly used, but their results are confined for the most part to molecules of equal size, that is, $n = 1$.

12.3.4 Ternary mixtures

It is clear that the complexity and variety of phase behaviour are almost limitless in ternary systems. Van der Waals did not, therefore, make any systematic study of them in the way he did for binary mixtures. He did, however, show how such investigations could be made [65]; as before, the method is the combination of thermodynamics with the equation of state. The latter is constructed in the same way as before with the parameters a and b being expressed as quadratic functions of mole fraction because only pair interactions are taken into account. The absence of a cubic term is important since it allows the parameters of a ternary mixture to be determined from those of the three binary mixtures. This circumstance is a great simplification and is to be contrasted with that for binary mixtures for which the parameters cannot be obtained from a knowledge of those of the pure substances; and so, he wrote, 'the theory of ternary mixtures does not therefore require any new data above those of the theory of binary mixtures' [66e, p. 235]. He demonstrated this conclusion by an example relating to the conditions for a minimum critical temperature in a ternary system [66].

12.3.5 The reception and evaluation of the theory of mixtures

In contrast to the equation of state and the principle of corresponding states, van der Waals's theory of mixtures was received with indifference by European scientists. An obvious reason for this was their inability to master it at a time when most chemists had little knowledge of mathematics and physics; it is probable that all but his own pupils were frightened by his cumbersome calculations. His style of writing added to these difficulties and even in the Netherlands it is doubtful, according to Kohnstamm, whether the *Molecular Theory* would have had much success if it had not been complemented with the author's oral explanations. As a result the appreciation of his work on mixtures in the rest of Europe was at least twenty years behind that in the Netherlands. Only after the publication of the second part of the *Textbook of Thermodynamics*, and some later reviews by his son [162] and by Kohnstamm [163] was its true value realized by the world at large. By then, however, it was too late for the centre of scientific interest had moved on and this field was buried under the succeeding 'culture layers'. Only towards the middle of this century were excavations stimulated by a revival of scientific and practical interest. Early in the twentieth century, things were quite different, as may be shown by this quotation from Alexander Findlay in 1913. 'The striking analogy which was shown by van 't Hoff to exist between dilute solutions and gases seemed to point quite naturally to the view that the van 't Hoff laws, when applied to concentrated solutions, would have to be modified in much the same way as the laws for gases under high pressures. In other words, it was thought that the quantitative relationships between osmotic pressure and concentration should be expressed by an equation similar to that of van der Waals for gases' [*42*, p. 27].

It was quite clear from van der Waals's theory of mixtures that this way of taking into account the non-ideality of solutions was wrong in principle, as van Laar repeatedly stressed, 'there is nothing less suitable for this case' [*43*]. It is unfortunate that a quarter of a century after the appearance of van der Waals's first paper a leading authority on the phase rule and, indeed many others, should think that the way to build a theory of non-ideal solutions was to modify van 't Hoff's equation 'in imitation of van der Waals'.

There was also direct opposition to his views, and some, in particular Duhem, refused to accept the idea of combining thermodynamics with an equation of state. Duhem's rejection was based principally on his opposition to atoms and on his wish to build a purely phenomenological picture of nature, but may also have been fired by feelings of competition or even jealousy. His first work on the theory of binary mixtures in the critical region had been published in 1888 [*19*], before the appearance of van der Waals's *Molecular Theory*, but the Dutch work in this field became the better known, particularly after the discovery of retrograde condensation. Duhem presented his views in English [*44*], and, in more detail, in French in his well-known *Treatise* [*20*, t. 4] which contained few references to van der Waals. He did, however, give a short description of the ψ-surface which concluded, 'It is through consideration of the ψ-surface that Mr van der Waals, and after him Mr Kuenen, have developed the theory of the continuity between the liquid and gaseous states of fluid mixtures. Although we have no intention of using this method we believe it will have been useful to have given the reader some information about its principles' [*20*, t. 4, p. 108].

In his turn, van der Waals was equally restrained in his mention of the 'interesting theoretical papers of Mr Duhem', saying that since 'Mr Duhem had not admitted the use of any definite form of the equation of state he could discuss only what followed from an application of thermodynamics, and some of his results, at any rate in their primary form, are different from ours' [58, p. 610]. The next year, 1901, Kohnstamm published a paper [*45*] which was, in part, a popularization of van der Waals's theory, and this time Duhem replied aggressively, basing some of his criticism on the probable differences between the theory and experimental results obtained in his own laboratory. He concluded, 'It does not follow that the theory of the ψ-surface proposed and developed by van der Waals must be wholly rejected. The theory has two distinct parts. On the one hand there are the laws of thermodynamics and some very general assertions that are undoubtedly correct, but on the other, there are some much more arbitrary assertions about the analytic form of the ψ-surface and, in particular, about its dependence on the concentration, x. It is clearly important to reject these latter assertions and to build the theory of binary mixtures only on the first set of foundations. This I have done, I believe, in my *Traité élémentaire de mécanique chimique*' [*46*, p. 606].

This attack drew no immediate response, if we except the tenth *Stelling* of Kohnstamm's thesis in which it is stated that Duhem's opinion that a theory of mixtures can be founded on thermodynamics alone is a delusion. Several years later the argument was resumed in the *Textbook of Thermodynamics* [91, p. 177–178]. Here is quoted a passage in French from Duhem's *Treatise* [*20*, t. 3, p. 42] in which it was stated that a theory of dissociation in a gas mixture could not be founded on thermodynamics alone. 'It is therefore hard to understand how Duhem who shows, in the sentences just quoted, such a clear understanding of the boundaries set to any thermodynamic investigation, could insist elsewhere in applying thermodynamics without supporting it with a definite form for the dependence of the free energy on its variables.' The conclusion of the chapter is van der Waals's favourite assertion that without this additional information any more detailed knowledge of equilibrium relations would be closed off.

These two extreme views, that is the naïve transfer of the equation of state directly to mixtures, and the total rejection of such an equation, cleaving only to the orthodoxy of pure thermodynamics[23] lay either side of the more rational way of studying non-ideal systems that was explored by van der Waals. By the early years of this century the pre-Gibbsian (or van 't Hoff's) school, with its emphasis on dilute and ideal systems, was running out of steam, and the more general examples of phase equilibria were outside their scope. It was at this time that Kohnstamm wrote, 'Physical chemistry in its present state reminds us strongly with regard to its quantitative part, of the navigation of a people who do not yet know of the compass. The coastal-trade is carried on with great vigour, the same limited region is traversed again and again;

but they do not dare to venture on the main sea far from the coast, and with reason, for great is the danger of ruin in the towering waves of random hypotheses. This can only be remedied by a trustworthy compass. Physical chemistry may obtain it if it will abandon the method of the osmotic pressure and adopt that of the thermodynamic potential in connection with a well grounded equation of state' [*47*, p. 751].

The research workers in the Netherlands who followed thorny paths mapped out by van der Waals had to contend with two obstacles. The first was the notorious inaccuracy of the equation of state, which could be compensated only by direct experiments, and the second, the extreme complexity of the calculations; as Tsiklis remarked later, 'the length of the equations can be measured in metres' [*38*, p. 121].

The demand for a simpler approach to the description of non-ideal systems was satisfied by the method of activities introduced by G. N. Lewis, and still widely used in physical chemistry. At first this was regarded as the antithesis to van der Waals's theory and so evoked sharp criticism from his pupils, especially van Laar who used every chance to attack 'American' thermodynamics, with its *activity, fugacity* and other 'words substituted in place of exact notions' [*48*]. This was echoed in the comment of Landau: 'Lewis's theory has a philological character; it really gives rise to no inferences but only introduces new words' [*49*].

In 1929 an important monograph [*50*] characterized the method of activities as 'purely formal' and the van der Waals theory of binary mixtures as 'the outstanding picture, van der Waals's equation being often able to give a useful starting point to the analytic description of experimental results, although it is difficult and even impossible to forecast values of parameters in the resulting formulae'. At the same time the authors remarked that the need to incorporate empirical coefficients, makes this method, 'at any rate, at present, scarcely less formal then the method of activities' [*50*, p. 296,364].

Meanwhile the work on activities[24] showed that they incorporated the very same principles that van der Waals had advocated, namely the division of a thermodynamic function into two parts, the first of which (the *ideal* term) is known *a priori* and depends explicitly on the concentration, and the second of which depends on the activity coefficient and must be measured experimentally or determined from a molecular theory. Van der Waals's conception is still viable and is reflected, in some cases, even in the methods by which the activity coefficients are determined, as, for example, in the Newton–Dodge plots and by chemical engineers to relate activity coefficients to reduced temperatures and pressures (see, e.g. [*22,55*]).

In passing from a phenomenological to a molecular approach we should like to direct attention to the scientific programme that began in the 1950s. One of us wrote in 1959: 'It is unfortunate that both the theory and practice of the study of the liquid mixtures have often been divorced from the study of pure liquids. I have tried, in this book, to put the two subjects together again' [169(1959), p. vii]. It was from these years that such a connection, so typical of the van der Waals school, came again to the fore in theories of solutions. One of its first manifestations was, perhaps, that of I. Prigogine who wrote in 1957: 'The molecular theory of mixtures was given a brilliant start by the fundamental work of Van der Waals and his school . . . Van der Waals built a detailed and consistent theory of multicomponent systems which even now we cannot but admire. . . . In a certain sense the average potential model widely used in the present book is a modern version of Van der Waals treatment' [*56*]. More closely related to the original ideas are methods in which theories of simple liquids are used for building models of mixtures based substantially on the van der Waals picture of a mixture, as, for example, the so-called *van der Waals one-fluid approximation* [*57,58*,169,181]. Here we add only that its roots were not only in the van der Waals quadratic approximation, eqn (12.15), but also in his idea of 'a mixture considered as a single substance', the idea displayed in a long series of attempts to define 'pseudocritical constants' for mixtures.

Other models in which the direct connection with the van der Waals model of a liquid (Chapter 11, Section 3) is proclaimed, are the 'augmented van der Waals theory of mixtures' of M. Rigby *et al.* [*59*] and 'the generalized van der Waals theory' of S. Nordholm and his colleagues [*60*]. Here, as well as in the physics of pure liquids [*61*], the van der Waals model occupies the important position of being a limiting case, or idealized standard, which is used to characterize the more complicated behaviour of real systems.

In phenomenology the same may be said about the van der Waals equation proper; as was shown in the work already mentioned [*40,41*], it has proved to be sufficient for the design of a qualitative picture of the great variety of phase equilibria in binary fluid mixtures. Nevertheless analytic investigations founded on a straightforward application of the van der Waals equation are probably less promising than geometric approaches after his manner, which, although qualitative, are more general.

It is characteristic of modern science that, in parallel with exact quantitative investigations, the importance and necessity of qualitative studies are becoming more apparent, probably because of the recognition of how unsatisfactory is the basis for 'exact' investigations. It is this general background that makes geometric approaches so interesting, as, for example, in the work of G. Feix [*62*]. These methods seem to be especially promising for experimental research in the field of phase equilibria under pressure which, even now, is still not sufficiently studied. Even leading laboratories (see reviews [*37,36*]) have not used fully the rich trove of van der Waals and we can expect more work on the development of his geometric methods.

Ostwald prophesied in 1913 that 'The famous author gives such an abundance of new and important results that their assimilation by experimental science will probably take not years but decades' [113, a_1]. His words have proved to be true.

Notes

1. It is characteristic that van der Waals nevertheless tried to see whether it is possible to substitute kinetic theory for thermodynamics, and his opinion was that this could be done in principle. In a paper *On the kinetic interpretation of the thermodynamic potential* [39] (a paper which he always valued highly), he had compared the thermodynamic conditions for phase equilibrium (that is, equality of temperatures, pressures and chemical potentials in the coexisting phases) with the condition determined by the kinetic theory, namely that the numbers of molecules of each kind passing during an arbitrary period of time from phase α to phase β and *vice versa* must be equal. For one- and two-component systems these numbers can be expressed through the parameters a and b of the equation of state, an equation in which temperature, pressure and composition are also variables. Moreover, chemical (or thermodynamic) potentials can also be expressed in terms of the same quantities, and so the two conditions are found to be equivalent. Later he confessed, however, that 'Although these properties [of coexisting phases] and hence the conditions of coexistence, as well as the equation of state itself, could be derived from mechanical (or kinetic) considerations, it is more common to derive them through thermodynamic considerations from the equation of state' [61, p. 90].
2. 'Geometric thermodynamics', as the field had been called by a Russian thermodynamicist A. B. Mlodzeevskii, creates 'so to say, "a geometric story" of transformations in nature; on the other hand, being based on the general laws of thermodynamics, it establishes particular physical laws by using geometric methods' [*7*, p. 5]. Solubility diagrams were the first geometric models of systems, and Gibbs's 'geometric illustrations' in his work on heterogeneous equilibria mark an important stage in the development of geometric thermodynamics. Here the synthesis of the model and thermodynamics led to the design of thermodynamic surfaces whose peculiarities were imposed by the model.
3. Although a general way of obtaining analytic forms for thermodynamic functions from structural models is given by statistical mechanics, this route is almost always impassable for most real systems, so that at some point in the derivation from first principles we must introduce assumptions on the values of parameters or on the forms of functions in the general equations of statistical mechanics.
4. He thought that 'experimentally the existence of such a phase would mean that two menisci would disappear simultaneously in a Cailletet tube' [113, p. 40], but it is now known that slightly more subtle experiments are needed to show the presence of such a state [*8*].
5. In a review [*8*] it is stated that a paper by G. Efremova, published in 1961, is the first in which tricritical phenomena have been described. This is a misunderstanding, however, because this paper [*10*] is an investigation of the ternary system acetic acid–water–*n*-butane up to 180 °C and it established the, then unusual, case of liquid–liquid equilibrium at temperatures above the critical point of one of the components, butane, the field of coexistence of two liquid phases being reduced as the temperature is increased. It is true that it was these peculiarities that suggested the existence of 'critical phenomena of higher order' in this system, but the statement proper, as well as evidence in favour of the existence of such a critical point at about 190 °C, appeared only in a later paper [*11*] in which there was a reference to the first paper [*10*]. It is interesting that these works became known in the West only a decade later when they were quoted by Griffiths and Widom [*12*].
6. We may consider not only tricritical points, but also even tricritical lines, the first

of which was experimentally described also in Kritschevskii's laboratory [*13*]; more detailed features, predicted by theory, were established experimentally by the work of Lang and Widom [*14*].

7. These assertions are undoubtedly right for phase transitions between liquids and ideal crystals. Discussions started again, however, when physics moved away from this ideal case. In 1935 Ya. Frenkel put forward a theory of fusion and crystallization based on the concept of continuity between the solid and liquid states [*15*]. Considering fusion to be a result of a continuous increase in the concentration of defects in the crystal, he wrote: 'The only difference between this transition and the continuous transition from the liquid to the gas phase on Van der Waals's theory consists in the fact that in the latter case one has to deal with only one independent variable – the pressure (the temperature being kept constant), whereas in the former case one has to take into account, along with the pressure, the degree of regularity – or the degree of amorphization as a function of the volume. We thus arrive at a theory of fusion or crystallization qualitatively quite similar to the Van der Waals theory of the boiling or condensation process' [*15*, p. 914]. Next year Landau criticized this idea 'which would require the existence of a miraculous state which is neither isotropic nor anisotropic' [*16*]. From his theory of phase transitions [*17*] he came to the opinion that, 'Any transition of a crystal into the liquid or into crystals of other symmetry is connected with the disappearance or the appearance of some symmetry elements. However, a symmetry element is either present or not present and there can exist no intermediate state whatever. Therefore a continuous transition . . . which is connected with changes of the symmetry is absolutely impossible' [*17*, p. 26–27]. Frenkel maintained his own opinion in his well-known *Kinetic Theory of Liquids*, 1946, as well as in his *Statistical Physics*, 1947. We can see three fields for which his intuition was probably correct: liquid crystals, transitions in polymers, especially under tension (that is, under a negative one-dimensional pressure) and small clusters.
8. Van der Waals returned to the question once again in a late publication [116] and here he assumed a possibility of 'the viscous-solid state' at infinite pressure and wrote that 'at the highest temperature for the existence of the solid state the difference between solid and viscous has probably disappeared under infinite pressure.'
9. There is a confusion in the literature as to which assertion is described as which law. Following the original publications, and their interpretation by Bakhuis Roozeboom, we may call the first law the assertion that the vapour contains a higher fraction than the solution of that component the increase of which makes the vapour pressure rise. The second law, which is more exactly named the Gibbs–Konovalov law, asserts that at the maximum or the minimum of the pressure (or of the temperature if isobars are considered) the compositions of coexisting phases are equal. A third law is often included which is implicit in Konovalov's derivation, namely that the compositions of both phases change in the same direction with change of pressure or temperature (see e.g. [*3*, p. 204; *22*, p. 550]).
10. This result is often ascribed to Schreinemakers but he used such equations only from 1900 onwards [*23*].
11. How much Storonkin was indebted to van der Waals's ideas one can see from this first paper which already contained many of his results published in the books mentioned above (see Chapter 9, Refs [*104*] and [*105*]).
12. Krichevskii's book [*26*], in particular, denounced 'a prejudice that volume effects in the gas phase are always greater than volume effects in the liquid phase' [*26*, p. 115] and concluded with the assertion that 'measurements of volumes of

solutions are urgent not only at various temperatures but also at various pressures' [*26*, p. 168].

13. This was the only really polemical paper that he wrote without provocation. The discussion with Boltzmann, considered in the preceding chapter, was an answer forced from him by Boltzmann's challenge.
14. Thus he wrote in 1900 on the subject of eqn (12.14): 'I have refuted this opinion some time ago, first because the ground which was alleged for concluding to this relation seemed incorrect to me, as it does still, but secondly because the great variety which critical phenomena of mixtures show, seemed to me to clash with the assumption of such a single relation between a_{12}, a_1, and a_2. Since I have learned to ascribe many complications which mixtures show to the anomaly of the components themselves, a great many objections, which I had against the relation $a_{12} = \sqrt{a_1 a_2}$ have lost their weight, and any rate I think it desirable to keep in view at every phenomena the possibility that this relation should be fullfilled' [60e, p. 170].
15. It is, therefore, an exaggeration in modern literature when this geometric mean approximation is ascribed to van der Waals (e.g. [*32*, p. 45]).
16. Later van der Waals pointed out that such a classification was convenient when the domains of the longitudinal and of the transverse plaits are separated by a wide temperature interval; when plaits overlapped with each other, then it would be better to distinguish the' main' and the 'branch' plaits [76e].
17. The etymology of the term stems from the Latin *spina* (a thorn): a section of the ψ-surface by the tangent plane at the plait point is a curve with a turning point, that is, it has a cusp or 'thorn' (cf. [61, p. 99]).
18. Krichevskii expressed it thus: 'The backbone of a phase diagram is the critical curve, the course of which determines the peculiarities of phase equilibrium' [*34*, p. 114].
19. Gas–gas equilibrium in a ternary system (see Section 2.3) occurs at a temperature below the critical point of the least volatile component, and when the temperature increases the phase gap contracts to a point and vanishes.
20. There was a dramatic episode in the history of this discovery. On 2 January 1940 the Editor of *Acta physicochimica URSS* received a letter which was published in No. 3 of vol. 12 in which the limited mutual solubility of gases in the system NH_3–N_2 at 140 °C and at pressures above 5000 kg cm^{-2} was claimed. It is that publication (the only one on the subject in a foreign language) that is usually referred to in the Western scientific literature. It turned out to be erroneous, however; Professor Krichevskii told one of the authors that he had been mistaken in these difficult measurements and it was only a year and a half later that the effect sought for had, in fact, been detected [*36*]. Neither Krichevskii himself nor his coworkers ever mentioned the first publication afterwards.
21. It has been shown by Tsiklis that a stepwise transition from the one type of separation to the other is observed in the series of binary systems with helium (from He–H_2 to He–Xe) and also in the series of pseudo-binary systems in the ternary system He–N_2–NH_3 [*38*, p. 117–119].
22. It is also of interest to point out the simple criterion, or the necessary condition, for the immiscibility of gases proposed by M. Temkin, since his analysis was based wholly on van der Waals's theory of mixtures. He searched for the possibility of there being a longitudinal plait on the ψ-surface near the critical temperature T_2 of the less volatile (the second) component. Using the condition $\partial^2\psi/\partial x^2 < 0$ for the plait, together with eqns (12.16) and (12.11)–(12.12) and the expression for the critical temperature $(T_c)_2 = 8a_2/27b_2$, he could show that 'gas–gas equilibria are

possible if the parameter a_{12} is significantly less than arithmetic mean of a_1 and a_2' [*39*, p. 2044].

23. In the Netherlands, Schreinemakers belonged to this school. He discussed his theoretical problems with Lorentz and Korteweg, but never with van der Waals, and one can guess that there was here some hidden disagreement.
24. It is appropriate to mention here the introduction of the 'activity coefficient' by N. Bjerrum [*51*], the combination of activity and the chemical potential by P. Debye [*52*], which was the second principal step in the development of the method, (since Lewis himself never did it, although later he used partial molar free energies), and finally the systematic presentation of the thermodynamics of non-ideal solutions on the basis of the method of chemical potentials, including activities, by F. H. McDougall [*53*] and especially E. A. Guggenheim [*54*].

References

1. Andrews, T., Address by the President, *BAAS Rep.*, 1876, v. 46, p. lxvii–lxxxvi.

2. Planck, M., Allgemeines zur neueren Entwicklung der Wärmetheorie, *Z. ph. Ch.*, 1891, Bd 7, S. 647–656.

3. Haase, R., *Thermodynamik der Mischphasen mit einer Einführung in die Grundlagen der Thermodynamik*, Berlin, 1956, 597 S.

4. Kurnakow, N. S., Singuläre Punkte chemischer Diagramme, *Z. anorg. allg. Chem.*, 1925, Bd 146, S. 69–102.

5. Gibbs, J. W., *Collected Works*, vol. I, New York, 1928, 434 p.

6. Maxwell, J. C., Tait's 'Thermodynamics', *Nature*, 1878, v. 17, p. 257–259, 278–280.

7. Mlodzeevskii, A.B. (А. Б. Млодзеевский), *Geometricheskaya termodinamika* (*Geometric Thermodynamics*), Moscow, 1956, 92 p.

8. Knobler, C. M. and Scott, R. L., Multicritical points in fluid mixtures. In *Phase transitions and critical phenomena*, ed. C. Domb and J. L. Lebowitz, vol. 9, London, 1984, p. 164–231.

9. Khazanova, N. E. (Н. Е. Хазанова), Critical phenomena in solutions, *ЖФХ*, 1976, t. 50, c. 1372–1380; *Russ. J. Phys. Chem.*, 1976, v. 50, p. 831–837.

10. Efremova, G. D. and Pryanikova, R. O., (Г. Д. Ефремова, Р. О. Пряникова), Liquid–liquid equilibrium in the system acetic acid–water–butane, *Chimicheskaya promishlennost*, 1961, c. 46–48 (564–566).

11. Krichevskii, I. R., Efremova, G. D., Pryanikova, R. O. and Serebryakova, A. V. (И. Р. Кричевский, А. В. Серебрякова), A possible case of critical phenomena, *ЖФХ*, 1963, t. 37, c. 1924–1925; *Russ. J. Phys. Chem.*, 1963, v. 37, p. 1046–1047.

12. Griffiths, R. B. and Widom, B., Multicomponent-Fluid Tricritical Points, *Phys. Rev.*, 1973, v. A8, p. 2173–2175.

13. Efremova, G. D. and Schwarz, A.V. (А. В. Шварц), Critical phenomena of higher order in four-component systems, *Dokl. Akad. Nauk USSR*, 1969, t. 188, c. 1108–1110.

14. Lang, J. C. and Widom, B., Equilibrium of three liquid phases and approach to the tricritical point in benzene–ethanol–water–ammonium sulfate mixtures, *Physica*, 1975, v. 81A, p. 190–213.

15. Frenkel, J., The liquid state and the theory of fusion. I. Continuity between the solid and liquid states. II. The theory of fusion and crystallization, *Acta physicochim. URSS*, 1935, v. 3, p. 633–648, 913–938.

16. Landau, L., The theory of phase transitions, *Nature*, 1936, v. 138, p. 840–841.

17. Landau, L., Zur Theorie der Phasenumwandlungen, I,II, *Physik. Zt. Sowjetunion*, 1937, Bd 11, S. 26–47, 545–555. English translation in *Collected Papers of L. D. Landau*, Oxford, 1965, p. 193–209, 209–216.

18. Guldberg, C. M., Beitrag zur Theorie der unbestimmten chemischen Verbindungen, *Ostwald's Klassiker, No. 139*, Leipzig, 1903, S. 27–70.

19. Duhem, P., Sur la liquéfaction de l'acide carbonique en présence de l'air, *J. Phys. (Paris)*, 1888, t. 7, p. 158–168.

20. Duhem, P., *Traité élémentaire de mécanique chimique fondée sur la thermodynamique*, Paris, t. I, 1897, t. II,III, 1898, t. IV, 1899.

21. Ostwald, W., Studien über die Bildung und Umwandlung fester Körper, *Z. ph. Ch.*, 1897, Bd 22, S. 289–330.

22. Dodge, B. F., *Chemical engineering thermodynamics*, New York, 1944.

23. Schreinemakers, F. A. H., Dampfdrucke ternärer Gemische, *Z. ph. Ch.*, 1901, Bd 36, S. 287–289, 413–489, 710–740; Bd 38, S. 227–255.

24. Storonkin, A., Über die Gleichgewichtsbedingungen in Zweiphasensystemen mit vielen Komponenten I., *Acta physicochim. URSS*, 1940, Bd 40, S. 505–530.

25. Haase, R., Einige allgemeine Beziehungen für Zustandsfunktionen bei Vielkomponentensystemen, *Z. Naturforsch.*, 1948, Bd 3a, N5, S. 285–290.

26. Krichevskii, I. R., *Fasovie ravnovesiya v rastvorach pri visokich davleniyach*, (*Phase equilibria in solutions at high pressures*), 2nd edn, Moscow, 1952, 168 p.

27. van der Waals, J. D., Jr., Gibbs' invloed op de mengseltheorie, *Ch. W.*, 1926, Jg. 23, b. 410–411.

28. Kay, W. B., Density of hydrocarbon gases and vapors, *Ind. Eng. Chem.*, 1936, v. 28, p. 1014–1019.

29. Berthelot, D., Sur le mélange des gaz, *C.R.*, 1898, t. 126, p. 1703–1706.

30. Berthelot, D., Sur le mélange des gaz, *C.R.*, 1898, t. 126, p. 1857–1858.

31. Van Dalfsen, B. M., On the function a/b for multiple mixtures, *Proc.*, 1904, v. 7, p. 94–107.

32. Prausnitz, J. M., *Molecular thermodynamics of fluid-phase equilibria*, Englewood Cliffs, New Jersey, 1969.

33. Korteweg, D. J., Ueber Faltenpunkte, *Wien. S.-Ber.*, 1889, Bd 98, Abt. 2a, S. 1154–1191.

34. Krichevskii, I. R., *Fasovie ravnovesiya v rastvorach pri visokich davleniyach*, (*Phase equilibria in solutions at high pressures*), Moscow, 1946, 120 p.

35. Krichevskii, I. R., and Bolshakov, P. E. (П. Е. Большаков), Heterogeneous equilibria in the ammonia-nitrogen system at high pressures, *ЖФХ*, 1941, v. 15, p. 184–192.

36. Krichevskii, I. R., and Tsiklis, D. S. (Д. С. Циклис), The limited mutual solubility of gases at high pressures, *ЖФХ*, 1941, v. 15, p. 1059.

37. Tsiklis, D. S., Phase equilibria in solutions at high pressures, *ЖФХ*, 1976, v. 50, p. 1361–1371.

38. Ciklis, D. S. (Tsiklis), *Phasentrennung in Gasgemischen. Phasengleichgewichte in binären und ternären Systemen im Gebiet überkritischer Parameter. Übersetszung aus dem Russisch, ergänzt und erweitert von der Verfasser*, VEB Deutscher Verlag für Grundstoff-industrie, Leipzig, 1972, 224 S.

39. Temkin, M. I. (М. И. Тёмкин), On conditions determining the possibility of gaseous phases coexistence, *ЖФХ*, 1959, v. 33, p. 2040–2044.

40. Scott, R. L. and van Konynenburg, P. H., Van der Waals's and related models for hydrocarbon mixtures, *Discuss. Faraday, Soc.*, 1970, v. 49, p. 87–97.

41. van Konynenburg, P. H. and Scott, R. L., Critical lines and phase equilibria in binary van der Waals mixtures, *Philos. Trans. R. Soc. A*, 1980, v. 298, p. 495–540.

42. Findlay, A., *Osmotic pressure*, London, 1913.

43. van Laar, J. J., *Sechs Vorträge über das thermodynamische Potential*, Braunschweig, 1906, S. 31, 86.

44. Duhem, P., On the liquefaction of a mixture of two gases, *J. Ph. Ch.*, 1897, v. 1, p. 273–297.

45. Kohnstamm, Ph., Über Dampfdrucke binärer Gemische, betrachtet im Lichte der Theorie von van der Waals, *Z. ph. Ch.*, 1901, Bd 36, S. 41–62.

46. Duhem, P., Ueber die Verdampfung binärer Gemische, *Z. ph. Ch.*, 1901, Bd 36, S. 605–606.

47. Kohnstamm, Ph., Osmotic pressure or thermodynamic potential, *Proc.*, 1905, v. 7, p. 741–751.

48. van Laar, J. J., Richard Lorenz, *Ch. W.*, 1929, Jg. 26, b. 406–407.

49. Landau, L. D. (Л. Д. Ландау), A comment after V. K. Semenchenko's lecture. In *Svoystva rastvorov electrolitov* (*Properties of electrolytic solutions*), Leningrad, 1930, p. 41.

50. Schottky, W., Ulich, H. and Wagner, C., *Thermodynamik. Die Lehre von der Kreisprozessen, den physikalischen und chemischen Veränderungen und Gleichgewichten*, Berlin, 1929.

51. Bjerrum, N., Der Aktivitatskoeffizient der Ionen, *Z. anorg. allg. Chem.*, 1920, Bd 109, S. 275–292.

52. Debye, P., Osmotische Zustandsgleichung und Aktivität verdünnter starker Elektrolyte, *Phys. Z.*, 1924, Bd 25, S. 97–107.

53. MacDougall, F. H., *Thermodynamics and Chemistry*, New York, 1923, 2nd edn, 1926.

54. Guggenheim, E. A., *Modern Thermodynamics by the methods of Willard Gibbs*, London, 1933.

55. Reid, R. C., Prausnitz, J. M. and Sherwood, T. K., *The properties of gases and liquids*, 3rd edn, New York, 1977.

56. Prigogine, I., *The molecular theory of solutions*, Amsterdam, 1957, p. xiii.

57. Leland, T. W., Rowlinson, J. S. and Sather, G. A., Statistical thermodynamics of mixtures of molecules of different sizes, *Trans. Faraday Soc.*, 1968, v. 64, p. 1447–1460.

58. Leland, T. W., Rowlinson, J. S., Sather, G. A. and Watson, I. D., Statistical thermodynamics of two-fluid models of mixtures, *Trans. Faraday Soc.*, 1969, v. 65, p. 2034–2043.

59. Rigby, M., Alder, B. J., Sapse, A.-M. and Hecht, C. E., Augmented van der Waals theory of mixtures, *J. Chem. Phys.*, 1970, v. 52, p. 3665–3670.

60. Nordholm, S. and Haymet, A. D. J., Generalized van der Waals theory. I Basic formulation and application to uniform fluids; Nordholm, S., Harrowel, P. R. and Cheung, K., Generalized van der Waals theory. IX Excess properties of simple fluid mixtures; *Aust. J. Chem.*, 1980, v. 33, p. 2013–2027; 1982, v. 35, p. 247–263.

61. Chandler, D., Weeks, J. D. and Andersen, H. C., Van der Waals picture of liquids and phase transformations, *Science*, 1983, v. 220, p. 787–794.

62. Feix, G., Die Theorie der Faltenpunkte und ihre Anwendung auf Phasengleichgewichte Flüssigkeit-Flüssigkeit-Dampf, *Z. ph. Ch.* (*Leipzig*), 1983, Bd 264, S. 369–378.

63. Schneider, G. M., Fluid mixtures at high pressures, *Pure Appl. Chem.*, 1983, v. 55, N3, p. 479–492.

CHAPTER 13

Conclusion

Three points of view have been implicit in this presentation of van der Waals's life and work. There is nothing new in the assertion that science has both a national and an international aspect and he was right to emphasize the particular character of Dutch science while recognizing always its place in the wider world. To these two aspects of his work we must add a third, the personal point of view that arises from the circumstances of his life and character. His whole work is a shining example of the fruitful synthesis of these three aspects.

His single-mindedness left its stamp on his work with its particular emphasis on its continuity. There are few if any other cases in the history of science where the whole of a scientist's life work has flowed so smoothly from his first piece of research. The idea of continuity carries with it, however, its opposite, that of discreteness, and both formed significant themes in his work. We recall Poincaré's saying that 'science is always doomed to periodic transitions from atomism to continuity and back again.'

The theme of continuity is to be found not only in the title of his thesis but also in his use of geometric methods and in its characteristic treatment of the interfacial layer between a liquid and its vapour. The theme of discreteness manifested itself mainly in his commitment to an atomic and molecular point of view, but it appeared also in his belief in a hierarchical design of nature. At a time when the theory of evolution was becoming accepted, Kohnstamm wrote that 'van der Waals retained the conviction that there were innate differences in the forms and essences even of inorganic nature, that everything did not merge continuously into everything else, but that there were impenetrable boundaries for both things and ideas' [218, p. 381].

The heart of his work remained his molecular model of fluids and phase transitions, but this original work branched into other fields in a way that makes it particularly interesting to follow his influence on science world-wide. This influence did not spread uniformly in either time or space and it made its way at different rates into different disciplines. After the First World War his work was generally thought to be old-fashioned and was often ignored, and its revival in the second half of the century started first with the work on the physics of fluids [183]. We might guess that this revival had its roots in the increasing recognition of the importance of qualitative studies in modern science, a recognition of how inexact is the basis of almost all 'exact' investigations.

Van der Waals's work provides also an interesting example of the interaction between physics and chemistry, sciences that often use quite different concepts but which yet blend seamlessly the one into the other. It is hard to say which science benefited more from his studies and it is not surprising that many obituarists described him as a physical chemist [207–209,211]. Here his legacy was the extension of the classical physical chemistry of ideal systems to that of non-ideal ones, a trend that is very much in keeping with current work.

We must remember also, however, the value of his work in reviving Dutch science. In 1873, the year of van der Waals's thesis, de Candolle wrote 'If my analysis [of the state of Dutch science in the 1860s] is true then Holland will rise again. The actual eclipse will be as temporary as that in England at the end of the eighteenth century. . . . There are still so many strong influences in Holland that one can reasonably hope for a return of the previous scientific lustre of this country' [*1*]. Van der Waals was born in this time of stagnation and fought his way through poverty and bureaucratic restrictions to become the first leader of the renaissance of Dutch science, a renaissance that was marked by a strong and fruitful interaction between mathematics, physics and chemistry.

Finally we must not ignore the high ethical standard he set for his contemporaries with his steadfast nature, loath to compromise but free from vanity. 'That life, full of labour, never looking back, shunning every superficial extravagance, is this not the pattern for all of us, not the way of life so characteristic of our times – restless, bustling, nervous and devoted to surface gloss and glory?' [186d′, p. 45]. Zeeman [198] quoted Carlyle to describe his great compatriot:

'The great *silent* man! Looking round on the noisy inanity of the world, words with little meaning, actions with little worth, one loves to reflect on the great Empire of *Silence*. The noble silent men, scattered here and there each in his department; silently thinking, silently working; whom no Morning Newspaper makes mention of! They are the salt of the Earth. A country that has none or few of these is in a bad way' [*2*].

References

1. de Candolle, A., *Histoire des Sciences et des Savants*, Genève, 1873, p. 211–22.
2. Carlyle, T., *On Heroes, Hero-Worship and the Heroic in History*, London, 1841, p. 351.

Bibliography

Abbreviations

A. N.	Archives Néerlandaises des sciences exactes et naturelles publiées par la Societé Hollandaise des sciences à Harlem
A. Ph.	Annalen der Physik
BAAS Rep.	Reports of British Association for Advancement of Science
Ber.	Berichte der Deutschen Chemischen Gesellschaft
Beibl.	Beiblätter zu Wiedemanns Annalen der Physik
C. R.	Comptes rendus hebdomadaires des séances de l'Academie des sciences (Paris)
Ch. Cb.	Chemisches Centralblatt
Ch. W.	Chemisch weekblad. Orgaan van de Nederlandsche Chemische Vereeniging (Amsterdam)
Ch. Ztg.	Chemiker-Zeitung
F. Ph.	Die Fortschritte der Physik im Jahre . . .
E. B.	Encyclopaedia Britannica
Jb. AU	Jaarboek van der Universiteit van Amsterdam
Jb. Ch.	Jahresberichte über Fortschritte der Chemie und der verwandten Teile anderer Wissenschaften.
Jb. KAW	Jaarboek van de Koninklijke Akademie van Wetenschappen gevestigt le Amsterdam
J. C. S.	Journal of the Chemical Society of London
J. Ph. Ch.	Journal of Physical Chemistry
Levensber.	Lebensberichten van professoren en lectoren: Het aanhangsel tot het Gedenkboek van de Universiteit van Amsterdam, uitgegeven bij de herdenking van haar 300-jarig bestaan. Amsterdam, 1932, 179 b.
NNGC	Handelingen van het Nederlandsch Natuur- en Geneskundige Congres
N. P. R. I.	Notices of the Proceedings at the meetings of the members of the Royal Institution of Great Britain
N. T. N.	Nederlandsch Tijdschrift voor Natuurkunde
N. V.	Maatschapij Diligentia ('s Gravenhage). Natuurkundige Voordrachten
P. C.	Amsterdamsch Studenten-Weekblad, Propria Cures
P. N. T. N.	Physica, Nederlandsch Tijdschrift voor Natuurkunde
Ph. Z.	Physikalische Zeitschrift
Proc.	Koninklijke Akademie van Wetenschappen te Amsterdam. Proceedings of the Section of Sciences
P. V.	Processen-Verbaal van de vergaderingen der Afdeling der natuurkunde
Rec.	Recueil de travaux chimiques de Pays-Bas
V. M.	Verslagen en Mededeelingen der Koninklijke Akademie van Wetenschappen. Afdeling der Natuurkunde
Verh.	Verhandelingen der Koninklijke Akademie van Wetenschappen te Amsterdam

Versl. Koninklijke Akademie van Wetenschappen te Amsterdam. Verslagen van de gewone vergaderingen der Wis- en Natuurkundige Afdeling

W. G. N. Werken van het Genootschap ter Bevordering van Natuur-, Genees- en Heelkunde te Amsterdam

Wien. S.-Ber. Sitzungsberichte der mathematisch-naturwissenschaftlichen Klasse der Königliche Akademie der Wissenschaften zu Wien

Z. El. Zeitschrift für Elektrochemie und angewandte physikalische Chemie

Z. ph. Ch. Zeitschrift für physikalische Chemie, Leipzig

ВУИ Варшавские Университетские Известия (Bulletin of Warsaw University)

ЖРФО Журнал Русского Физико-химического Общества, часть Физическая (Journal of the Russian Physico-Chemical Society, Physical Section)

ЖРХО Журнал Русского Физико-химического Общества, часть Хизическая (Journal of the Russian Physico-Chemical Society, Chemical Section)

ЖФХ Журнал Физической Химии (Journal of Physical Chemistry)

КУИ Киевские Университетские Известия (Bulletin of Kiev University)

ТИИЕТ Труды Института Истории Естествознания и Техники АН СССР, Москва (Transactions of the Institute for the History of Natural Sciences and Technology of the USSR Academy of Sciences, Moscow)

УАН Успехи Физических Наук (Advances in Physical Sciences)

Notation

a, review; c, abstract; ac, author's abstract; d,e,f,h,r, translations into German, English, French, Dutch, and Russian respectively; a prime (′) on the letter indicates changes in the translation or in a later edition; a subscript indicates the ordinal number of the edition, or of the abstract, or of the review. Thus, $1d'_2$ means the second (revised) German edition of the work [1]; $1c_3$ means the third abstract of the work [1].

The titles of van der Waals's works which have been translated into English, German or French are given only in one of these languages. The titles of van der Waals's communications which have not been published in full or as author's abstracts are marked with one asterisk (*). In the few cases where the titles are given by the compilers these are placed in square brackets. Publications which have not been checked in the original form are marked with two **; in such cases the source of the bibliographic description is given in brackets. Letters to the authors are designated as L/A.

Publications of van der Waals[1]

1. *Over de Continuiteit van den Gas- en Vloeistoftoestand (On the Continuity of the Gaseous and Liquid States)*, Thesis, Leiden, 1873. viii, 134 b.; a_1, *Nature*, 1874, v. 10, p. 477–480; c_1, *F. Ph.*, 1874, Bd 30, S. 691–693; c_2, *Beibl.*, 1877, Bd 1, S. 10–21; d'_1, Leipzig, 1881. viii, 168 S.; c_3, *F. Ph.*, 1881, Bd 37, S. 637–679; e'_1, in *Phys. Memoirs, Phys. Soc. London*, 1890, v. 1, p. vi, 333–496; f', Paris, 1894. xvi, 280 p.; a_2, *Arch. Sci. Phys.*, 1894, t. 32, p. 216–217; d'_2, Leipzig, 1899. viii, 182 S.; c_4, *F. Ph.*, 1899, Bd 55, Abt. 2, S. 367–368; a_3, *J. Ph. Ch.*, 1899, v. 3, p. 164; a_4, *Z. ph. Ch.*, 1899, Bd 30, S. 184; e_2, *Stud. Stat. Mech.*, Amsterdam, 1988, v. 14, p. 121–239.

[1] This list does not include the reports made in his capacity of General Secretary of the Academy of Sciences. See Jb. KAW, 1897–1912.

2. De invloed van magneten op electrische stroomen en in 't bijzonder op electrisch licht, *N. V.*, 1874–75, Ser. 3, b. 85–93.
3. Het lijnen-spectrum van koolstof, *P. V.*, 1876, N 4, b. 2–3.
4. Sur le nombre relatif des chocs que subit une molécule suivant qu'elle se meut au milieu de molécules en mouvement ou au milieu de molécules supposées en repos, et sur l'influence que les dimensions des molécules, dans la direction du mouvement relatif, exercent sur le nombre de ces chocs, *V. M.*, 1876, d. 10, b. 321–336; f, *A. N.*, 1877, t. 12, p. 201–216; c, *F. Ph.* 1877, Bd 33, S. 682–685.
5. Sur le nombre des chocs et la distance de choc moyenne dans les mélanges gazeux, *V. M.*, 1876, d. 10, b. 337–348; f, *A. N.*, 1877, t. 12, p. 217–228; c, *F. Ph.* 1877, Bd 33, S. 682–685.
6. [Mitteilung zu E. Wiedemann, betreffend der Formel von Andrews (Recknagel)], *Beibl.*, 1877, Bd 1, S. 27–28.
7. L'influence de la pression sur la température du maximum de densité de l'eau, *V. M.*, 1877, d. 11, b. 119–131; c_1, *Beibl.*, 1877, Bd 1, S. 511–513; c_2, *F. Ph.*, 1877, Bd 33, S. 711–713; c_3, *Mondes*, 1877, t. 44, p. 471; f, *A. N.*, 1877, t. 12, p. 457–469; c_4, *Jb. Ch.*, 1877, S. 46.
8. Over de specifieke warmte van den verzadigden damp, *V. M.*, 1878, d. 12, b. 169–183; ac, *P. V.*, 1878, N 7, b. 4–5; c_1, *Beibl.*, 1878, Bd 2, S. 328–332; c_2, *F. Ph.* 1878, Bd 34, S. 625–626.
9. Een meetkundige voorstelling voor de voortplanting van het geluid in de lucht, *P. V.*, 1879, N 6, b. 8–11; c_1, *Beibl.*, 1880, Bd 4, S. 531–532; c_2, *F. Ph.*, 1880, Bd 36, S. 1108–1110.
10. De betrekking tusschen spanning, volumen en temperatuur bij dissociatie, *V. M.*, 1880, d. 15, b. 199–217; ac, *P. V.*, 1880, N 7, b. 3; c_1, *Beibl.*, 1880, Bd 4, S. 749–753; c_2, *F. Ph.*, 1880, Bd 36, S. 551–555, 1069–1073.
11. Over de samendrukbaarheid van ethyleengas, *V. M.*, 1880, d. 15, b. 426–432; ac, *P. V.*, 1880, N 1, b. 5; c, *Beibl.*, 1880, Bd 4, S. 704.
12. Untersuchungen über die übereinstimmenden Eigenschaften der Zustandsgleichungen der verschiedenen Körper, *Verh.*, 1880, d. 20, N 5, b. 1–32; ac_1, *P. V.*, 1880, N 3, b. 4–5, N 5, b. 2–3; c_1, *Beibl.*, 1881, Bd 5, S. 27–28; c_2, *Beibl.*, 1881, Bd 5, S. 250–259; c_3 *F. Ph.* 1880, Bd 36, S. 564–569; ac_2, 1d′; ac_3, 1e′; ac_4, 1f′; ac_5, $1d_2'$.
13. Electrische verschijnselen in ruimten bij hoogen graad van verdunning der stof, of zoogenaamde stralende stof, *N. V.*, 1880–81, Ser. 9, b. 42–51.
14. Eenige opmerkingen naar aanleiding van de algemeene theorie der ponderomotorischan krachten van Dr. D. J. Korteweg, *Verh.*, 1880, d. 20, N 1, b. 1–12.
15. Over de coefficienten van uitzetting en van samendrukking in overeen-stemmende toestanden der verschillende vloeistoffen, *Verh.*, 1880, d. 20, N 6, b. 1–11; c_1, *Beibl.*, 1881, Bd 5, S. 250–259; c_2, *F. Ph.*, 1880, Bd 36, S. 564–569; ac_1, $1d_1'$; ac_2, 1e′; ac_3, 1f′; ac_4, $1d_2'$.
16. Bijdrage tot de kennis van de wet der overeenstemmende toestanden, *Verh.*, 1881, d. 21, N 5, b. 1–10; c, *Beibl.*, 1881, Bd 5, S. 567–569.
17. Over de physische constanten van diaethylamin, *P. V.*, 1881, N 7, b. 4–5.
18. Over het stroomen van vloeistoffen en gassen door capillare buizen met elliptische doorsneden, *P. V.*, 1881, N 2, b. 2.
19. Over een mogelijken temperatuursfactor van den molekulairdruk bij vloeistoffen en gassen, *P. V.*, 1881, N 6, b. 3.
20. Een methode bij de terugkaatsing van gepolarised licht op de poolvlakte van een electromagnet (proef van Kerr) de phase en de intensiteit van den magnetischen lichtocomponent te bepalen, *P. V.*, 1884, N 9, b. 7–8.

21. Influence de la température sur la richesse en gaz d'une solution et sur l'equilibre entre des solutions gazeuses et des hydrates solides, *V. M.*, 1885, d. 1, b. 377–380; f, *Rec.*, 1885, t. 4, p. 135–138; c_1, *Jb. Ch.*, 1885, S. 88–89; c_2, *F. Ph.*, 1885, Bd 41, Abt. 1, S. 484.

*22. Over de plaats de buigspunten in de isothermische lijnen, *V. M.*, 1886, d. 2, b. 165.

*23. Over de thermodynamisch oppervlak van Gibbs, *V. M.*, 1886, d. 2, b. 199.

*24. Over het stabiel evenwicht van heterogene stoffen, *V. M.*, 1886, d. 2, b. 199, 231.

*25. Over de grenzen van het gebied van den vasten toestand voor enkelvoudige en samengestelde stoffen, *V. M.*, 1886, d. 2, b. 240.

*26. Over de oplosbaarheid der zouten bij verschillende temperaturen, *V. M.*, 1886, d. 2, b. 376.

27. De samenhang tusschen de wijze waarop de dichtheid verandert in der overgangslaag van vloeistof op damp en de wijze der werking der molekulaire krachten, *P. V.*, 1888, 26 Mai, b. 13–14; *V. M.*, 1889, d. 5, b. 171–172; c_1, *F. Ph.*, 1888, Bd 44, Abt. 2, S. 331–332; c_2, *Beibl.*, 1889, Bd 13, S. 134–135.

28. De voorwaarden, wooraan voldaan moet worden als een gegeven stof zich bij gegeven temperatuur T_K in een gegeven ruimte in evenwicht schikt. *P. V.*, 1888, 30 Juni, b. 1–3; *V. M.*, 1889, d. 5, b. 199–202; c, *Beibl.*, 1889, Bd 13, S. 140–141.

29. Eenige uitkomsten verkregen bij het zoeken naar een moleculair-theorie voor een mengsel van twee stoffen, *V. M.*, 1889. d. 6, b. 163–166; c, *Beibl.*, 1889, Bd 13, S. 465–466.

30. Over het evenwicht van vasle verbindingen tegenover vloeistofen dampmengsels, *V. M.*, 1890, d. 7, b. 4.

31. Théorie moléculaire d'une substance composée de deux matières différentes, *A. N.*, 1891, t. 24, p. 1–56; d, *Z. ph. Ch.*, 1890, Bd 5, S. 133–173; c_1, *F. Ph.*, 1890, Bd 46, S. 136; c_2, *Ch. Cb.*, 1890, T. 1, S. 789; c_3, *Ber.*, 1890 Bd 23, Abt. 2, S. 265; c_4, *Jb. Ch.*, 1890, S. 115; e, $1e_2$, p. 241–282.

32. Die Grösse des Druckes bei koéxistierenden Phasen von Mischungen, besonders bei Salz- und Säurelösungen, *V. M.*, 1891, d. 8, b. 409–447; ac, *V. M.*, 1891, d. 8, b. 397–399; d, *Z. ph. Ch.*, 1891, Bd 8, S. 189–214; c_1, *Ber.*, 1891, Bd 24, Abt. 2, S. 163; c_2, *J. C. S.*, 1891, v. 60, p. 1319; c_3, *F. Ph.*, 1891, Bd 47, Abt. 1, S. 343, Abt. 2, S. 241–242; c_4, *Jb. Ch.*, 1891, S. 185; f, *A. N.*, 1893, t. 26, p. 91–95.

33. Die Formel der elektrolytischen Dissociation, *V. M.*, 1891, d. 8, b. 448–459; ac, *V. M.*, 1892, d. 9, b. 3; d′, *Z. ph. Ch.*, 1891, Bd 8, S. 215–222; c_1, *F. Ph.*, 1891, Bd 47, Abt. 2, S. 559–560; c_2, *Ber.*, 1891, Bd 24, Abt. 2, S. 613–614; f′, *A. N.*, 1893, t. 26, p. 126–136; c_3 [also for [32]], *Beibl.*, 1892, Bd 16, S. 137–139.

34. Over een toepassing van de wet der overeenstemmende toestanden bij opgeloste stoffen, *V. M.*, 1892, d. 9, b. 346–349; c, *F. Ph.*, 1892, Bd 48, Abt. 1, S. 377–378; c_2, *Beibl.*, 1893, Bd 17, S. 307–308.

35. Opmerking, betreffend het onderzoek van Dr. de Vries, *Versl.*, 1892, d. 1, b. 158–160; c, *Beibl.*, 1894, Bd 18, S. 525.

36. De formule voor de wet van moleculaire attractie, *Versl.*, 1893, d. 2, b. 20–21; c, *Beibl.*, 1894, Bd 18, S. 734.

37. The thermodynamic theory of capillarity under the hypothesis of a continuous variation of density, *Verh.*, 1893, Sect. 1, d. 1, N 8, b. 1–56; d′, *Z. ph. Ch.*, 1894, Bd 13, S. 657–725; c_1, *F. Ph.*, 1893, Bd 49, Abt. 1, S. 463–467; c_2, *Beibl.*, 1894, Bd 18, S. 170–172; c_3, *Beibl.*, 1895, Bd 19, S. 608–611; f′, *A. N.*, 1895, t. 28, p. 121–209; e′, *J. Stat. Phys.*, 1979, v. 20, p. 200–244.

38. Over de afwezigheid van of wijziging in de kritische verschijnselen voor een mengsel, ten gevolge van het bestaan der lengteplooi op het ψ-vlak bij hoogere

temperaturen, *Versl.*, 1894, d. 3, b. 133–137; c_1, *Beibl.*, 1895, Bd 19, S. 557; c_2, *F. Ph.*, 1894, Bd 50, Abt. 2, S. 205; c_3, *F. Ph.*, 1895, Bd 51, Abt. 2, S. 229.

39. L'interprétation cinétique du potential thermodynamique, *Versl.*, 1895, d. 3, b. 205–219; c, *Beibl.*, 1895, Bd 19, S. 552–553; f, *A. N.*, 1897, t. 30, p. 137–153.

40. Sur les caractères qui décident de l'allure de la courbe de plissement dans le cas d'un mélange de deux substances, *Versl.*, 1895, d. 4, b. 20–30; c, *Beibl.*, 1895, Bd 19, S. 864; f, *A. N.*, 1897, t. 30, p. 266–277.

41. Sur les conditions critiques, ou de plissement d'un mélange, *Versl.*, 1895, d. 4, b. 82–93; c, *F. Ph.*, 1895, Bd 51, Abt. 2, S. 226; f, *A. N.*, 1897, t. 30, p. 278–290.

42. Over de wijze van uitstraling de X-stralen, *Versl.*, 1896, d. 4, b. 293–294; c, *F. Ph.*, 1896, Bd 52, Abt. 2, S. 632–633.

43. Contribution à la connaissance de l'équation d'état, *Versl.*, 1896, d. 5, b. 150–153; c_1, *F. Ph.*, 1896, Bd 52, Abt. 2, S. 199–200; c_2, *Nature*, 1896, v. 55, p. 112; f′, *A. N.*, 1901, t. 4, p. 299–313; c_3, *F. Ph.*, 1901, Bd 57, Abt. 2, S. 201.

44. L'état moléculaire du dissolvant a-t-il une influence sur la diminution des tensions de vapeur produite par des sels dissous?, *Versl.*, 1897, d. 5, b. 342–350; c_1, *Beibl.*, 1897, Bd 21, S. 570; c_2, *J. Ph. Ch.*, 1897, v. 1, p. 687; c_3, *F. Ph.*, 1897, Bd 53, Abt. 1, S. 473; f, *A. N.*, 1901, t. 4, p. 332–340.

45. Bijzonderheden in den loop der smeltkromme, *Versl.*, 1897, d. 5, b. 385–388; c_1, *Nature* 1897, v. 55, p. 551–552; c_2, *F. Ph.*, 1897, Bd 53, Abt. 2, S. 291–292.

46. De l'équilibre d'un corps solide complexe en présence de gaz et de liquide, *Versl.*, 1897, d. 5, b. 482–491; c_1, *Beibl.*, 1897, Bd 21, S. 715–716; c_2, *F. Ph.*, 1897, Bd 53, Abt. 2, S. 180–181; f, *A. N.*, 1898, t. 1, p. 78–88; c_3, *F. Ph.*, 1897, Bd 53, Abt. 1, S. 236–237.

47. Sur la représentation graphique des équilibres à l'aide de fonction ζ, *Versl.*, 1897, d. 6, b. 209–218; c_1, *Beibl.*, 1898, Bd 22, S. 87; c_2, *F. Ph.*, 1897, Bd 53, Abt. 2, S. 181; f, *A. N.*, 1899, t. 2, p. 68–78; c_3, *F. Ph.*, 1898, Bd 54, Abt. 2, S. 217.

48. Une règle approchée relative à la forme de la courbe de plissement d'un mélange, *Versl.*, 1897, d. 6, b. 279–303; c_1, *Beibl.*, 1898, Bd 22, S. 87–89; c_2, *F. Ph.*, 1897, Bd 53, Abt. 2, S. 182–183; c_3, *Z. ph. Ch.*, 1899, Bd 30, S. 157; f, *A. N.*, 1899, t. 2, p. 79–102; c_4, *F. Ph.*, 1898, Bd 54, Abt. 2, S. 217–218.

49. Sur le mélange des gaz: Extrait d'une lettre à M. H. Becquerel, *C. R.*, 1898, t. 126, p. 1856–1857; c_1, *J. Ph. Ch.*, 1899, v. 3, p. 416; c_2, *Z. ph. Ch.*, 1899, Bd 28, S. 565; c_3, *F. Ph.*, 1898, Bd 54, Abt. 2, S. 487–488.

50. Simple deduction of the characteristic equation for substances with extended and composite molecules, *Versl.*, 1898, d. 7, b. 160–165; e, *Proc.*, 1898, v. 1, p. 138–143; c_1, *Z. ph. Ch.*, 1899, Bd 30, S. 157–158; c_2, *F. Ph.*, 1898, Bd 54, Abt. 2, S. 229; f′, *A. N.*, 1901, t. 4, p. 299–313; c_3, *F. Ph.*, 1901, Bd 57, Abt. 2, S. 201.

51. Variation of volume and of pressure in mixing, *Versl.*, 1898, d. 7, b. 239–250; e, *Proc.*, 1898, v. 1, p. 179–191.

52. On the accurate determination of the molecular weight of gases from their density, *Versl.*, 1898, d. 7, b. 258–261; e, *Proc.*, 1898, v. 1, p. 198–201; c_1, *Z. ph. Ch.*, 1899, Bd 30, S. 158–159; c_2, *F. Ph.*, 1898, Bd 54, Abt. 1, S. 163–164.

53. Variation of volume and of pressure in mixing, II, *Versl.*, 1898, d. 7, b. 270–280, 297; e, *Proc.*, 1898, v. 1, p. 232–243; c, *F. Ph.*, 1898, Bd 54, Abt. 2, S. 214–215.

54. An anomaly in the course of the plaitpoint curve in a mixture of anomalous substances, *Versl.*, 1899, d. 7, b. 464–469; e, *Proc.*, 1899, v. 1, p. 385–390; c, *F. Ph.*, 1899, Bd 55, Abt. 2, S. 258.

55. On variation of volume and pressure. III, *Versl.*, 1899, d. 7, b. 469–477; e, *Proc.*,

1899, v. 1, p. 390–397; c_1, *F. Ph.*, 1899, Bd 55, Abt. 2, S. 249; c_2, *Z. ph. Ch.*, 1899, Bd 30, S. 159–160.

56. On the deduction of the characteristic equation – discussion with prof. Boltzmann, *Versl.*, 1899, d. 7, b. 537–542; e, *Proc.*, 1899, v. 1, p. 468–473; c_1, *Z. ph. Ch.*, 1900, Bd 35, S. 504; f′, *A. N.*, 1901, t. 4, p. 299–313; c_2, *F. Ph.*, 1901, Bd 57, Abt. 2, S. 201.
57. The cooling of a current of gas by sudden change of pressure, *Versl.*, 1900, d. 8, b. 441–451; e, *Proc.*, 1900, v. 2, p. 379–389; c, *F. Ph.*, 1900, Bd 56, Abt. 2, S. 207–209.
58. Statique des fluides (mélanges), *Rapports présentés au Congrès Internationale de Physique réuni à Paris en 1900*, Paris, 1900, t. 1, p. 583–614; c_1, *F. Ph.*, 1900, Bd 56, Abt. 2, S. 199; a, *Ph. Z.*, 1900, Bd 1, S. 608–610.
59. Sur la relation entre les modifications, subiés par le volume spécifique de la vapeur saturée et celui de liquide coexistant sous l'influence des variations de température, *A. N.*, 1900, t. 5, p. 407–416; c_1, *F. Ph.*, 1900, Bd 56, Abt. 2, S. 199–200; d, *Z. ph. Ch.*, 1901, Bd 36, S. 460–468; c_2, *Ch. Cb.*, 1901, T. 1, S. 871; c_3, *Jb. Ch.*, 1901, S. 21.
60. The properties of the pressure curves for coexisting phases of mixtures, *Versl.*, 1900, d. 9, b. 166–180; e, *Proc.*, 1900, v. 3, p. 163–177; c, *F. Ph.*, 1900, Bd 56, Abt. 2, S. 200.
61. Die Continuität des gasförmigen und flüssigen Zustandes, T. 2. Binäre Gemische, Leipzig, 1900. iv, 192 S.; c, *F. Ph.*, 1900, Bd 56, Abt. 2, S. 198; a_1, *Z. ph. Ch.*, 1900, Bd 35, S. 507; a_2, *J. Ph. Ch.*, 1901, v. 5, p. 141.
62. The equation of state and the theory of cyclic motion, *Versl.*, 1901, d. 9, b. 499, 556–559, 614–627, 701–712; e, *Proc.*, 1901, v. 3, p. 515–528, 571–584, 643–654; f, *A. N.*, 1901, t. 4, p. 231–269; c_1, *F. Ph.*, 1901, Bd 57, Abt. 2, S. 199–201; d, *Z. ph. Ch.*, 1901, Bd 38, S. 257–288; c_2, *Ch. Cb.*, 1901, T. 1, S. 871; c_3, *F. Ph.*, 1901, Bd 57, Abt. 1, S. 112–113; c_4, *Jb. Ch.*, 1901, S. 18.
63. Sur une formule exacte exprimant la variation de *b* avec volume, *A. N.*, 1901, t. 6, p. 47–54; c, *F. Ph.*, 1901, Bd 57, Abt. 2, S. 201.
64. Condensation of gases, *E. B.*, 10th edn, London, 1902, v. 27, p. 183–188; 64_2 *E. B.*, 11th edn, Cambridge, 1910, v. 6, p. 844–849.
65. Ternary systems. I–V, *Versl.*, 1902, d. 10, b. 544–560, 665–686, 862–876; d. 11, b. 88–109, 224–243; e, *Proc.*, 1902, v. 4, p. 448–463, 533–560, 681–694; v. 5, p. 1–21, 121–139; c, *F. Ph.*, 1902, Bd 58, Abt. 1, S. 182–83; f, *A. N.*, 1902, t. 7, p. 343–442.
66. On the conditions for the occurrence of a minimum critical temperature for a ternary system, *Versl.*, 1902, d. 11, b. 285–294; e, *Proc.*, 1902, v. 5, p. 225–235; c_1, *F. Ph.*, 1902, Bd 58, Abt. 1, S. 123; f, *A. N.*, 1903, t. 8, p. 69–81; c_2, *F. Ph.*, 1903, Bd 59, Abt. 2, S. 537.
67. Some observations on the course of the molecular transformation, *Versl.*, 1902, d. 11, p. 391–395; e, *Proc.*, 1902, v. 5, b. 303–307; c_1, *F. Ph.*, 1902, Bd 58, Abt. 2, S. 255; f, *A. N.*, 1903, t. 8, p. 104–108; c_2, *F. Ph.*, 1903, Bd 59, Abt. 2, S. 539.
68. Critical phenomena in partially miscible liquids, *Versl.*, 1902, d. 11, b. 396–400; e, *Proc.*, 1902, v. 5, p. 307–311; f, *A. N.*, 1903, t. 8, p. 109–113; c, *F. Ph.*, 1902, Bd 58, Abt. 2, S. 251.
69. Het Zeeman-verschijnsel, *De Gids*, 1903, d. 21, b. 493–512.
70. The liquid state and the equation of condition, *Versl.*, 1903, d. 12, b. 82–110; e, *Proc.*, 1903, v. 6, p. 123–151; c_1, *F. Ph.*, 1903, Bd 59, Abt. 2, S. 525; f_1, *A. N.*, 1904, t. 9, p. 1–33; f_2, *J. Chim. Phys.*, 1904, t. 2, p. 7–46; c_2, *F. Ph.*, 1904, Bd 60, Abt. 2, S. 664–665.

71. The equilibrium between a solid body and a fluid phase, especially in the neighbourhood of the critical state, *Versl.*, 1903, d. 12, b. 439–454, 606–615; e, *Proc.*, 1903, v. 6, p. 230–244, 357–366; c_1, *F. Ph.*, 1903, Bd 59, Abt. 2, S. 531; f, *A. N.*, 1904, t. 9, p. 158–185; c_2, *F. Ph.*, 1904, Bd 60, Abt. 2, S. 664.
72. Variation de la grandeur *b* de l'equation d'état considèrée comme une diminution apparente de la molécule, in *Boltzmann-Festschrift*, Leipzig, 1904, S. 305–312; f, *A. N.*, 1904 t. 9, p. 381–388; c_1, *Ch. Cb.*, 1905, T. 2, S. 734; c_2, *F. Ph.*, 1904, Bd 60, Abt. 2, S. 666; c_3, *Jb. Ch.*, 1904, S. 31.
73. Das Prinzip der Korrespondenz in der Physik der Materie, *Deutsche Rev.*, 1904, Jg. 29, Bd 1, S. 275–287; h, *Wet. bladen*, 1904, d. 2, b. 161–177.
74. De arbeid van Kamerlingh Onnes voor de Vaterlandsche Natuurkunde, in *Het Natuurkunding laboratorium des Rijks-universiteit te Leiden in de jaren 1882–1904: Gedenkboek aangeboden aan . . . H. Kamerlingh Onnes . . .* Leiden, 1904, b. 71–87.
75. The derivation of the formula which gives the relation between the concentration of coexisting phases for binary mixtures, *Versl.*, 1904, d. 13, b. 145–151; e, *Proc.*, 1904, v. 7, p. 156–162; f, *A. N.*, 1905, t. 10, p. 113–120.
76. The transformation of a branch plait into a main plait and vice versa, *Versl.*, 1905, d. 13, b. 625–630; e, *Proc.*, 1905, v. 7, p. 621–626; f, *A. N.*, 1905, t. 10, p. 284–290.
77. The shape of sections of the surface of saturation normal to the *x*-axis, in the case of a three phase pressure between two temperatures, *Versl.*, 1905, d. 14, b. 176–185; e, *Proc.*, 1905, v. 8, p. 184–193; f, *A. N.*, 1905, t. 10, p. 483–485.
78. The (T, x)-equilibria of solid and fluid phases for variable values of the pressure, *Versl.*, 1905, d. 14, b. 185–187; e, *Proc.*, 1905, v. 8, p. 193–195.
79. Properties of the critical line (plaitpoint line) on the side of the component, *Versl.*, 1905, d. 14, b. 230–240; e, *Proc.*, 1905, v. 8, p. 271–280; f, *A. N.*, 1906, t. 11, p. 115–148; c *Ch. Cb.*, 1906. T. 1, S. 1592.
80. The properties of the sections of the surface of saturation of a binary mixture on the side of the components, *Versl.*, 1905, d. 14, b. 240–249; e, *Proc.*, 1905, v. 8, p. 280–289; f′, *A. N.*, 1906, t. 11, p. 115–148; c, *Ch. Cb.*, 1906, T. 1, S. 1592.
81. The exact numerical values for the properties of the plaitpoint line on the side of the components, *Versl.*, 1905, d. 14, b. 249–258; e, *Proc.*, 1905, v. 8, p. 289–298; f′, *A. N.*, 1906, t. 11, p. 115–148; c_1, *Ch. Cb.*, 1906, T. 1, S. 1592; c_2, *Jb. Ch.*, 1905–8, S. 162.
*82. Een onstandigheid welke in acht genomen moet worden bij de toepassing der toestandsvergelijking voor groote dichtheid der materie, *Versl.*, 1906, d. 14, b. 623.
83. A remark on the theory of the ψ-surface for binary mixtures, *Versl.*, 1907, d. 14, b. 540–555; e, *Proc.*, 1907, v. 9, p. 524–528; c, *F. Ph.*, 1907, Bd 63, Abt. 1, S. 290.
84. Contribution to the theory of binary mixtures, *Versl.*, 1907, d. 15, b. 686–704; e, *Proc.*, 1907, v. 9, p. 621–638.
85. Ibid., II, *Versl.*, 1907, d. 15, b. 823–847; e, *Proc.*, 1907, v. 9, p. 727–749.
86. Ibid., III, *Versl.*, 1907, d. 15, b. 915–939; e, *Proc.*, 1907, v. 9, p. 826–849.
87. Ibid., IV, *Versl.*, 1907, d. 16, b. 12–31; e, *Proc.*, 1907, v. 10, p. 56–74.
88. Ibid., V, *Versl.*, 1907, d. 16, b. 143–161; e, *Proc.*, 1907, v. 10, p. 123–141.
89. Ibid., VI, The plaitpoint line, *Versl.*, 1907, d. 16, b. 216–232; e, *Proc.*, 1907, v. 10, p. 183–199, 84–89; f, *A. N.*, 1908, t. 13, p. 17–90, 199–259; c, *Ch Cb.*, 1908, T. 2, S. 1758; c_2, *Jb. Ch.*, 1905–1908, S. 87.
90. The shape of the empiric isotherm for the condensation of a binary mixture,

Versl., 1907, d. 15, b. 847–851; e, *Proc.*, 1907, v. 9, p. 750–753; f, *A. N.*, 1907, t. 12, p. 418–422.

91. *Lehrbuch der Thermodynamik in ihrer Anwendung auf das Gleichgewicht von Systemen mit gasförmig-flüssigen Phasen. Nach Vorlesungen von Van der Waals ... bearbeitet von Dr. Ph. Kohnstamm.* T. 1. Leipzig; Amsterdam, 1908. xii, 287 S.; a_1, *F. Ph.*, 1908, Bd 64, Abt. 2, S. 501; a_2, *Z. ph. Ch.*, 1908, Bd 66, S. 381–382; a_3, *Физ обозр* (Physical Review Kiev), 1919, т. 10, с. 164–165; a_4, *Ph. Z.*, 1910. Bd 11, S. 511; **91_2, Leipzig, 1923. xii, 289 S.
92. Contribution to the theory of binary mixtures. XII, *Versl.*, 1908, d. 17, b. 140–151; e, *Proc.*, 1908, v. 11, p. 146–157.
93. Ibid., VIII, *Versl.*, 1908, d. 17, b. 148–161; e, *Proc.*, 1908, v. 11, p. 187–200.
94. Ibid., IX, *Versl.*, 1908, d. 17, b. 262–273; e, *Proc.*, 1908, v. 11, p. 201–212.
95. Ibid., X, *Versl.*, 1908, d. 17, b. 375–391; e, *Proc.*, 1908, v. 11, p. 317–333.
96. Ibid., XI, *Versl.*, 1908, d. 17, b. 439–457; e, *Proc.*, 1908, v. 11, p. 426–444.
97. Ibid., XII, *Versl.*, 1908, d. 17, b. 504–521; e, *Proc.*, 1908, v. 11, p. 477–493.
98. Ibid., XIII, *Versl.*, 1909, d. 17, b. 790–807; e, *Proc.*, 1909, v. 11, p. 698–715.
99. Ibid., XIV, Double retrograde condensation. *Versl.*, 1909, d. 17, b. 856–872; e, *Proc.*, 1909, v. 11, p. 816–831.
100. Ibid., XV, Splitting of the spinodal line, *Versl.*, 1909, d. 17, b. 1013–1029; e, *Proc.*, 1909, v. 11, p. 890–905; f, *A. N.*, 1909, t. 14, p. 181–221, 389–465; 1911, t. 15, p. 126–160; c, *Ch. Cb.*, 1909, T. 2, S. 1188; 1910, T. 1, S. 127.
101. Quasi-association or molecule-complexes, *Versl.*, 1910, d. 9, b. 78–105, 519–561; e, *Proc.*, 1910, v. 13, p. 107–134, 494–503; c_1, *Ch. Cb.*, 1911, T. 1, S. 110; f, *A. N.*, 1912, t. 1A, p. 90–135; c_2, *F. Ph.*, 1911, Bd 57, Abt. 2, S. 607.
102. [Le discours sur le principe des états correspondantes]. In *Les Prix Nobel en 1910*, Stockholm, 1911, p. 43–44.
103. Die Zustandsgleichung: Nobel-Vortrag ... — In *Les Prix Nobel en 1910*, Stockholm, 1911, p. 1–14; 103_2, Leipzig, 1911, 24 S.; r, *ЖРХО* 1911, t. 43, отд. II, c. 175–185; a_1, *Z. ph. Ch.*, 1911, Bd 76, S. 510; a_2, *Z. Elektrochem.*, 1911, Jg. 17, S. 544; a_3, *Ph. Z.*, 1912, Bd 13, S. 790–791; h_1', *Ch. W.*, 1911, Jg. 8, b. 69–82; a_4, *Ch. W.*, 1911, Jg. 8, b. 483; c, *Ch. Cb.*, 1911, T. 1, S. 1338; h_2', *W. G. N.*, 1912, d. 7, Afl. 1, b. 123–133; e, in *Nobel Lectures in Physics, 1901–1921*, Amsterdam, 1967, p. 254–265.
104. On the value of the critical quantities, *Versl.*, 1911, d. 19, b. 1310–1330; e, *Proc.*, 1911, v. 13, p. 1211–1231; c_1, *Ch. Cb.*, 1911, T. 1, S. 1536; f, *A. N.*, 1912, t. 1A, p. 136–158; c_2, *F. Ph.*, 1911, Bd 67, Abt. 2, S. 606–607.
105. Some remarks on the value of the volumes of the coexisting phases of a single substance, *Versl.*, 1911, d. 19, b. 1458–1467; e, *Proc.*, 1911, v. 13, p. 1253–1262; c_1, *Cb. Ch.*, 1911, T. 2, S. 333; f, *A. N.*, 1912, t. 1A, p. 292–303; c_2, *F. Ph.*, 1912, Bd 68, Abt. 2, S. 650.
106. Contribution to the theory of the binary mixtures. XVI, *Versl.*, 1911, d. 20, b. 597–603; e, *Proc.*, 1911, v. 14, p. 503–509.
107. Ibid., XVII, The concentration of the gas phase between that of two coexisting liquid phases, *Versl.*, 1911, d. 20, b. 654–673; e, *Proc.*, 1912, v. 14, p. 655–673.
108. Ibid., XVIII, *Versl.*, 1912, d. 20, b. 937–955; e, *Proc.*, 1912, v. 14, p. 875–892, N 106–108; f, *A. N.*, 1912, t. 2A, p. 96–138; 1913, t. 3A, p. 196–227.
109. Ibid., XIX, *Versl.*, 1912, d. 20, b. 1215–1229; e, *Proc.*, 1912, v. 14, p. 1049–1063.
110. Ibid., XX, *Versl.*, 1912, d. 20, b. 1366–1379; e, *Proc.*, 1912, v. 14, p. 1217–1230.
111. Ibid., XXI, The conditions for the existence of minimum critical temperature, *Versl.*, 1912, d. 21, b. 615–627; e, *Proc.*, 1912, v. 15, p. 602–614; c, *F. Ph.*, 1912, Bd 68, Abt. 2, S. 641–642.

112. Some remarkable relations either accurate or approximative for different substances, *Versl.*, 1912, d. 21, b. 800–807; e, *Proc.*, 1913, v. 15, p. 903–910.
113. *Lehrbuch der Thermodynamik in ihrer Anwendung auf das Gleichgewicht von Systemen mit gasförmig-flüssigen Phasen ... bearbeitet von Dr. Ph. Kohnstamm*, Leipzig, 1912, T. 2, xvi, 646 S.; a_1, *Z. ph. Ch.*, 1913, Bd 83, S. 761–762; a_2, *Ch. W.*, 1913, Jg. 10, b. 350–351; a_3, *J. Ph. Ch.*, 1914, v. 18, p. 460.
114. The law of corresponding states for different substances, *Versl.*, 1912, d. 21, b. 947–957; e, *Proc.*, 1913, v. 15, p. 971–981; c, *F. Ph.*, 1913, Bd 69, Abt. 2, S. 646–647.
115. Some remarks on the course of the variability of the quantity *b* of the equation of state, *Versl.*, 1913, d. 21, b. 1074–1088; e, *Proc.*, 1913, v. 15, p. 1131–1144.
116. On the point in which the solid state disappears as an answer to the question in how far this point can be compared to the critical point of a liquid: The easiest way to do this is by means of the ψ-curve, *Versl.*, 1913, d. 21, b. 1187–1188; e, *Proc.*, 1913, v. 16, p. 39–40.
117. Over de formule die aangeeft hoe de waarde van de grootheid *b* der toestandsvergelijking vermindert met verkleining van het volume, *Ch. W.*, 1913, Jg. 10, b. 628–634; c_1, *Ch. Cb.*, 1913, T. 2, S. 1450; c_2, *F. Ph.*, 1913, Bd 69, Abt. 2, S. 649.
118. Weiteres zur Zustandsgleichung, Leipzig, 1913. 55 S.; a_1, *Z. El.*, 1914, Bd 20, S. 142; a_2, *Ch. W.*, 1914, Jg. 11, b. 29–30.
119. The volume of molecules and the volume of the component atoms, *Versl.*, 1914, d. 22, b. 782–792; e, *Proc.*, 1914, v. 16, p. 880–890; c, *F. Ph.*, 1914, Bd 70, Abt. 2, S. 543.
120. On the critical density for associating substances, *Versl.*, 1914, d. 22, b. 1125–1131; e, *Proc.*, 1914, v. 16, p. 1076 – 1081; c, *Ch. W.*, 1914, Jg. 11, b. 426.
121. The increase of the quantity *a* of the equation of state for densities greater than the critical density, *Versl.*, 1916, d. 25, b. 275–284; e, *Proc.*, 1916, v. 19, p. 359–367; c, *F. Ph.*, 1916, Bd 72, Abt. 2, S. 402–403.
122. *Lehrbuch der Thermostatik, das heisst des thermischen Gleichgewichtes materieller Systeme*, T. 1, *Allgemeine Thermostatik ... bearbeitet von Dr. Ph. Kohnstamm*, Leipzig, 1927. xvi, 390 S.; r, Moscow, 1936 Ч. 1, 462 c.; a_1, *Ph. Z.*, 1928, Bd 29, S. 913–914; a_2, *УФН*, 1928, t. 8, c. 420–422.
123. Ibid., T. 2, *Binäre Gemische*, Leipzig, 1927. viii, 402 S.; r, Moscow, 1936, 439 c.

Reports on Papers

124. Van H. A. Lorentz, *V. M.*, 1878, d. 13, b. 139–143.
125. Van H. A. Lorentz, *V. M.*, 1880, d. 15, b. 345–349.
126. Van D. J. Korteweg, *V. M.*, 1879, d. 14, b. 389–393.
127. Van H. Kamerlingh Onnes, *V. M.*, 1881, d. 16, b. 214–243, 294–296.
128. Van H. Haga, *V. M.*, 1882, d. 17, b. 207–209.
129. Van J. W. Giltay, *V. M.*, 1884, d. 20, b. 71–77.
130. Van P. H. Dojes, *V. M.*, 1889, d. 5, b. 223–225.
131. Van P. H. Dojes, *Versl.*, 1895, d. 3, b. 242–244.
132. Van A. C. van Rijn van Alkemade, *V. M.*, 1892, d. 9, b. 333–336.
133. Van L. H. Siertsema, *Versl.*, 1896, d. 4, b. 208–209.
134. Van C. H. Wind, *Versl.*, 1896, d. 4, b. 210–215.

Reports and Accounts of Commissions

135. Betreffend bliksemafleiders, *V. M.*, 1880, d. 15, b. 33–37.
136. Ibid., *V. M.*, 1881, d. 16, b. 215–222.
137. Ibid., *V. M.*, 1882, d. 17, b. 254–259.
138. Ibid., *V. M.*, 1888, d. 4, b. 121–125.
139. Ibid., *V. M.*, 1889, d. 5, b. 6–8.
140. Ibid., *Versl.*, 1895, d. 4, b. 2–3.
141. Ibid., *Versl.*, 1895, d. 4, b. 158–164.
142. Ibid., *Versl.*, 1896, d. 5, b. 196–199.
143. Ibid., *Versl.*, 1897, d. 6, b. 360–361.
144. Ibid., *Versl.*, 1902, d. 10, b. 641–645.
145. Over het gevaar van de aanwezigheid van gecomprimeerde gassen in het Natuurkundig Laboratorium te Leiden, *Versl.*, 1896, d. 5, b. 3–17.
146. Betreffend de gehoorigheid gevangenissen, *Versl.*, 1897, d. 6, b. 103–172.
147. Ibid., *Versl.*, 1897, d. 6, b. 361–370, 538–544.
148. Betreffend Laboratorium op den Col d'Olen, *Versl.*, 1909, d. 17, b. 855–856.
149. Betreffend subsidieaanvrag van het 'Comité Internationale des Tables physico-chimiques' *Versl.*, 1910, d. 19, b. 613–615.

Comments and Acknowledgments

150. De toekomste bewaring der standaarden, *V. M.*, 1887, d. 3, b. 266–267.
151. Prof. D. Bierens de Haan, *Versl.*, 1895, d. 4, b. 95–96.
152. M. C. Ver Loren van Theemat, *Versl.*, 1900, d. 8, b. 558.
153. Missive van de Geologische Commissie, *Versl.*, 1906, d. 15, b. 312–313.
154. Letter to Royal Institution of Great Britain, *N. P. R. I.*, 1891, v. 13, p. 477.
155. Toespraak bij uitreikening de gouden medaille, *W. G. N.*, 1909, d. 6, Afl. 2, b. 138–139.
156. Toespraak bij neerlegging van het Secretariaat, *Jb. KAW.*, 1912, b. cxxxvi–cxxxvii.

Letters

157. Unpublished letters by van der Waals: (a) to H. A. Lorentz, 8/iii/1878, 13/iii/1881, Algemeen Rijksarchief, Den Haag; (b) to R. Threlfall, 29/iii/1888, 9/ii/1889, 25/ii/1889, Archives of Sydney University; (c) to N. A. Umov, 20/vi/1898, Archive of Moscow Naturalists' Society, inv. N727; (d) to F. Yu. Levinson-Lessing, 20/vi/1898, Archive of the Russian Academy of Sciences, fund 347, inv. N3, dossier 88; (e) to H. Kamerlingh Onnes, 6/xi/1893 etc., Museum Boerhaave, Leiden; (f) to W. Ostwald, 14/i/90 to 2/v/91, Ostwald Nachlass 3401, Deutsche Akad. Wiss., Berlin.

Publications relating to van der Waals's scientific work

158. Kuenen, J. P., *Theorie der Verdampfung und Verflüssigung von Gemischen und der fraktionierten Destillation*, Leipzig, 1906, xii, 244 S.
159. Kuenen, J. P., *Die Zustandsgleichung der Gase und Flüssigkeiten und die Kontinuitätstheorie*, Braunschweig, 1907, x, 241 S.

160. Kamerlingh Onnes, H. and Keesom, W. H., Die Zustandsgleichung, *Leiden Comm.*, 1912, Suppl. N 23, 352 S.
161. van Laar, J. J., *Die Zustandsgleichung von Gasen und Flüssigkeiten: Mit besonderen Berücksichtigung der veränderlichkeit der Werte von a und b, des kritischen Zustandes und der Theorie der Dampfspannungskurven*, Leipzig, 1924, x, 368 S.
162. van der Waals, J. D. jr., Die Zustandsvergleichung der gasförmigen und flüssigen Körper. In *Handbuch der Physik*, Berlin, 1926, Bd 10, S. 126–222.
163. Kohnstamm, Ph., Thermodynamik der Gemische, Ibid., S. 223–274.
164. Bakker, G., Kapillarität und Oberflächenspannung, *Handbuch der Experimentalphysik*, Leipzig, 1928, Bd 11, S. xv, 458.
165. van Laar, J. J., *Die Thermodynamik einheitlicher Stoffe und binärer Gemische, mit Anwendungen auf verschiedenen physikalisch-chemische Probleme*, Groningen, Batavia, 1935, 379 S.
166. Internationale Natuurkundige Congres toe Van der Waals 100-jarig jubileum, Van der Waals Centenary Number, *P. N. T. N.*, 1937, v. 5, p. 915–1180.
167. van Laar, J. J., Het bestaansrecht der theoretische en toegepaste thermodynamica naast de moderne atoomtheorie, *Ch. W.*, 1938, Jg. 35, b. 351–363.
168. Vukalovich, M. P. and Novikov, I. I. (М. П. Вукалович, И. И. Новиков), *Uravenie sostoyaniya real'nikh gazov (The equation of state of real gases)*, Moscow, 1948, 339 p.
169. Rowlinson, J. S., *Liquids and Liquid Mixtures*, London, 1959, 2nd edn 1969, 3rd edn with F. L. Swinton, 1982, viii, 328 p.
170. Brush, S. G., Development of the kinetic theory of gases, 5. The equation of state, *Am. J. Phys.*, 1961, v. 29, p. 593–605.
171. Rowlinson, J. S., Thomas Andrews and the critical point, *Nature*, 1969, v. 224, p. 541–544.
172. Rigby, M., The van der Waals fluid: a renaissance, *Chem. Soc. Q. Rev.*, 1970, v. 24, p. 416–432.
173. Brush, S. G., J. D. Van der Waals and the states of matter, *Phys. Teacher*. 1973, v. 11, p. 261–270.
174. Rowlinson, J. S., Legacy of Van der Waals, *Nature*, 1973, v. 244, p. 414–417.
175. de Boer, J., Van der Waals in his time and the present revival: Opening address, *Physica*, 1974, v. 73, p. 1–27.
176. Klein, M., The historical origins of the Van der Waals equation, Ibid., p. 28–47.
177. Levelt Sengers, J. M. H., From Van der Waals equation to the scaling laws, Ibid., p. 73–106.
178. Novikov, I. I. (ed.), *Uravneniya sostoyaniya gazov i zhidkostei: K stoletiyu uravneniya Van-der-Waalsa (Equation of state for gases and liquids; on the centenary of van der Waals's equation)*, Moscow, 1975, 260 p.
179. Brush, S. G., The kind of motion we call heat, Amsterdam, 1976, v. 1, p. 249–273; v. 2, p. 386–421.
180. Rowlinson, J. S., Van der Waals revisited, *Chem. Br.*, 1980, v. 16, p. 32–35.
181. Rowlinson, J. S. and Widom B., *Molecular theory of capillarity*, Oxford, 1982, xii, 327 p.
182. 'Johannes Diderik van der Waals', *Studievereiniging voor technische natuurkunde, 1960–1985*, Eindhoven, 1985, 96 + 102 b.
183. Rowlinson, J. S., Van der Waals and the physics of liquids. In $1e_2$, p. 1–119.
184. Gavroglu, K., The reaction of British physicists and chemists to van der Waals' early work and to the law of corresponding states, *Hist. Stud. Phys. Biol. Sci.*, 1990, v. 20, p. 199–237.

Biography

185. Lorentz, H., Prof. J. D. van der Waals, *Eigen Haard*, 1898, N 26, b. 407–408.
186. van Laar J. J., J. D. van der Waals, *Mannen en vrouwen van beteekenis in onze dagen*, 1900, Bd 31, b. 87–138, d′, Leipzig, 1900, 51 S.; c, *Ph. Z.*, 1900, Bd 2, S. 96.
187. Kohnstamm, Ph., J. D. van der Waals, 25 October 1877–1902, *P. C.*, 1902, 25 Oct., Jg. 14, N5, b. 56–58.
188. Anon., J. D. van der Waals, *Ch. W.*, 1907, Jg. 4, b. 800–801.
189. Anon., Twee hollandsche geleerde, *Nieuws van den Tag*, 1908, 2 Apr.
190. Van Ploeg, J., Prof. J. D. van der Waals, *P. C.*, 1908, 11 Juli, Jg. 19, N33, b. 328–329.
191. Anon., Gedenksteen – Van der Waals, *Algemeen Handelsblad*, 1908, 13 juli.
192. Traube J., J. D. Van der Waals und sein universelles Gesetz: Die Zustandsgleichung, *Ch. Ztg.*, 1908, 23 Nov., S. 1166–1169.
193. K[ohnstamm] Ph., Prof. Dr. van der Waals, *Amsterdamsche Studentenalmanak voor 1909*, 1909, b. 157–163.
194. Anon., Professor J. D. van der Waals, die verleden week den Nobelprijs outving, *Weekbl. Het Leven*, 1910, 18 Nov., b. 938.
195. Lorentz, H. A., Prof. van der Waals bekroning met den Nobelprijs. (1) *Nieuwe Rotterdam. Courant*, 1910, 22 Nov.; (2) Lorentz, H. A., *Collected Papers*, Den Haag, 1939, v. 9, p. 308–314.
196. Kamerlingh Onnes, H., De onderzoekingen van van der Waals, *De Amsterdammer*, 1910, 4 Dec., b. 1–2.
197. Kohnstamm Ph., J. D. van der Waals, *Ch. W.*, 1912, Jg. 9, b. 560–562.
198. Zeeman P., J. D. Van der Waals zu zeinem 80. Geburtstag, *Die Naturwiss.*, 1917, Bd 5, S. 701–703.
199. Kohnstamm Ph., Van der Waals, *Algemeen Handelsblad*, 1923, 11 Maart; reprinted in [206], b. 1–6.
200. Anon., Begrafenis van professor dr. J. D. van der Waals, *Algemeen Handelsblad*, 1932, 12 maart.
201. Kamerlingh Onnes H., Ter herinnering aan prof. dr. J. D. van der Waals, *De Telegraaf*, 1923, 11 Maart.
202. Went, F. A. F. C., J. D. van der Waals, *Versl.*, 1923, d. 32, b. 212–217.
203. Anon., J. D. van der Waals, *Ch. W.*, 1923, Jg. 20, b. 141, 167.
204. Leendertz, W., J. D. van der Waals, *De Gids*, 1932, Jg. 87, b. 151–155.
205. Zeeman, P., Prof. Dr. J. D. van der Waals Sr.: In memoriam, *P. N. T. N.*, 1923, Jg. 3, N 4, p. 101–113.
206. Kohnstamm Ph., Het lebenswerk van J. D. van der Waals, *Onze Eeuw*, Haarlem, 1932, Jg. xxiii, N 10, b. 1–32.
207. Scheffer, F. E. C., Over de beteekenis van het werk van van der Waals voor de chemie, *Ch. W.*, 1923, Jg. 20, b. 222–225.
208. Kamerlingh Onnes, H., Prof. J. D. van der Waals, *Nature*, 1923, v. 111, p. 609–610; (b) *Great Chemists*, ed. E. Farber, New York, 1961, p. 749–755.
209. Haller, A., Van der Waals, *C. R.*, 1923, t. 176, p. 793–794.
210. Haller, A., Van der Waals, *C. R.*, 1923, t. 177, p. 1350.
211. Jeans, J. H., Van der Waals memorial lecture, (1) *J. C. S.*, 1923, p. 3398–3414; (2) Memorial lectures delivered before the Chemical Society 1914–1932, London, 1933, v. 3, p. 75–90.
212. Mathias, E., Nécrologie J. D. van der Waals, *Rev. Gén. Sci. Pures Appl.*, 1923, t. 34, p. 354–355.

213. Maggi, G. A., Commemorazione del Socio Straniero Van der Waals, *Rend. Accad. Nat. Lincei. Cl. fis.*, 1924, t. 33, p. 152–159.
214. Schwab, G. M., J. D. Van der Waals, *Allgem. Ztg.*, 1923, 18 März (Bibl. d. Deutsch. Ztschr. Literatur, 1923, Leipzig, 1924, Bd 2, S. 240).
215. Lint, J. G. de., Medallic illustrations of the history of science, No 6: J. D. Van der Waals, *Isis*, 1927, v. 9, p. 258, 359.
216. Johan Dirk van der Waals, *Levensber.*, b. 169–170.
217. Voorwoord, *Faraday*, Groningen, 1937, Jg. 8, N 2 (Van der Waals-Nummer), b. 33.
218. Kohnstamm Ph., Van der Waals als docent en leermeester, Ibid., b. 34–41.
219. Keesom, W. H., Beteekenis van der Waals voor het onderzoek bij lage temperaturen, Ibid., b. 42–49.
220. Smits, A., De beteekenis van het werk van J. D. van der Waals Sr. voor de chemische wetenschap., Ibid., b. 50–71.
221. Feltkamp, T. E. W., Artikel., Ibid., b. 71.
222. Michels, A. M. J. F., Tegenwoordige onderzoekinen in verband met de theorie van van der Waals, Ibid., b. 72–77.
223. Bachinskii, A.I. (А. И. Бачинский), Johannes Diderik van der Waals (in Russian), *Fizika v shkol'e*, 1937, No. 1, p. 27–30.
224. Gerrits, G. C., Van der Waals als docent, *Algemeen Handelsblad*, 1937, 26 nov.
225. Debye, P., J. D. van der Waals: Rede, *N. T. N.*, 1937, Jg. 4, b. 257–267.
226. Sevensma, T. R., *Nederlandsche Helden der Wetenschap*, Amsterdam, 1946, b. 125–165.
227. Gerrits, G. C., *Grote Nederlanders bij de opbouw der Natuurwetenschappen*, Leiden, 1948, b. 275–315.
228. Heathcote N. H. de., *Nobel prize winners in physics (1901–1950)* New York, 1953, p. 87–93.
229. Oesper, R. E., J. D. van der Waals, *J. Chem. Educ.*, 1954, v. 31, p. 599.
230. *Nobel lectures: Including presentation speeches and laureates' biographies: Physics 1901–1921*, Amsterdam, 1967, p. 266–268.
231. Kox, A. J., J. D. van der Waals: De realiteit van atomen, (a) *Intermediar*, 1980, d. 16, N 32, b. 35–45; (b) In *Van Stevin tot Lorentz: Portretten van Nederlandse natuurwetenschappers*, Red. A. J. Kox and M. Chamalaun, Amsterdam, 1980, b. 149–161.
232. van der Ent H. and Kramer-Vreugdenhil, J., *Jacqueline E. van der Waals: Haar leven en haar werk*, Nijkerk, 1982, 189 b.
233. Kamerlingh Onnes, H., Letters to N. P. Kasterin of 8 and 10 Oct. 1898, The M. V. Lomonosov Moscow State University Archives, fund 219, inventory 1, dossier 299.
234. Collection Municipal archives, The Hague.
235. Processen-Verbaal van Buitengewone Vergaderingen Wis- en Natuurkundige Afd. gehouden 27 Febr., 27 Maart en 23 Apr. 1875, Rijksarchief in Noord-Holland, Haarlem.
236. Downer, W., L/A, 7/xi/1983.
237. Tabak, J., L/A, 7/vi/1983.
238. Bordewijk, H., L/A, 17/vi/1983.
239. Bordewijk, H., L/A, 5/iii/1984.
240. Valenbreder, J., L/A, 5/xii/1983.
241. Kohnstamm, G. A., L/A, 28/vii, 22/xii/1983.
242. van der Waals J. C. and van der Waals, J., L/A, 1/iii/1984.
243. Wurfbain, M. L., L/A, 7/vi/1984.

244. J. D. van der Waals' archief, Van der Waals Laboratory, Amsterdam.
245. Archief Nederlandse Akademie van Wetenschappen, Inv. vis 26, 27, Rijksarchief in Noord-Holland, Haarlem.
246. Maxwell, J. C., Van der Waals' equation, Manuscript in Cambridge University Library.
247. Beek, L., *Dutch Pioneers of Science*, Assen, van Goreum, 1985, p. 100–107.
248. Yavelov, B. E. and Kipnis, A. Ya. (Б. Е. Явелов, А. Я. Кипнис), Maxwell and Van der Waals (in Russian). In *Maksvell i razvitie fiziki xix–xx vekov*, Moscow, 1985, p. 140–157.
249. Yavelov, B. E., On van der Waals's 150th Anniversary (in Russian), *Fizika Nizkich Temperatur*, 1987, t. 13, N12, c. 1310–1312.
250. van Emmerik, E. P., Zoals het nog is; het laboratorium voor Van der Waals, *Laboratorium Praktijk*, Mei 1988, b. 276–280.
251. van Emmerik, E. P., De invloed van Van der Waals op de vacature Bakhuis Roozeboom in 1907, Ibid., Augustus 1988, b. 384–387.
252. Anon. Prof. van der Waals. *Weekbl. Het Leven*. 1910, 7 Nov.

Name index

This index covers the main text and the notes, but not the Preface nor the references. The diphthong ij is indexed as two letters and not, as sometimes in Dutch, as a y. Page numbers in bold type are entries with significant biographical information. This information is not given for the better-known scientists covered by such works as the *Dictionary of Scientific Biography*. Relationship to van der Waals is shown in brackets.

Subject index

Breinigsville, PA USA
15 December 2010
251440BV00003B/50/A
9 780198 552109